BEYOND THE BATAAN DEATH MARCH

WILLIAMS-FORD
TEXAS A&M UNIVERSITY
MILITARY HISTORY
SERIES

BEYOND THE BATAAN DEATH MARCH

The Life and Times of K. L. Berry

DANA BERRY FRAZEE

Texas A&M University Press
College Station

Copyright © 2026 by Dana Berry Frazee
All rights reserved
First edition

∞ This paper meets the requirements of ANSI/NISO Z39.48–1992 (Permanence of Paper). Binding materials have been chosen for durability.

Library of Congress Cataloging-in-Publication Data
Names: Frazee, Dana author
Title: Beyond the Bataan Death March: the life and times of K. L. Berry / Dana Berry Frazee.
Other titles: Williams-Ford Texas A&M University military history series
Description: First edition. | College Station: Texas A&M University Press, [2026] | Series: Williams-Ford Texas A&M University military history series | Includes bibliographical references and index.
Identifiers: LCCN 2025031743 (print) | LCCN 2025031744 (ebook) | ISBN 9781648433269 paperback | ISBN 9781648433276 ebook
Subjects: LCSH: Berry, Kearie Lee, 1893-1965 | United States. Army–Biography | Texas. Adjutant General's Department–Biography | Generals–United States–Biography | Adjutants–Texas–Biography | Ex-prisoners of war–United States–Biography | Death march Survivors–Texas–Biography | Male college athletes–Texas–Biography | World War, 1939-1945–Prisoners and prisons, Japanese | Bataan Death March, Philippines, 1942 | LCGFT: Biographies
Classification: LCC U53.B3 F73 2026 (print) | LCC U53.B3 (ebook)
LC record available at https://lccn.loc.gov/2025031743
LC ebook record available at https://lccn.loc.gov/2025031744

Cover concept by Rick Berry.

For my family

CONTENTS

Preface ix
Acknowledgments xiii

PART I. EARLY YEARS

Chapter 1. K. L. Berry: Becoming an Athlete *3*

Chapter 2. Military Training and Football on the Border *12*

PART II. MILITARY CAREER

Chapter 3. Lieutenant Berry Begins His Military Career *23*

Chapter 4. New Posts, Athletics, and Promotions *33*

Chapter 5. War is on the Horizon *47*

PART III. WAR

Chapter 6. Lieutenant Colonel Berry in the Battle of Bataan: Part 1 *59*

Chapter 7. Colonel Berry in the Battle of Bataan: Part 2 *76*

Chapter 8. Surrender and the Bataan Death March *94*

PART IV. PRISONER OF WAR

Chapter 9. Luzon POW Camps and Hell Ships *109*

Chapter 10. Taiwan POW Camp Karenko: Part 1 *122*

Chapter 11. Taiwan POW Camp Karenko: Part 2 *136*

Chapter 12. Taiwan POW Camp Shirakawa: Part 1 *153*

Chapter 13. Taiwan POW Camp Shirakawa: Part 2 *168*

Chapter 14. En Route to Manchuria *185*

Chapter 15. Manchuria POW Camps *194*

Chapter 16. Rescue and Freedom *208*

PART V. A NEW LEASE ON LIFE

Chapter 17. Recognition *219*

Chapter 18. The Adjutant General Years *228*

Chapter 19. An Honorable Life *250*

Epilogue *255*

Appendix A. Birthday Poems to K. L. Berry *259*

Appendix B. Books Read by K. L. Berry While Interned as a Prisoner of War *261*

Appendix C. Letter from Capt. Jack Walker to K. L. Berry *265*

Notes 267

Bibliography 303

Index 317

PREFACE

This book has its origin in an incident that occurred on a sweltering August day in Austin, Texas, in 1964. I was sixteen years old and had traveled to Texas with my mother and younger brother to visit my grandfather Gen. K. L. Berry and his wife, Alice. My grandfather had recently suffered a heart attack, and although I was too young then to fully understand the purpose of our visit, I now know that my mother had arranged it knowing that it would likely be the last time we would see General Berry.

My grandfather was sitting outside in his pajamas and bathrobe when my mother told me to go sit with him and wait for her, saying she would be only a few minutes. Under a bit of shade, facing one another, each a stranger to the other, my grandfather and I made small talk. I had no idea what to say to the tall, gaunt, tired-looking man, my grandfather—a figure both imposing and remote in my teenage eyes. He had no idea what to say to me, either. After all, I was a teenage girl while my grandfather was seventy-one—as it happened, the same age I was when I began writing this book.

On previous visits, my grandfather had never spent any time with me; he had always been in the company of my father and my older brother, doing "man" things. On this visit, however, my father and older brother were not with us, and here I was alone with him for the first time in my life with no idea what I should be talking with him about.

The minutes dragged on as my impatience at my mother's tardiness—and my own awkwardness—grew. Sweat trickled down my back. I was bored with the whole Texas visit and looked forward to doing something, anything, with my mother that day. After twenty minutes or so, I excused myself and went inside to find her.

K. L. Berry Diary. This is the first page of K. L. Berry's diary, dated February 19, 1943, at POW Camp Karenko. Source: Author's private collection.

When I found my mother, her reaction to my leaving the conversation with my grandfather startled me. She was visibly upset. Why did I come inside and leave the conversation with my grandfather? He was important, special. Going to the store we could do anytime, but a chance to talk with him, ask him questions, get to know him would most likely never happen again.

As it turned out, my mother was right—I would never have the opportunity to speak with my grandfather again. We left Texas two days later, and by the next April, he was gone. The strange feeling of having missed something important that day with him in the yard never left me, even though I had not understood at the time what I had been supposed to talk about, or the questions I was supposed to ask.

In 2004, while cleaning out our grandparents' house in Austin, my cousin Bruce Reilly discovered a diary our grandfather had written when he was a prisoner of war (POW) of the Japanese. I decided to make it my mission to transcribe the diary, and in so doing, I would get to know my grandfather. The transcription turned out to be very difficult to do, and it took me years to complete it.

Through those years, as I transcribed, I realized how much research I needed to do to understand the diary. As I researched the Battle of Bataan, which led to my grandfather's imprisonment in World War II, I became deeply interested in how and why my grandfather ended up in the Philippines, what his role in that battle was, and how he survived imprisonment. Questions I was supposed to ask as a sixteen-year-old came into view. I followed the questions. How did a man from Denton, Texas, end up a prisoner of war of the Japanese? How did he survive the Bataan Death March when thousands of others did not? What set him apart? What gave him strength? I wanted to understand the man who had been a mystery to me as a teenager so long ago. My quest to know him led me to write about his entire life, not just his POW life. I have found that my grandfather Kearie Lee Berry lived a remarkable life as an athlete, a soldier, a war hero, a prisoner of war, and adjutant general of Texas. I am proud to tell his story.

ACKNOWLEDGMENTS

As with any endeavor of this type, many people deserve my thanks. My mother, Phyllis Anne Berry, was the first to make me aware that K. L. Berry was an important Texan and that I needed to appreciate him. Eager for me to know him, she gave me a beautifully framed copy of Texas House Resolution No. 110, which outlined his life. I am grateful that she lived long enough to know that I had finished the book. I thank her for her love and encouragement, without which I may never have taken this journey.

In pursuit of more information about my grandfather's life in the military during World War II, I wrote the periodical *The Quan: Official Publication of The American Defenders of Bataan and Corregidor* in 2004 and asked if anyone could tell me anything about my grandfather's role in the battles. To my surprise, several people responded, including William Bartsch, author of *December 8, 1941: MacArthur's Pearl Harbor*. Bartsch pointed me to several books that cited my grandfather in the battles and another book that chronicled his life as a prisoner of war.

My cousins Bruce and King Reilly have been extraordinarily helpful to me. They have graciously shared all of our grandfather's documents, letters, the diary, and their precious memories of our grandfather. Even when I thought I had all of the materials, Bruce sent me four more boxes of important letters, many of which are referenced in the book. He and his wife, Kathy Clevenger, hosted a Berry reunion for many of us in Texas in 2021, which included Alice Berry, Jim Berry, DeDe Berry, Kearie Lee Berry IV, and Janet Berry Dundas.

In addition to sharing his memories, King made me aware of Desmond Power's and Hal Leith's books, and he also gave me the Gerte Bolte story and his interview of Col. Harrison Browne. King accompanied me to the Philippines in 2016 for Araw ng Kagitingan (Day of Valor) where we met the Rio and Vitriolo families. Eliseo D. Rio had been assigned to my grandfather as an aide-de-camp from February to April 1942 during the Philippine battles on the Bataan Peninsula in Luzon. Rio had written a book, *Rays of a Setting Sun: Recollections of World War II*, which was published just before his death in 2004. Rio had stories about my grandfather in his book that none of us in the family had ever heard before. His daughter Elisea Rio Vitriolo edited a new edition of her father's book, which led her to me when she sought a photo of General Berry for the new edition. With the Vitriolo and Rio families, King and I traced our grandfather's steps on the infamous Bataan Death March. Elisea Rio Vitriolo and her brothers Eliseo Rio Jr. and Ephraim Rio and her sister-in-law Judy Rio were our teachers and guides. I don't want to forget to mention that Lito Vitriolo (Elisea's husband), a college executive, acquired housing for us in college dormitories all over Luzon as we traveled. At night he taught us karaoke! We cannot thank those families enough for our momentous trip to the Philippines.

I want to thank some friends who have assisted me. Transcribing the POW diary got me started down this road. My dear friend Maureen Keller assisted me in the transcription of the diary. She typed as I read each sentence slowly trying to decipher handwriting and meaning, and she became as caught up in my grandfather's life as a prisoner of war as I was. We both cried each time K. L. did not get mail. In addition, Maureen proofread the captions for the photos, maps, and caricatures as well as some chapters of the book. My childhood friend Beth Quinn read my first draft. Her advice and encouragement have been steady throughout this process. I loved her for saying that she had already chosen a Hollywood star to play my grandfather in a movie of the book. Whitney Galbraith made a copy of his father's diary and notes for me. His father, Nicoll Galbraith, was my grandfather's roommate at the Shirakawa POW camp. Sharon O'Connor and Rebecca Sieve read one of the last drafts of the book and cheered me on.

I want to thank two new Texas friends, John Keck and John Adams. John Keck contacted me out of the blue about a book he commissioned to be written about the 1914 University of Texas football team. His grandfather Ray Keck was a teammate of my grandfather on that famous team. He gave me great encouragement along the way as I worked on my book. Very importantly, John also connected me with John Adams, a renowned writer of Texas and military history. John Adams took me to Texas A&M University's Cushing Memorial Library & Archives so I could read Col. Ed Aldridge's diary. Then, John took me to Texas A&M University Press where he introduced me to Jay Dew, the press director. John also read one of the drafts of the book and gave me valuable pointers on how to put it in publishable form.

Family members have encouraged me along the way. My brother K. L. Berry shared some of his memories with me about our grandfather. His daughter Lisa Berry Shearrer and his granddaughter Grace Shearrer encouraged me to go see the 1914 University of Texas football team exhibit at the H. J. Lutcher Stark Center for Physical Culture and Sports at the University of Texas. His grandson Kearie John Berry assisted me on a couple of occasions with the transcription of Berry's private notebook that he wrote as a prisoner of war.

My brother Rick Berry came with me to Texas in 2004 to go through boxes of our grandfather's material at Bruce's house, while I got my first look at the diary. Then he and I met with Lisa Sharik, who at that time was the deputy director of the Texas Military Forces Museum at Camp Mabry. Over the years, Lisa assisted me with copying pages of the private notebook and answering many questions by email. I gave her copies of the transcribed K. L. Berry diary and private notebook to put with Berry's artifacts at the museum. Years later when I realized the biography of my grandfather would be published, Rick stepped back into the process and improved the cover photo, designed the book cover, and, best of all, came up with the title for the book.

My daughters Olivia Navarro, Samantha Frazee, and Caroline Frazee, and my daughter-in-law Vanessa Connery, heard many stories about my grandfather as I researched and wrote the book. I thank them for their love and support.

Most of all, I want to thank my love, my husband, John Frazee. I had initially thought of my project as a book only for family members. When John read my first draft, he gave me sound advice on how to structure the manuscript into a biography that might be read by people outside of the family. He also told me I should seek a publisher for the book when it was ready. Seeking a publisher was a "heady" thought for me to hear so many drafts ago. In order to be ready, I enlisted John's editing expertise. Tirelessly, he read through many different drafts of the book, noting everything from sentence structure to grammar, parts that needed clarification, parts that could be deleted, and parts that were good. Thank you, John, for your love and for believing I could do this.

BEYOND THE BATAAN DEATH MARCH

PART I

Early Years

"Our blue-eyed boy raised his voice."
—*Viola Berry from* The Alamo and Other Poems

CHAPTER 1

K. L. Berry
Becoming an Athlete

DENTON, TEXAS

The United States was in the middle of a severe economic crisis in 1893. President Grover Cleveland blamed the crisis primarily on the Sherman Silver Purchase Act, which was enacted in 1890 during Benjamin Harrison's term as president. As a result of the act, production of silver increased, and the supply of gold decreased, thus making gold more expensive. As Cleveland focused on the gold crisis, a depression grew. Unemployment rose, banks shut their doors, and railroad construction fell by 50 percent. Labor unrest grew and railroad workers went on strike. People felt President Cleveland was not dealing directly with the depression by establishing public works programs, and many in his own Democratic Party lost confidence in his actions.[1]

In Denton, Texas, however, where future Texas athlete, soldier, and war hero Kearie Lee Berry was born on July 6, 1893, the Democratic Party and Cleveland were still supported. Cleveland had returned captured Confederate battle flags to the South, a gesture that made people think of him as a friend of the former Confederacy.[2] Citizens of Texas held onto memories of the Republic of Texas, the Mexican War, the Civil War, and the institution of slavery. Stories of those past events lived on in the Berry family when Kearie Lee was born to Thomas Eugene Berry and Viola Eugenia Riley. He was their second of seven children born between the years 1892 and 1905. The name Kearie began in the family with Thomas's brother Kearie Albert Berry, who lived in Quanah, Texas. The middle name Lee was a nod to the famous Confederate general, Robert E. Lee, whom the family revered.

Kearie's paternal grandfather, William David Berry, had been an overseer on a cotton plantation in Wetumpka, Alabama, in the 1850s and 1860s. Overseers generally ran the day-to-day operations of cotton farming and managed and disciplined the slave workers. After the Civil War, when former slaves were free, William became a cotton farmer in Wetumpka, no longer an overseer. Because Wetumpka declined in population and prosperity after the war, William eventually moved his family to Collin County, Texas, and continued cotton farming. His son Thomas Berry followed in his footsteps and became a well-to-do farmer in northwestern Denton County, Texas, in an area called Bolivar about thirty to thirty-five miles from Collin County where William lived. When William Berry died in 1899, Kearie Lee was too young to really know his grandfather, but as he grew up, he became aware of William's livelihood as an overseer and cotton farmer.[3]

On the other hand, Kearie Lee did have the time and opportunity to know his maternal grandfather, John Sleychk Riley. John was born in Pennsylvania in 1812 and lived a long, remarkable life as a physician, soldier-physician, and plantation owner. He died in Denton in 1915 when Kearie Lee was in college. Later in life, Berry spoke about this grandfather, "the general," and proudly recounted details about Riley's life to a reporter. Riley had studied medicine in Ohio, had fought in the Mexican War in and around Burleson, Texas, had owned a plantation near Burleson, and had been a surgeon in the Confederate Army. After the Civil War, he moved his family to Cooke County, Texas, where he and his wife adopted eight children, adding to the seven they already had. Eventually, John Riley moved to Denton to put his children in school.[4]

John Riley was never a general, but he must have seemed like one to his grandson, who grew up listening to Riley's stories about Texans' roles in the Republic of Texas when Texas declared itself a separate Republic from Mexico (1836–1846); about the war with Mexico after the United States annexed Texas (1846–1848); and about the military heroics of Confederate soldiers in the Civil War. Riley, who lived a long life, had participated in all those conflicts, and enjoyed talking about them.

Viola, Kearie's mother, had also grown up listening to those stories from her father. She had absorbed them and studied their history. In 1898, when the United States and Spain fought over Cuba and the Philippines, she paid close attention to all the details of the Spanish-American War as they made headlines in the newspapers. Comm. George Dewey of the US Navy was praised widely for wiping out the Spanish fleet in Manila Bay in the Philippines in one day. Viola was so captivated by the subject of war—the history, the people, the military chain of events—that she began writing long poems romanticizing historical conflicts. Among her published poems is "Manila, Our War with Spain," in which she depicted George Dewey as a hero. Viola's poetic subjects came to include her children. In a poem entitled, "Our Boys," she described young Kearie Lee's future as a "soldier boy" who would become a hero, like Dewey, in some future war.[5]

> When July smiled on land and sea,
> And Independence day came 'round,
> Filling our land with joy and glee
> And patriotic shouts and sounds,
>
> Our blue-eyed boy raised his voice
> And joined the resounding revelry,
> Showing thus early that his choice
> To defend his country's flag would be.
>
> And Hope once more spread visions fair,
> Filling our throbbing hearts with joy,
> As we bowed us once again in prayer
> By the cradle of our soldier boy.
>
> We forward look with faith and pride
> Adown Time's vista dim and far,
> And see him sail the ocean wide,
> The Dewey of some future war.[6]

Kearie Lee was greatly influenced by his grandfather and mother and did become a soldier as soon as the opportunity came his way. Eventually, he even would even "sail the ocean wide" and fight a war in the Philippines, not as a naval officer like George Dewey, but as a soldier like Robert E. Lee.

Seemingly unaffected by the depression of 1893, Denton became a fast-growing agricultural trade center north of Dallas and Fort Worth. The railroads had come to Denton County in the 1880s, a development that enabled light agricultural manufacturing businesses like flour mills and cottonseed oil factories to grow and thrive. Corn and vegetable farming decreased as cotton and wheat farming increased. The county was either first or second in wheat production for the state from 1890–1920.[7] Thomas Berry, young Kearie's father, farmed both wheat and cotton and was wealthy enough to hire help for his farm.

In addition to being an agricultural area, Denton was becoming a college town—the Texas Normal College (now the University of North Texas) was established in Denton in 1890, and the Girls Industrial College (now Texas Woman's University) put down roots in 1903. These colleges ensured that Denton would be a destination spot within the state.[8]

Joining the hustle and bustle around Denton, Thomas Berry moved his family into town in 1900 to put the children in school. Kearie Lee attended North Ward school and at age seven became a proud paper boy for the *Denton Record-Chronicle* newspaper for "quite a long while," covering his "route in the southwest part of Denton on my Shetland pony." When he was not delivering newspapers, he enjoyed hunting in the woods north of town for quail, possums, and rabbits. "I believe I knew every possum hollow for miles around. At times my two brothers, Eugene and John with the Skiles, Stroud, Forrester and Taliaferro boys used to make longer excursions to Clear and Hickory Creeks to gather pecans and to hunt."[9] Kearie Lee's early love of hunting was to continue throughout his life.

As young Kearie Lee grew up, Vice President Theodore Roosevelt became president of the United States after President William McKinley was assassinated in 1901. Texans already knew Teddy Roosevelt as the man who trained the famous "Rough Riders" near San Antonio before heading to Cuba to fight with them against Spain in the Spanish-American War in 1898. The United States won that war, and Spain left Cuba and ceded Guam, Puerto Rico, and the Philippines to the United States.[10] During his time as president, Roosevelt expanded the US influence in the Western Hemisphere by establishing the construction of the Panama Canal and by formulating the Roosevelt Corollary to the Monroe Doctrine, which stated that the United States would intervene in Central and South American countries' internal affairs to keep European countries from colonizing in the Western Hemisphere. And in the Eastern Hemisphere, Roosevelt became a peacemaker between Russia and Japan, helping negotiate a peace settlement for their international disagreements in 1904, for which he won the Nobel Peace Prize in 1909.[11] Because young Berry's mother and grandfather always read the newspapers, it can be inferred that Kearie Lee heard about Teddy Roosevelt, the Rough Riders, and some of these world events as he grew into a teenager.

In 1908, fifteen-year-old Kearie enrolled in Denton High School, which had been established as a public school in 1884 in a two-story building near

Denton High School. Kearie Lee Berry's high school senior year photo. Source: Author's private collection.

downtown. By 1904, the school received "the equivalent of accreditation by becoming affiliated with the University of Texas."[12] Perhaps this affiliation is the reason brothers Eugene (Gene) and Kearie Berry ultimately attended the University of Texas. The Denton High School 1911 yearbook, *The Bronco*, provides some interesting details about Kearie Lee. Under his senior photo, the unusual nickname "Yochum" was printed after his name. No one in the Berry family today knows where that nickname came from. The description went on to say, "Owing to his reserved disposition, it has been difficult to ascertain anything very definite about Kearie. He says but little, though we believe he thinks much." It was not until Kearie was a senior that he tried out for the football team. His older brother, Gene, had played football and encouraged Kearie to play. Kearie made the team and later recalled, "I did not distinguish myself as a student, but being 'kinda large' for my age, I managed to make guard on the football team. I was not a star."[13]

Even though he did not think of himself as a star, the sportswriter for *The Bronco* thought he was. *The Bronco* included a football team photo with a handsome, tall, blue-eyed Kearie Lee standing in the back row. At that time, he was five foot eleven and 159 pounds. The writer called Kearie "Big" Berry and "Yock." He continued, "'Big' Berry started the season at guard. He was afterward placed at tackle, where he remained during the season. 'Yock' often distinguished himself by tackling, but his long suit was breaking up interference. He was one of the few who really played in the Fort Worth game. The way he tore up their interference was a real caution. Berry will be greatly missed next year."[14]

If Berry did not excel as a student, it was clear from the yearbook that he had a field where he could excel—the football field. He graduated from high school in 1911 but did not join his brother Eugene at the University of Texas (UT) to play football until the fall of 1912. In the interval, Kearie Lee said, he "became a horny-handed son of toil" on his father's farm for a year. By the time Kearie Lee entered the university, the nicknames "Yock" and "Yochum" were gone, replaced by "K" or "K. L."[15]

FOOTBALL, TRACK, AND WRESTLING AT THE UNIVERSITY OF TEXAS

Toiling on his father's farm came to a close during the unusual presidential election of 1912, which saw Woodrow Wilson defeat incumbent President Howard Taft and former President Theodore Roosevelt, who was a third-party candidate. Roosevelt's candidacy split the Republican Party, making it possible for Democrat Wilson, born in Virginia before the Civil War, to become president of the United States. As president, Wilson would eventually lead the United States into the war in Europe in 1917 to "make the world safe for democracy."[16]

Presidential politics dominated the news as K. L. matriculated at the University of Texas in the fall of 1912. But politics was not on K. L.'s mind as he became a student. Playing sports, not academics, became his focus. He excelled at sports but needed "much prodding by various and sundry ones, to stay eligible" academically so he could play football and participate in wrestling and track.[17]

1914 University of Texas Football Team. K. L. is top row, far right, in the 1914 UT football team photo. Source: University of Texas 1915 *Cactus* yearbook.

Several editions of the University of Texas yearbook, the *Cactus*, featured K. L. Berry's athletic prowess. The 1913 edition summarized the 1912 football team, noting that older brother Gene Berry was one of the reserves and second men of the 1911 football team to return in 1912. The summary continued, "And the freshmen! They were plentiful. Among the most promising were Littlefield, Higginbotham, Berry and Brown." The Berry brothers were featured in the football team photos and were listed as "Wearers of the 'T,'" meaning they both lettered in football in 1912.[18]

However, in the 1913 football season, there was no mention of K. L in the yearbook. Gene appeared in photos, on rosters, and in summaries of team games. Where had the "promising" player from 1912 gone? Clearly, K. L. had not played that season. As noted earlier, K. L. needed prodding to keep his grades up so he could play. Apparently, he was not prodded enough to be eligible for the 1913 season.[19]

K. L. came back for the 1914 football season stronger and better than ever. He was now 6 feet and 180 pounds playing right tackle. The *Cactus* noted that K. L. "pushed opponents right and left and opened up wagon roads for the backfield men." Apparently, K. L. was also "one of the few men in the line who could receive forward passes."[20]

This 1914 football team that K. L. played on became one of the most notable teams in the university's history and was featured after the turn of the century in an exceptional exhibit called *A Perfect Season* at the H. J. Lutcher Stark Center for Physical Culture and Sport at the University of Texas. Visitors to the exhibit learned that the 1914 football team set the stage for future UT football "excellence and prestige" by achieving a "points per game record that stood for 91 years."[21]

For his part, Berry "played a consistent game and was a sure tackler. He was alert, active, and strong on offense. When it came to running interference, he could not be beaten. His ability as a tackler was recognized when the officials selected him for a place on the All-State team."[22] K. L. had a terrific year, winning both All-State and All-Southwestern honors for the right tackle position and being chosen to lead the Longhorns as captain for the 1915 season.

As captain of the team in 1915, K. L. was described as "probably the most powerful man on the team . . . charging and holding the line."[23] Even though an injury kept K. L. out of a few games that season, he was thought of as a good leader by his teammates, one who could instill fighting spirit into the team. They had great confidence in him throughout the season. Those qualities would be abundantly evident in K. L.'s future life and contribute to his successful career in the military.

As if football were not enough, Berry joined the new collegiate sport of wrestling in 1915. This first team consisted of five men who played in only one match and won. The *Cactus* reported, "Berry downed his opponent with a cradle hold after two minutes and forty-five seconds of fast work."[24] The wrestling team grew to forty members the next year.

In spring of 1915, K. L. participated in track. Following in his brother Gene's footsteps, he threw the discus, shot put, and hammer. "One state record was broken and other high marks made in the 1915 class track meet, one of the most successful ever held here—K. L. Berry won the greatest number of points—K. L. Berry Shot Put 35.3 feet, Hammer Throw 97.8 feet."[25] In the Southwestern Intercollegiate meet later in the season, he established a conference record in the shot put of forty-two feet and five inches.[26]

Continuing sports the next year, K. L. was inducted into the prestigious Sigma Delta Psi Honorary Athletic Fraternity along with his good football and track buddy Clyde Littlefield. The story in the *Cactus* noted, "Berry, Captain of the 1915 Longhorns, leading heavyweight on the wrestling squad and holder of the State and Southwestern records for the shot put, is the third man to win his 'T' in three different sports."[27] K. L. was having the time of his life as an athlete. All the coaches were happy that he had one more year of college to play on their teams.

However, as K. L. Berry enjoyed his renowned career as a college athlete, world events loomed larger and larger. The war in Europe between the Central powers of Germany, Austria-Hungary, the Ottoman Empire, and Bulgaria and the Entente Nations of France, the United Kingdom, Russia, Italy, and Japan had been dragging on since August 1914. President Woodrow Wilson feared the United States would eventually be caught up in it. But closer to home, a Mexican outlaw named Pancho Villa had been making headlines for several months raiding towns along the Texas border and, in some cases, killing Americans. The Texas National Guard was called to active duty to protect the border. With family stories of Texans fighting Mexicans ringing in his ears, the college athlete Berry answered the call of the Texas Guard at age twenty-three. Football, track, and wrestling coaches at the university lamented Berry's departure. But, at long last, Berry was going to become a "soldier boy" as his mother had predicted he would be.

CHAPTER 2

Military Training and Football on the Border

ON THE BORDER

The National Guard enlistment program cost the University of Texas football team several of its stars. In May 1916, K. L., along with other football friends from the university, eagerly enlisted in the 2nd Texas Infantry to protect their state along the border. Border problems between Mexico and United States had been increasing since the Mexican Revolution began in 1910. Many Mexicans who had opposed President Porfirio Diaz's presidency had been fleeing Mexico for years to join Tejanos, Texas-born-people of Mexican or Spanish descent. The community of immigrants grew in many South Texas counties. However, their loyalties remained with Mexico. With revolution still on the minds of the new immigrants, it was not long before the Mexican communities began to protest about the problems the Mexican community had been facing for years in Texas after the Mexican-American War. They decried inequalities in education, land ownership, labor rights, and discrimination. Many disgruntled immigrants joined an armed movement called The Plan of San Diego. The plan urged people from South Texas and northern Mexico to join a movement to reclaim Mexican lands from Texas and the United States that had been lost in battles in 1836 and 1848.[1] This call to reclaim lost Mexican lands stirred up ire in the citizenry in the states along the border. The young K. L. Berry was similarly moved.

In 1911, Mexican President Diaz was ousted, succeeded by Francisco Madero. Madero's term was short-lived, however, as Victoriano Huerta, an opponent, forced him to step down in February 1913. Madero was later assassinated. Huerta declared himself Mexico's president, but the United States under President Woodrow Wilson refused to recognize Huerta's presidency. The Mexican community in South Texas aligned itself with the leaders of one of three factions—Victoriano Huerta, Venustiano Carranza, or Francisco (Pancho) Villa. Villa and Carranza joined forces against Huerta, forcing him to step down in 1914. The United States then supported Carranza's faction, recognizing his presidency in 1915. As a result, Francisco "Pancho" Villa, formerly an ally of Carranza, developed animosity toward the United States for supporting Carranza and not him. Villa and his followers began to attack and murder people along the border.[2]

When Villa and his men attacked the town of Columbus, New Mexico, in March 1916, President Wilson quickly reacted by charging Gen. John Pershing with the task of leading a "punitive" expedition from Fort Bliss into Mexico to subdue Villa's soldiers and capture Villa himself. After Villa's men attacked Glenn Springs, Texas, on May 5, President Wilson federalized the Texas, Arizona, and New Mexico National Guards on May 8 and ordered the guard to increase its forces and station them on the border. Troop strength at Fort Bliss, the military post near El Paso, Texas, increased to fifty thousand troops, the majority of whom were from the National Guard.[3]

Even though Berry joined the guardsmen too late to be with Pershing's soldiers, he was by June 1916 among the sixty thousand National Guard troops serving to protect the border. That number rose to an estimated 112,000 guardsmen by the end of July. Berry and the 2nd Texas Infantry were stationed, after several moves, at Camp Scurry in Corpus Christi under the command of Brig. Gen. John A. Hulen.[4]

2ND TEXAS INFANTRY

Writing to Harold Ratliff of the Associated Press in 1955, Berry, with twenty-twenty hindsight, mused that the large buildup of troops was motivated by more than a desire to keep peace on the Mexican border: "The 2nd Texas went to the border along with all the National Guard troops in the U.S. The border trouble was only a lame excuse to my way of thinking. President Wilson saw the big war coming and decided to get the Guard in as good shape as he could early and as cheaply as possible.... We lived in tents, drew 50 cents a day and our ration cost about 20 cents a day."[5] Wilson knew the war in Europe would entangle the United States sooner rather than later, and he was, in fact, building up the readiness of the guard.

Serving with the 2nd Texas Infantry, Private Berry spent long weeks in the humid summer heat of Texas marching in full gear to and from firing ranges. In September, Berry was part of a multi-states training exercise. Guardsmen from Texas, Kansas, Illinois, Missouri, and Wisconsin went on a fifteen-day, 166-mile march from San Antonio to New Braunfels. By the end of the summer 1916, these former civilians had become conditioned soldiers.[6]

Continuing his narrative to Ratliff in 1955, Berry said, "By sending us to the border he [Wilson] hardened us up and taught us something."[7] This early training with the National Guard during the summer of 1916 was an important touchstone for Berry. It demonstrated to him that he had the strength, perseverance, and fortitude as a soldier to undergo hardships, including those he would encounter on the infamous Bataan Death March twenty-six years later. As noted in the *History of the Texas Guard*, serving was "a great opportunity for these units, as it enabled them to perfect their training, and before their tour of duty was over, they were well disciplined and trained."[8] Indeed, they were prepared for the war Wilson saw coming.

As they trained, life on the border could be boring for the guardsmen. Berry wrote Ratliff, "We were far removed from Pancho Villa who operated around El Paso and Douglas, Arizona. We did exchange a few shots across the river, but this was caused by nervousness or plain cussedness."[9] While waiting for bandits to strike, the young soldiers' enthusiasm for service in the guard waned. Life in the fields was hard, and the drills, rifle practice, and patrols grew monotonous. They had signed up to fight an enemy at the border, but, at this point, there was no enemy near Corpus Christi. General Hulen experienced firsthand how monotony and boredom affected his men when 150 members of the 2nd Texas Infantry left camp on January 17, 1917, to protest in the streets of Corpus Christi, shouting that they wanted to go home.[10]

The monotony and boredom soldiers experienced when they were not training were not new problems for the military. Military leaders were beginning to recognize that sports of any kind relieved the monotony of daily life for soldiers. Consequently, sports were intentionally linked to the military's training curriculum after the Spanish-American War, explained historian Steven Pope, when "a younger, reformist generation of uniformed officers assumed a moral commitment to the soldiers' welfare and used sport initially to combat desertion, alcohol, and the lure of prostitution."[11] The military incorporated sports fully in 1916 during the border problems with Mexico, because otherwise the saloons and red-light districts were all that the one hundred thousand American troops had for entertainment. US Secretary of War Newton Baker believed that soldier athletic programs would "keep their bodies busy with wholesome, healthful, and attractive things."[12] More importantly, as more athletics were introduced into military training, the military could see that many athletic skills were similar to the skills soldiers needed in war. For example, wrestling skills helped men with hand-to-hand combat exercises; throwing baseballs gave men skills for throwing grenades; the gymnastic exercises of jumping, vaulting, and scaling readied men more ably for the similar movements in trench warfare; and boxing maneuvers were similar to the skills needed in bayonet fighting.[13]

Integrating athletics and organized sports into military training proved to be wildly popular; baseball, football, and basketball were considered to be the "big three." In 1917, football became a popular sport among soldiers on the border because many of them had played in college or even coached college teams. Berry remembered fondly, "When the Second Texas Infantry was ordered to the Mexican Border in May 1916 a great many college students enlisted for the 'War' and among them were quite a number of football players from Texas, A&M, and Baylor. Along about October the first, the football urge got the upper hand, so a team was organized."[14] Not only did the 2nd Texas organize a team, but almost all the other National Guard units on the border did as well. Competition among these teams grew in intensity with each passing month.

FOOTBALL

It is highly unlikely that Private Berry was shouting in the streets that he wanted to go home, as 150 of his fellow guardsmen had done on January 17. On the contrary, he was a soldier—training, watching for bandits, *and* playing football. Two of Berry's friends from the University of Texas football team, Bill Birge and Charlie Turner, played alongside him on the 2nd Texas Infantry team. In addition, star players from A&M were on the team. With such a lineup, this team became the best National Guard infantry team in the country.[15]

The 2nd Infantry scored 432 points to their opponents' 6 points. They dominated the army league. Those lonely 6 points were scored against the 2nd Texas in Austin on January 16, 1917—the day before the protest in the streets in Corpus Christi—by the 12th Division all-star team from Camp Wilson, near San Antonio, which was coached by future President Dwight D. Eisenhower. The 2nd Texas defeated Eisenhower's team 34 to 6. This game was the first time that Berry and Eisenhower crossed paths, although they did not meet. In the future, they would work together and become friends.[16]

The Reveille newsletter, which was "devoted to the Interests of the Texas Brigade and Other Troops Stationed at Corpus Christi," contained an article about a football game between the 2nd Texas and the Wisconsin Infantry. Texas won the game 60–0. Private Berry was the standout, as reported in this humorous anecdote:

> The man deserving mention as the star of the game, however, is K. L. Berry.... He distinguished himself especially by two long runs for touchdowns through a large number of tacklers and without interference. On both these runs, Berry showed the prettiest stiff-arm work we have ever seen. About to be tackled, he merely shot his open hand out to his would-be-tackler's head and ran along for yards holding the man off while he was making every effort to get to him. Then, both times, just as this man fell Berry successfully evaded two more men who were waiting for him and carried the ball across. One person remarked about Berry's stiff-arming, "he seems to do that with the same ease that he would eat a good meal."[17]

RETURNING TO CIVILIAN LIFE, ALMOST

Although Pershing's troops never captured Pancho Villa, the presence of the National Guard on the border deterred further raids by the Mexican insurgents. Pershing's expedition left Mexico in February 1917. President Wilson needed Pershing back in the United States to prepare for the war in Europe that Wilson feared the United States would be drawn into. The war had been ongoing since 1914 between the Central powers (Germany, Austria-Hungary, the Ottoman Empire, and Bulgaria) and the Entente Nations (France, the United Kingdom, Russia, Italy, and Japan). Wilson's foreboding came to pass when the Germans began unrestricted submarine warfare against the United States. It became impossible for the United States to remain a neutral nation.[18]

By early 1917, with the border problems with Mexico having subsided, all the Texas National Guard units were moved to Fort Sam Houston, where they trained until they were deactivated. Private Berry and the 2nd Texas Infantry were deactivated on March 23, and Berry, along with other guardsmen, left by train for Austin the next day. When the soldiers arrived in Austin, hundreds of citizens, family, and friends greeted them. As band music played, the guardsmen "marched up Congress Avenue to the Driskill Hotel where they stacked arms."[19] Waiting for Berry was Alice Celeste Fleming, a woman one year younger than he, whom he met during his college years when he boarded at her mother's rooming house. A happy Berry reentered the University of Texas to finish his degree right after he got home.

Wedding Day. K. L. Berry and Alice Fleming on their wedding day, May 7, 1917. Source: Author's private collection.

Around the same time that the Mexican border incursions were dying down, a provocative telegram from German Foreign Minister Arthur Zimmermann to Mexico was intercepted in late February. The Germans were seeking Mexico's alliance against the United States if the United States joined the war against Germany. In return Germany promised that Mexico would get territory in the United States. When the news of this telegram reached the public on March 1, there was intense scrutiny about the authenticity of the telegram. The Germans denied sending the telegram. Then, without prompting, Zimmermann admitted that he wrote the telegram. Once Americans understood that the telegram was genuine, public sentiment about getting involved in the war increased. A State Department official in El Paso, Texas, reported that Pancho Villa was amassing soldiers to assist the Germans, proclaiming that he would get Texas, Arizona, and California back for Mexico. Men like K. L. Berry, who had previously been protecting the border, became understandably incensed when this news broke. By April, Woodrow Wilson asked a special joint session of the US Congress to declare war on the Central powers.[20]

With war declared, highly qualified young officers were needed as soon as possible to train volunteers and draftees entering the army. For the action-oriented Berry, an opportunity presented itself that was too hard to resist. Recruitment for the US Army was taking place, and the noncommissioned personnel from the ranks of the Texas National Guard who had served on the Mexican border were offered the chance to attend the First Officers Training Camp (FOTC) at Camp Funston in Leon Springs, Texas, beginning May 8, 1917. Private Berry applied and was one of those selected to attend the training camp. He withdrew from the university the morning of May 7, rushed to marry Alice Fleming that afternoon at the University of Texas YMCA, and departed for FOTC in Leon Springs the very next day to become an officer in the Regular Army.[21]

FIRST OFFICERS TRAINING CAMP AT LEON SPRINGS

On May 8, several weeks after the United States declared war on the Central powers, the First Officers Training Camp opened at Camp Funston, later called Camp Stanley, in Leon Springs, Texas, located about twenty miles northwest of San Antonio. This first class of officer trainees became known as the "Ninety-Day Wonders," a name the commanders at Leon Springs had adopted from a training model developed by Maj. Gen. Leonard Wood in Plattsburg, New York. General Wood believed in turning civilians or noncommissioned men into junior officers in ninety days. Three thousand young men applied to become officers by the middle of August. Private Berry was ready for a grueling ninety days.[22]

Berry's good friend and fellow 1914 UT football star, Gus C. Dittmar, attended the camp at Leon Springs with him that summer. Dittmar recalled the apprehension of the men who accepted the challenge: "No group of men ever assembled under greater uncertainties. They had no knowledge of whether they would actually be commissioned should they complete the camp; whether they would be paid anything for their efforts; or in what capacity they might be used should they win a commission."[23]

Private Berry was assigned to 7th Company, while Dittmar found himself in 2nd Company. Another UT friend, Beauford Jester, was in 1st Company. Jester, a future Texas governor, would play a pivotal role in Berry's life after World War II. (Coincidentally, his daughter Joann would marry Berry's son, Thomas.)

In anticipation of a well-trained officer corps, President Wilson drafted the entire National Guard into US Army service for World War I on August 5, 1917, just two weeks before the Leon Springs candidates would graduate ready to train draftees. Out of the three thousand men who entered the camp at Camp Funston in May 1917, only 1,846 men completed the grueling ninety-day training and graduated. Berry was one of them. His athleticism and previous assignment with the 2nd Infantry had prepared him well for the rigors of the camp training. He was commissioned as a 2nd lieutenant in the Infantry Section of the Officers Reserve Corps (ORC) on August 15, 1917.[24]

On commission day, there were no formal ceremonies for becoming an officer. The orders listing the names and ranks of those commissioned were simply handed out to each graduate. The Leon Springs First Officers Training Camp bonded these young men for life. Years later, in 1931, they organized the First Officers Training Camp Association, and members gathered for annual meetings for more than fifty years. They called themselves and one another "First Camp Men." The experience for Berry was pivotal—he had chosen to be a soldier and was ready to make his mark as an officer.

PART II

Military Career

"It is a proud privilege to be a soldier—a good soldier . . . [with] discipline, self-respect, pride in his unit and his country, a high sense of duty and obligation to comrades and to his superiors, and a self-confidence born of demonstrated ability."

—*George S. Patton Jr.*

CHAPTER 3

Lieutenant Berry Begins His Military Career

PREPARING FOR WAR

When the United States declared war on the Central powers in April 1917, a larger national army was needed immediately. Voluntary enlistment programs did not bring in the numbers of men needed for the war in Europe. Consequently, the Selective Service Act was enacted by Congress in May 1917. The act temporarily allowed the federal government to expand the military by conscription. All American males between the ages of twenty-one and thirty—later expanded to age forty-five on August 31, 1918—had to register for the draft.[1]

Volunteers and draftees, who came from all walks of life, lacked military training. In Texas, Camp Travis near Austin became a designated training center for the new recruits and draftees. Newly minted lieutenants from officer candidate schools would do the training. Almost all of the new Leon Springs FOTC officers went to Camp Travis to train soldiers for war. Surprisingly, 2nd Lieutenant Berry was not among the new instructors at Travis. Records indicate that Berry was sent to San Diego, California, to join the 21st Infantry instead of going to Camp Travis.[2]

No records indicate why Berry was sent to Camp Kearney in San Diego. Perhaps the 21st Infantry, which had been deployed along the Mexican border in 1916 and 1917, knew of Berry's military and athletic prowess and recruited him to come to Camp Kearney where the 21st had been "given the mission of training Army units for deployment to France." Berry served with the 21st from 1917 to 1919, training draftees in California at Balboa Park, Camp Kearney, and Palm

City. In addition, he traveled to Fort D. A. Russell in Wyoming for a while to train troops there. Ultimately, the 21st Infantry trained thousands of replacement soldiers for units fighting in France during the war. The regiment itself never saw combat during the war even though it was ready for deployment at the time of the armistice.[3]

In October 1918, while at Camp Kearney, Berry's first son, Kearie Lee Berry Jr. was born. Family lore tells the story that Berry and Alice were disappointed that they could not get back to Texas in time to have their son born a Texan. Two weeks after Kearie's birth, the war ended. In early 1919, Berry was promoted to 1st Lieutenant (P—permanent) Regular Army. Proudly, he took his young family back to Texas to await new orders. He was unassigned for a few months, but by June his orders came. They were a surprise: 1st Lieutenant Berry was headed to Siberia.[4]

POST-WORLD WAR I IN SIBERIA

Lieutenant Berry's first overseas duty began on June 5, 1919, when he was assigned to the Siberian Presidio Replacement Detachment No. 1 with the 27th Infantry. The first deployment of the 27th Infantry departed from the Philippines to Siberia in 1918. Berry was sent in 1919 as one of the replacement troops. He arrived in Vladivostok, Siberia, on June 28 and joined one of two battalions of the 27th Infantry at Verkhne-Udinsk about 2,100 miles west of Vladivostok near Lake Baikal.[5]

Why part of the US Army was sent to Siberia is a part of US history rarely taught in schools. In February 1917, Russian Tsar Nicolas II, leader of the Russian Empire, which was allied with France and Great Britain in World War I against the Central powers, was overthrown by a Russian faction called the Mensheviks. The United States believed that the new government headed by Alexander Kerensky was capable of fighting and winning against the Central powers; consequently, the United States sent $1 billion of American military equipment to Siberia in the spring of 1917 to support Kerensky's government.[6]

But in November 1917, a second revolution took place in Russia, and the Bolsheviks or Red Russians, who would later become the Communist Party, overthrew the Mensheviks. Russia was thrown into turmoil. The Germans put pressure on the inexperienced, shaky Bolshevik government to sign the Treaty of Brest-Litovsk, which took Russia out of the war. However, within Russia the Bolsheviks were fighting to maintain control of the government against a faction called the White Russians, or Cossacks, who were not identified as socialists or communists. The international community, not wanting to see the Bolsheviks

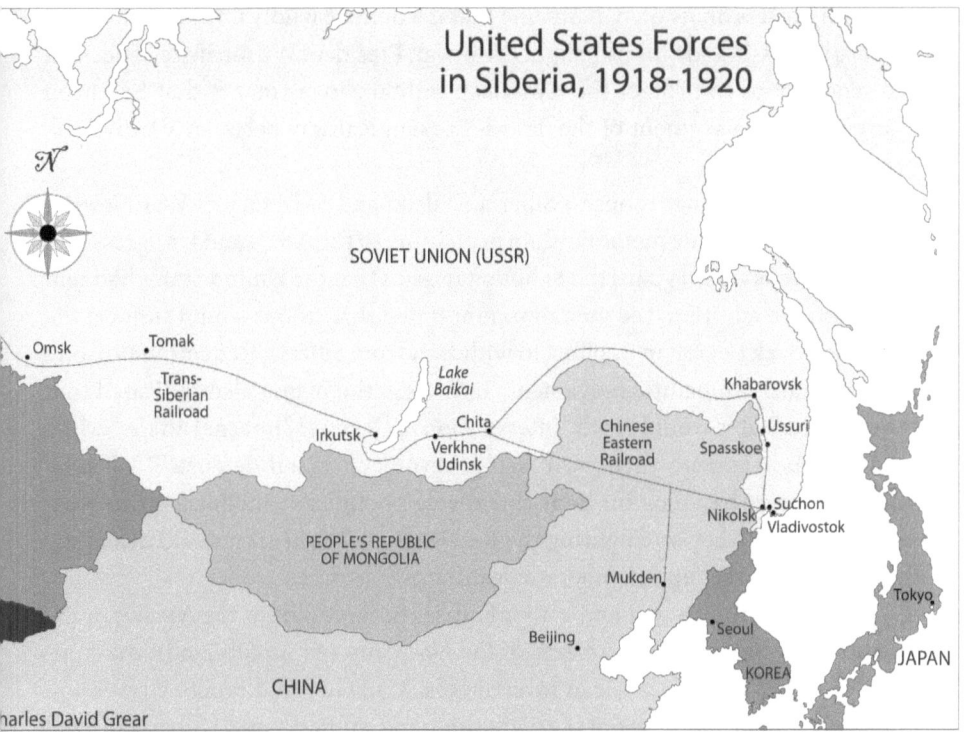

Map 1. Map of Siberia. The map shows the port city of Vladivostok and the city of Verkhne-Udinsk, where K. L. Berry spent most of his time while stationed with the 27th Infantry. Source: Map by Charles David Grear, 2024.

solidify their control in Russia, favored the White Russians and sent British, French, and Japanese troops to Russia to intervene. The Bolsheviks then found themselves fighting both the Cossacks and Allied forces. Meanwhile Germany, with no Russian enemy to fight to the east, moved numerous divisions to France, and by the spring of 1918 they were within striking distance of Paris. Allied nations then faced the entire German army closing in on Paris.[7]

As a result, France and Great Britain wanted to create a diversion that would force Germany to fight a two-front war once again. They pushed the United States to send troops to Russia and Siberia to assist the Czechoslovak Legion, a large force of Czechs and Slovaks living in Russia who had been fighting against the Central powers. When Russia withdrew from the war, the legion wanted to join an autonomous Czechoslovak army headquartered in Paris that had been formed in February 1918. But if the legion moved west, it would be fighting the

Central powers on its own. Knowing that it would be wildly unpopular to send troops to Russia and Siberia to fight in the war, President Wilson instead decided to send troops to protect the American military investments that had been "strewn along a segment of the Trans-Siberian Railway between Vladivostok and Nikolsk."[8]

To justify moving troops to Siberia, Wilson and Secretary of War Newton Baker crafted a vague memorandum explaining to the Allies and Congress that the US troops would guard the $1 billion in arms that the United States had sent to Russia. In addition, the memorandum stated that troops would support the Czechoslovak Legion in its effort to withdraw from Siberia. Reflecting Wilson's desire to aid democratic movements in Russia, the memo also indicated that the United States would avoid "intervention in [Russia's] internal affairs" while aiding Russians with their own "self-government or self-defense."[9] Wilson's decisions were based on his hope that the democratically inclined White Russian Cossacks who were fighting the Red Russian Bolsheviks would defeat the Bolsheviks and set up a democratic-leaning government.

Consequently, the 31st and 27th infantries became part of the American Expeditionary Force (AEF), an arm of the US Army sent to Siberia in the summer of 1918 to guard American investments. Maj. Gen. William S. Graves was hurriedly put in command of the eight thousand soldiers sent to Siberia instead of being sent to Europe to fight against the Central powers. Secretary of War Baker handed General Graves his orders, saying, "This contains the policy of the United States in Russia which you are to follow. Watch your step; you will be walking on eggs loaded with dynamite. God bless you and good-bye."[10]

Graves recognized very quickly that there was not a real strategic plan in place for him and his troops, so he decided to keep his troops as neutral as possible, protecting the United States' assets. Adding to the complexities faced by Graves and his troops, an unstated goal of Wilson's came to light. He was not happy with the Japanese expansion into Manchuria and western Russia after the Japanese defeated the Russians in 1905. Wilson hoped that the presence of American troops would stop the Japanese from gaining any more ground in the area. Unclear about his mission, Graves decided to focus on protecting access to the railroad to ensure that the Czech Legion could withdraw safely from Siberia and that American military assets would not fall into the hands of the Bolsheviks.[11]

Upon his arrival in Siberia in the fall of 1918, Graves deployed his two regiments in different locales. The 27th Infantry was split up—half of the 27th went to guard railroad depots and equipment caches along the stretch of railroad from

Wolfhound Friends. Lieutenant Berry is sitting in the front row, second from left, with his new Wolfhound friends in 1919. Source: Author's private collection.

the Trans-Baikal region approximately 2,100 miles west of Vladivostok (in the Verkhne-Udinsk and Irkutsk areas); the other half was sent to Khabarovsk on the Amur River, about 150 miles north of Vladivostok.[12]

In another area, north and east of Vladivostok, the 31st Infantry was assigned to keep the coal mines open and working. Stationed in small towns, the 31st guarded the roads and the railroad to the Suchan Valley coal mines. In June 1919, the Bolsheviks, moving to seize control of the railroad near the coal region, ambushed members of the 31st, killing twenty-nine Americans. With hostilities initiated by the Bolsheviks, General Graves dropped his stance of neutrality and ordered the 31st into combat against the Red Russians. In the monthlong battle, members of the 31st killed approximately five hundred of the enemy.[13]

As the 31st was engaged in battle with the Bolsheviks near the Suchan Valley, Lieutenant Berry arrived in Vladivostok, Siberia. He was part of the replacement contingent for the 27th Infantry troops who were going home after a year in

Siberia. Berry was posted with the 27th troops at the Verkhne-Udinsk sector (a sector about one hundred miles long) near beautiful Lake Baikal, the largest and deepest freshwater lake in the world. The 27th Infantry was supposed to impose tighter control of the railroad so that local merchants and railroad personnel would have protection from the Bolsheviks and Cossacks who were impeding the trains as they fought each other for control of Russia.

Berry learned that the 27th Infantry had become famous during their first year in Siberia for their pursuit tactics of the Bolsheviks as they carried out their assignment to protect US stockpiles at the railroad depots. The Bolsheviks, impressed with the Americans, nicknamed them the "Wolfhounds," a name that became a symbol and emblem of the 27th Infantry henceforth. Berry became a Wolfhound with a reputation to keep.[14]

On his journey to Verkhne-Udinsk in a dilapidated and drafty train car, Berry took photos of herds of Siberian camels that he saw in Manchuria and Siberia. He wrote on the backs of the photos for Alice: "We cross the Gobi Desert on the trip out here and they [camels] are the chief means of transportation in that section. Verkhne-Udinsk is one terminal for a large caravan route to Peking across the Gobi and camels frequently visit out here." Another photo showed Manchurian men riding "bull carts" for their transportation—wooden carts pulled by bulls. Berry noted, "Everything here is 100 years behind the times."[15]

When he arrived in Verkhne-Udinsk, Berry took photos of Russian refugees at the railroad station and wrote Alice on the backs of the photos, noting "there are thousands of these [refugees] living in box cars along the Trans-Siberian RR." The refugees had fled the war zones and were also fleeing from the Bolsheviks. Local peddlers were everywhere in the towns and around the train station, and Berry interacted with them, learning Russian at the same time. The landscape around Verkhne-Udinsk captivated young Berry, who wrote home to Alice, "I believe that some of the prettiest scenery in the world is found around the lake [Lake Baikal]." He dreamed of taking Alice on a train ride from Vladivostok to Petrograd "after things settle down." In his free time, Berry made his way to Irkutsk, a city west of Verkhne-Udinsk, and took photos of the second oldest Orthodox church in Siberia. He wrote on the back of that photo that it "is some classy looking place from the outside but like most Russian buildings it hasn't been taken care of."[16] Berry's interest in and efforts to learn about Siberia suggest that his curiosity transcended a narrow view of his role as a soldier.

Quickly, Berry noted that the White Russian Cossacks whom the Americans were supporting were far from unified in their goals. Adm. Alexander Kolchak had seized power from Cossack leader Grigori Semenov in Omsk (map 1) and

Вице-адмиралъ А. В. КОЛЧАКЪ.

Russian Adm. Alexander Kolchak. Source: Author's private collection.

declared himself supreme ruler of Russia. For a time, international leaders recognized Kolchak as the White Russian leader instead of Semenov, because Semenov was too closely aligned with the Japanese who were moving into Manchuria. Despite international support enjoyed by Kolchak, Semenov continued to fight for leadership. In his quest for leadership, Semenov's troops gained the grim reputation for rape and murder of civilians.[17]

Interestingly, Berry had the opportunity to meet Kolchak at his headquarters in Omsk, which was not far from Berry's station at Verkhne-Udinsk. Berry sent a photo of Kolchak home to Alice, writing on the back of the photo: "And this is Admiral Kolchak on whom the Allies depended so much. He is having a hard time of it. This isn't a good picture of him however as he is a stronger looking man than this picture shows him to be."[18]

Berry also met Kolchak's rival Semenov, the other Cossack leader. On the back of a photo of a train in Adrianovka, near Omsk, with an arrow pointing to a train car, Berry wrote with surprising nonchalance, "This is a view of Semenoff's armored train showing his tea garden. I have seen this several times and on my first trip up here had the commander [Semenov] over in my car for dinner. He seemed a real nice fellow but kinda bloodthirsty. He had just killed 43 Red prisoners." In another photo, Berry wrote that the people lined up against the railcars were Bolshevik prisoners, "targets for bayonets or rifles."[19] Perhaps these were the forty-three prisoners Semenov killed? Semenov's bloodthirstiness was in part to blame for the downfall of the international community's recognition of Kolchak as the Russian leader.

Both the White Russian Cossacks and Bolsheviks, locked in internal war, had early on recognized that the American stockpiles of military equipment, which included "barbed wire, cars, trucks, tools, weapons and ammunition could help them in their struggle against each other." Berry and his Wolfhound comrades fought off small Bolshevik bands along the railroad all that summer in the vicinity of Lake Baikal near Verkhne-Udinsk. But by fall 1919, in one of the ironies of military intervention in the complex politics after the Russian Revolution, the 27th found itself also defending American military assets from *both* the Bolsheviks and the Cossack bands ruled by Semenov. In the end, those assets had dwindled severely from looting by the Bolsheviks and Cossacks.[20]

Kolchak could not control Semenov. As a result, the Allies had become disenchanted with Kolchak's military and political potential. Relations between the American troops and the White Russians deteriorated. The Cossacks under Semenov were murdering the local population, giving all White Russians a bad reputation. General Graves wrote that they "were roaming the country like wild animals, killing and robbing the people."[21] Locals were running to the Bolsheviks for safety, and Americans were portrayed as the enemy because they were aligned with the White Russians. Morale among US soldiers was low. They did not know what they were doing there. Subsequently, soldiers from the 27th Infantry, out of frustration with Semenov's Cossacks, attacked and captured one of the heavily armored trains held by Semenov's men, thus straining relations between so-called allies to the breaking point. Four of Berry's comrades from the 27th Infantry died in the raid.[22]

By November, Kolchak's weak government collapsed. He sought assistance from the Czech Legion, but they handed him over to the French General Janin, who was the officer in command of the Allied troops at Irkutsk. Janin ended up turning Kolchak over to the Bolsheviks. At about the same time, President

Above, Semenov's Tea Car. Source: Author's private collection.

Left, Bolshevik prisoners loading onto the train. Source: Author's private collection.

Wilson ordered all American troops out of Siberia. Berry and his fellow infantrymen at Verkhne-Udinsk started moving toward Vladivostok to go home in February 1920. That same month, the Bolsheviks executed Kolchak by firing squad in Irkutsk.[23]

Berry shipped out from Vladivostok on the USS *Thomas* on March 10, 1920, bound for the Philippines, where the 27th Infantry had been at the outset of World War I. General Graves sent the last of the 27th to the Philippines on April 1, 1920. The expedition in Siberia cost the army 189 men who died from sickness and fighting. Graves later wrote, "I was in command of the United States troops sent to Siberia and I must admit, I do not know what the United States was trying to accomplish by military intervention."[24]

If the Siberian expedition had been a failure, the experience of chasing off bands of Bolsheviks in the vastness of Siberia provided Berry with some fighting experience. In July 1920, while posted in the Philippines with the 27th Infantry, 1st Lt. K. L. Berry, age 27, was promoted to captain, Regular Army, because of his service in Siberia as a "Wolfhound."

CHAPTER 4

New Posts, Athletics, and Promotions

PHILIPPINES, HAWAII, AND TEXAS

With the passage of the National Defense Act in 1920, the US Army was organized into three components—the professional Regular Army, the civilian National Guard, and the civilian Organized Reserves.[1] Captain Berry was part of the Regular Army, which was responsible for training the civilian components of the army in peacetime. The period between World War I and World War II saw Berry taking on several assignments with increasing responsibility to train Regular Army and civilian troops in the guard and reserves. For Berry, directing military athletic training programs was also part of those jobs. Since the time of the Spanish-American War, according to historian S. W. Pope, military training that included athletic training had grown "from a tentative experimentation with sports and athletics to its unqualified acceptance of them as essential elements of soldier training."[2] Military leaders saw sports and athletics as a way to instill "soldier" values of "obedience, citizenship, and combat," and they also believed sports would "repair class schisms and restore social order and patriotism to the nation."[3]

Berry had been an exceptional athlete both in college and with the 2nd Infantry on the border in 1916, and he had trained men for war; consequently, he was the right choice to lead military athletic training programs and build esprit de corps among officers and enlisted men in the period between wars. For Berry, a post that included both athletic training and military tactics was an ideal assignment.

Schofield Barracks. Captain Berry is fourth from the left in this 1921 photo of the Schofield Barracks championship basketball team. Source: Author's private collection.

When he was sent to the Philippines with the 27th Infantry after his post in Siberia, Berry became the athletic officer for the regiment from March 1920 until March 1921. His wife, Alice, and his young son joined him there. Their life was one of luxury in Manila because they had servants and cooks to do the mundane chores of life. Young Kearie had an amah to look after him, so Alice and Captain Berry could enjoy the many social events, including a ball at the Filipino president's Malacañang Palace in honor of visiting US congressmen and their families.[4]

After just one year, the life of luxury came to an end as Berry and family moved with the 27th to the Hawaiian Division. The division was stationed at Schofield Barracks, a base established in 1908 for the army's mobile defense of Pearl Harbor and the entire island of Oahu.[5] As in the Philippines, Berry was in charge of athletic training and participated in athletics as well. Berry's posting in Hawaii was relatively short. In July 1921, he returned to Texas and joined the 23rd Infantry at Fort Sam Houston, where he became commander of Company G of the infantry and the regimental athletic officer for the next three years. During his assignment at Fort Sam Houston, the Berrys added a daughter to the family.

Celeste Viola Berry, born March 5, 1922, was given her mother's middle name and her grandmother's first name. In July of the following year, Berry's father, Thomas Eugene, died from malaria at his home in Tennessee Colony, Anderson County, Texas, where he and Viola had lived since 1918. To honor Berry's father, they named their second son, born July 26, 1923, Thomas Eugene.

BACK TO COLLEGE 1924–25

In 1923, President Warren Harding died in office, and his vice president, Calvin Coolidge, became president. Coolidge won the office outright in the 1924 presidential election. During Coolidge's tenure as president, the country experienced a period of prosperity with "robust growth, rising wages, declining unemployment and inflation, and a bull market."[6] However, the period of prosperity was not built on a solid economic foundation and would not last. In the meantime, for several years, for many Americans life was good and lighthearted.

Berry experienced some of the lightheartedness as well, when he had the opportunity to return to the University of Texas as a War Department student in 1924 to finish his degree. He had not forgotten that when he departed from the university in 1916, he still had one year of eligibility left to play football, so perhaps with a twinkle in his eye at the unlikely age of thirty-one, he tried out for the football team upon entering the university. As a writer for *Longhorn T* observed, "Hundreds of 'T' men are watching with great interest the remarkable 'come-back' of K. L. Berry, captain of the Longhorn grid eleven in 1915.... K. L. has always taken excellent care of himself, and, now at age 32 [sic, he was 31], he is a marvel of endurance and stamina."[7] Not surprisingly, Berry made the team.

As it turned out, Berry did more than make the team that year. He made All-Southwest Conference, again, the oldest football player to earn that distinction. Memorabilia at the H. J. Lutcher Stark Center for Physical Culture and Sport in Austin reflect Berry's impact on the 1924 UT football team. Clippings from unidentified and undated newspapers, found in Alice Berry's scrapbooks at the Stark Center, are filled with accolades about Berry's prowess on the football field at age thirty-one. A few of those accolades were: "That K. L. Berry is my idea of a He-Man. We don't see many like him in these days," or, "Throughout his service in the army both during and after the war, he is still the same Berry, a man who can deal misery to any lineman."

By October, Berry found his photo on the front page of the *Fort Worth Record Sports* under the headline, "Five of 'Doc' Stewarts's Longhorn Football Aces." Once again, he was recognized as one of the stars of the university's football team. If anyone had any doubt about his ability to add value to the Longhorns,

Berry at Age 31. Berry played football on the University of Texas football team in 1924 and garnered All-Southwest Conference honors. Source: Author's private collection.

The Austin American-Statesman chose Berry and teammates Sprague and Marley to be on its "Honor Team" to be considered for the "All-Southwestern Eleven" football team. Sportswriter Lloyd Gregory wrote, "Berry is the outstanding guard in the conference. If there is a smarter football player in the conference, the writer has not seen him in action. Because of his football smartness and his great fighting heart and his ability to inspire his fellows, the writer has named "Cap" Berry captain of the mythical team."[8] As had been the case years earlier, Berry was recognized once again for his ability to inspire others.

To round out his last year of eligibility, Berry also participated in track, throwing the discus. As a mark of the years that had passed since Berry's previous matriculation at the university, Clyde Littlefield, Berry's buddy from the 1914 UT football team, was now coach of the track team. The new relationship between the two men must have worked fine because they remained friends for life. At the end of the year, Berry was awarded an athletic certificate from the university, which noted his success as an athlete in three different sports—football, track, and wrestling—for the years 1912, 1914–1916, and 1924–25.[9]

Berry enjoyed his return to college sports. Finishing his degree, however, was not in the cards. Berry's "Academic Certification" document from the UT registrar's office indicated that he attended the university from 1912 to 1916 and again from 1924 to 1925. The certification also stated, "This individual has no record of receiving any degrees at the University of Texas at Austin."[10] Berry had been studying engineering, but apparently obtaining the degree in engineering was not in his heart. On the other hand, his engineering knowledge would be put to important military use in World War II when during the Battle of Bataan, he taught his troops how to build strong fortifications for fending off the enemy.

FORT BENNING, GEORGIA, AND VERMONT

At the end of the 1925 academic year, Captain Berry served as Fort Sam Houston's Summer Training Camp athletic director before moving to Georgia to attend the Company Officers Course at Fort Benning Infantry School in the fall. Even at Fort Benning, football beckoned him. He played on the Infantry School football team that fall semester.

Upon completion of the Company Officers Course that year, Berry enrolled for the summer in a football coaching workshop at the University of Notre Dame. Becoming a football coach rather than a player appealed to him at this stage of his life. Returning to Fort Benning that fall, he served as second in command of the School Detachment and as the head football coach for one year. As head coach of the football team in 1926, Captain Berry was assisted by Maj. Dwight D. Eisenhower. The two men, whose teams had met on the football field in 1917, finally met each other. Although Eisenhower had a bad knee from playing football at West Point eleven years before the two met, Berry expressed his admiration that Eisenhower "suited out and demonstrated everything he was teaching his ball carriers." These two young officers became friends "during the long afternoons of practice and during a five-week road trip the team made to other Army posts."[11] Their assignment together as coaches was the beginning of a lifelong association that included football, the military, and politics.

The next year at Fort Benning, Berry continued his work as second in command of the School Detachment and became the line coach for the football team. At the same time, he turned to another sport and became a member of the infantry rifle team. He participated in national matches for three years, earning the rating of distinguished marksman.

STATE OF MISSISSIPPI
MILITARY DEPARTMENT
OFFICE OF THE ADJUTANT GENERAL
JACKSON

WILLIAM P. WILSON
MAJOR GENERAL
THE ADJUTANT GENERAL

18 March 1958

Major General Kearie L. Berry
The Adjutant General of Texas
PO Box 5218, West Austin Station
Austin 31, Texas

Dear General Berry:

    Of course, from the write-up under this picture which is attached we know you know who the third person is, as it is given in the first sentence. There will not be much trouble for those that know you in naming you as the number 2 man, we having heard you state that while on tour of duty at Fort Benning at one time you were with General Eisenhower, who helped to coach the football team.

    Will see you in Phoenix the middle of April.

    I am

Sincerely yours,

WILLIAM P. WILSON

1 Incl
Clipping from
Army Times, March 19, 1958.

KNOW THESE MEN? Second from right was Ft. Benning's football coach in 1922, then Maj. Dwight D. Eisenhower, of the 24th Infantry, now President of the United States. The other three men are unidentified. If anyone knows who they are contact The Journal or Sfc. David W. Chase, U. S. Army Infantry Center Museum curator, at Ft. Benning. This picture was among a historical collection about Ft. Benning owned by the late Maj. Gen. Charles S. Farnsworth, who commanded the post in 1919. The collection has been presented to the Infantry Museum by Gen. Farnsworth's son, Robert J. Farnsworth of Altadena, Calif. Anyone who wishes to contribute or lend items to the museum that portray the history of the Infantry should contact Sergeant Chase.

I wrote Sgt Chase, identifying "K"

Fort Benning Football Coaches. Captain Berry is the tall man with the white shirt and dark pants standing next to Maj. Dwight Eisenhower, who is wearing an army sweater and a cap. They were coaches together at Fort Benning, Georgia, in 1926. Berry's friend William Wilson sent him the clipping in 1958, telling Berry he wrote the *Army Times* and identified Berry. Source: Author's private collection.

Col. George C. Marshall arrived at Fort Benning that same year and revamped the instructional system to include research into other fields, which led to the development of machine gun techniques, especially antiaircraft firing.[12] In 1928–29, Berry took the Field Officer Course revised by Marshall, and on completion of the course, he was sent to the University of Vermont as assistant professor of military science and tactics (1929–30) for the Reserve Officers Training Corps. As with many of his other military assignments, Berry was involved in sports as the freshman football coach as well as the varsity basketball coach for the university. During the summer months, he served as assistant coach of the infantry rifle team at Camp Perry in Ohio. He was happy and doing well in his chosen career.

In 1928, surprising most people around him, President Coolidge decided not to run for president again. Republican Herbert Hoover and Democrat Al Smith faced each other, with Hoover winning the election. Hoover had promised to continue the prosperous times with lower taxes. But that was not to be. Several months into the Hoover presidency, the stock market collapsed, marking an end to the 1920s era of prosperity and the beginning of the Great Depression. There were many factors that led to a frail economy in the late 1920s, including overproduction in the farming and industrial sectors; an international economy that was not strong after World War I; and excessive speculation in the American stock market and other financial sectors. Whatever the causes, Hoover's policies failed to stop the slide of the stock market and the growing depression. Investors lost billions of dollars, and unemployment surged. Unemployment rose from seven million in 1931 to eleven million in 1933.[13] The Great Depression was the central issue in the 1932 presidential election. Hoover and his policies lost to Franklin D. Roosevelt and his New Deal.

As a thirty-six-year-old army officer, Berry was not affected by the rise in unemployment and was insulated from the worst effects of the Great Depression. In 1930, after his assignment in Vermont with the ROTC ended, Berry was transferred back to Fort Benning to be an instructor for the men of the 24th Infantry, an African American unit. Only part of the 24th Infantry—the Regimental Headquarters Company, Service Company, and Company D—were sent to Fort Benning in 1922.

At that time, the army had been granted congressional funds to build Camp Benning into a true military training fort. When the 24th Regiment arrived, its men became largely responsible for converting the camp into a fort. Aside from building the fort, they became famous for their band, which played concerts for all citizens at Fort Benning. They even used their band to cheer on their fellow infantrymen who were constructing the famous Doughboy Stadium at night.

With segregation still the official policy of the army, the African American soldiers who played such an important part in the construction of Fort Benning were excluded from using many of the facilities they helped build. Construction of Fort Benning continued into the 1930s as the army received funds for construction under federal works projects that were started during the Great Depression. Little is known about Berry's three years as the unit's instructor. The record is also silent about his thoughts on the discrimination his trainees faced. Nevertheless, Berry became the coach of the 24th Regiment's infantry rifle team in 1931 and captain of the team in 1932. Under his coaching, the team performed extremely well in matches against other infantry rifle teams.

THE 15TH INFANTRY IN CHINA

When the rifle team was no longer funded in 1933, Berry was posted to an exotic assignment in Tientsin (Tianjin, today), China, with the 15th Regiment. Military service in China had become a preferred overseas assignment in the 1920s,

24th Infantry Small Bore Rifle Team, 1933, Fort Benning, Georgia. Captain Berry is sitting in the middle of the front row. Source: Author's private collection.

so the Berry family was understandably excited for the transfer to Tientsin, where they would be exposed to Chinese culture and artifacts, far removed from the economic woes of the United States. The Berrys left on the USS *Grant* from San Francisco on May 9, 1933, and arrived in Chinwangtao (Qinhuangdao, today), China, a port city near Tientsin, on July 6, 1933, Captain Berry's fortieth birthday. Berry became the commander of the machine gun unit assigned to the regiment. He was well-suited for the job, having taken Colonel Marshall's class at Fort Benning in the latest techniques of machine gun warfare. In addition to those duties, he served as the athletic and recreation officer. His three-year posting with the 15th Infantry Regiment would afford the family many luxuries and adventures. It would also take them into the heart of Japanese aggression, which had been growing in the Pacific sphere.[14]

As in Siberia, Berry found himself in Asia on a military assignment that was relatively unknown to the American public. Some background is warranted. Throughout its history in China, the 15th Infantry, nicknamed the "Can Do" regiment, had helped keep access to Peking (Beijing, today) open by protecting the railroads. Tientsin, with its strategically placed railroad center, was the gateway to Peking because of its close proximity to the China Sea and the port city Chinwangtao. All major nations wanted to keep their access to Peking, which was seventy-five miles inland from Tientsin. Through a series of treaties in the late 1800s and early 1900s, several nations had established concessions—including privileges, offices, and housing—in Tientsin and other Chinese cities, ensuring access to the Peking market.[15]

In 1900, a group of Chinese calling themselves "the Boxers" rebelled against foreigners, imperialists, and Christians in the Tientsin-Peking area. They threatened to shut off access to Peking for all the concessions. The 15th Infantry was sent to China from the Philippines to assist with the defense of Peking and Tientsin and to keep the railroad open to Peking. The Eight Nation Alliance, made up of Russia, Britain, the German Empire, France, Japan, United States, Austria-Hungary, and Italy, recaptured Tientsin a month after the rebellion and stabilized the city.[16]

After the rebellion, many alliance nations established military garrisons within the city to safeguard their concessions and to ensure access to Peking. The United States, however, opted to withdraw the 15th Infantry. Several years later, the regiment was recalled to duty. In 1911 and 1912, after the reign of Qing dynasty ended, access to Peking was again threatened by several Chinese factions. In response to this internal upheaval, the 15th was again sent to Tientsin.

Guarding the railroad and protecting access to Peking became the infantry's responsibility once more.[17]

Over the course of the previous four decades, Japan had succeeded in gaining territory from both China and Russia, boosting its status as a major force in Asia. After the Sino-Japanese War of 1894–1895, China could not maintain its domination of Korea, and Japan gained control of Taiwan. After the Russo-Japanese War of 1904–1906, Japan solidified its control of Korea and much of South Manchuria, including Port Arthur and the railway that connected it with the rest of the region. Then, in 1915, China signed an agreement with Japan called the Twenty-One Demands, which granted Japan the former German railway and mining claims in Shandong province and extended Japanese leases in southern Manchuria for ninety-nine years. Japan thus succeeded in negotiating and intimidating its way into a powerful position in China. Recognizing the threat to American interests posed by the Twenty-One Demands treaty, President Wilson feared further penetration into China by Japan. He hoped that the presence of US troops in Siberia from 1918 to 1920 would keep the Japanese from gaining any more ground in China. But this effort did nothing to stop the Japanese from moving farther into China.[18] By 1931, Japan had spread its control in China by seizing Mukden in Manchuria, setting up a "puppet state," and renaming Manchuria Manchukuo.[19]

When Berry and the rest of the 27th Infantry left Vladivostok for the Philippines in 1920, the 15th Infantry was well established in Tientsin. At that time, the 15th consisted of one thousand officers and men organized in three battalions. In 1929, some of the rifle companies had been deactivated because of lack of funds. When Berry arrived in 1933, the 15th was two companies smaller than the original one thousand men.

Captain Berry and family were among the last of the "Can-Do" regiment in Tientsin. When they arrived in Tientsin in 1933, Japan and China had signed a truce by which Chinese troops withdrew south of the Great Wall and Japan moved troops into the area thus vacated. By 1934 Japan had forced Chinese troops out of the city of Tientsin. By this time, the executive officers of the 15th Infantry were questioning the value and purpose of the regiment, which had become too small to do anything more than protect American lives and property in the garrison.[20]

For the Berrys, though, regimental life was good. They lived in a rented home in the western quarter that had been arranged by the regiment. All of the houses were red brick with three bedrooms and two bathrooms on the second floor. The first floor had living and dining spaces, a kitchen, and quarters for eight servants. The Berrys appreciated and valued Chinese artifacts and rugs, which they collected during shopping expeditions in and around Tientsin and Peking. A favorite spot for the children was the German ice cream parlor, Keislings, in Tientsin.[21]

Kearie Lee was 14, Celeste 11, and Tom 9 when the Berrys moved to China for three years. They attended the Tientsin Grammar School, governed by the Anglican Church of England, where they took Chinese language classes. Within the concession areas, the Berry children were shielded from the political turmoil between China and Japan. Tom's friend Desmond Power remembered: "The Chinese and Japanese might be locked in mortal combat, but for foreigners in the sanctuary of their concessions, it was life as usual. In school, the same old morning assembly, the same thumping piano, the same march to the high-vaulted hall, the same formations facing the Headmaster—girls to the right—boys to the left."[22]

Each summer, the Berry children and their mother, Alice, would retreat to Chinwangtao to enjoy the cooler temperatures and beach life. Tom Berry, in a letter to a friend, reminisced about summers at Chinwangtao, saying, "All families of all officers and married non-commissioned officers would leave Tientsin as soon as school was out and move to Chinwangtao for the summer. There were little cabin houses well screened in for the families to stay."[23] He went on to explain that the weather in Tientsin during the summer was insufferably hot, up to 105 degrees with no air conditioning, when the weather in Chinwangtao stayed in the "balmy 75–80 degree" range. All the fathers worked in Tientsin during the summer suffering from the heat while the children and their mothers lived happily by the sea.

It is interesting to note that young Tom remembered Chinwangtao as a nice place to summer. Travel bureaus, too, popularized Chinwangtao as a magical place to visit with fanciful descriptions, advertising that "tiny maidens play moon guitars under a great moon pouring silver on tantalizing images and pigmy pavilions."[24] However, a writer for *The Sentinel*, the weekly newspaper for the 15th Infantry, apparently disagreed, describing Chinwangtao as grimy, filthy, and smelly. Seventy thousand people were "crowded into vermin infested and unsanitary hovels, with dope dens" lining the main street.[25]

But for the army children like Tom, life in the army camp on the beach was safe and carefree. Jean Speece, a classmate of the Berry children at the grammar school, remembered, "Right after breakfast each morning, about 10 to 15 teenagers would go to the stables where members of the Cavalry assigned saddle horses to each of us. Then we would be off for a ride through the Chinese countryside and villages." She continued, "From early morning until the setting of the sun, the beach offered endless hours of entertainment [as] a soldier life-guard in his tower on shore watched us as we braved the undertow, jellyfish and deep water." On rainy days, the teenagers would ride into Chinwangtao and back to their camp with the mail truck just for the adventure.[26]

During these carefree summers, families took trips to resort towns where they could buy handmade linens and other Chinese artifacts. On occasion, they visited the Great Wall of China, where the young teenagers were alarmed to see "ragged, sore-ridden, crippled and dirty beggars" asking for money.[27] At summer's end, the Berry family traveled by train back to their lives with Captain Berry in the 15th Infantry garrison in Tientsin.

Like his children at school, Berry had to learn Chinese; in fact, everyone associated with the 15th was mandated to learn Chinese. Classes were held several times a week, and troops had to pass a language test. The majority of officers learned the language well enough not to have to use interpreters. Berry, with his additional duty as advisor to the Chinese military, was able to interact in Chinese when he met with dignitaries. As the regimental athletic officer, Berry proved to be excellent. He succeeded in what no other athletic officer had done in Tientsin. Under his supervision and coaching, the track and field squad won the annual North China International competition, which was a notable achievement for the 15th Infantry. Berry even participated in the discus throw, coming in third. Col. R. J. Burt, the commanding officer of the 15th Infantry, wrote that this victory was "the greatest accomplishment of any such US Army team at any time in North China."[28] News of the win was even heard in Washington, DC, where Col. Karl Truesdell, former executive officer of the command, radioed Berry from the War College "congratulations" for the tug of war team victory.[29] As he had so many times before, Berry inspired his men to do their best.

As it had been in previous postings, athletics was a powerful means of maintaining fitness and discipline while also breaking the monotony of daily regimental life. Shortly after his big win, Berry was promoted to major. In 1935, Berry's oldest son, Kearie Lee, age 16, participated in the army track events alongside his father. Major Berry threw the discus, and his son ran hurdles and sprints. Several issues of *The Sentinel* mentioned young Berry placing in several events. People were intrigued by the Berrys because they were the first father-son duo to compete in the same army track meets.

In his leisure time, Major Berry liked to go on hunting expeditions in Mongolia. On one such expedition to the province of Shansi, he and his party rode

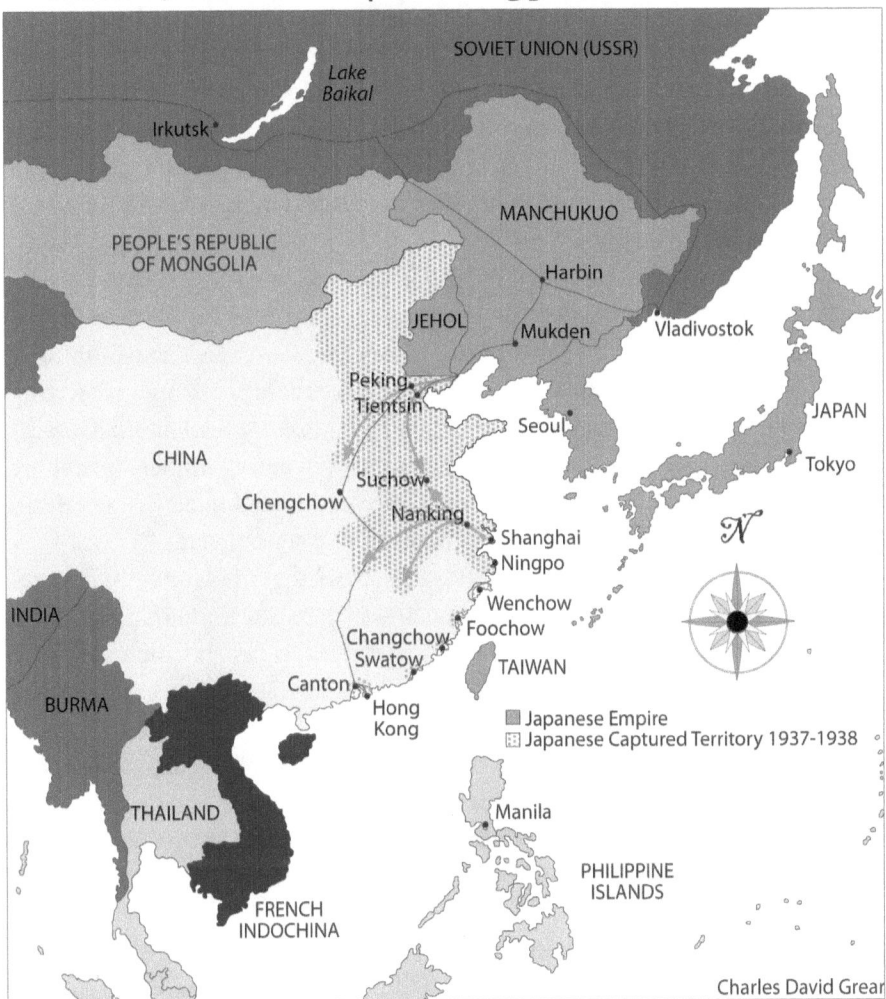

Map 2. Japanese Aggression. In the map, it is clear to see that Tientsin was a gateway city from the sea to Peking (Beijing). The railroads also met in Tientsin, and the Japanese needed these as they went north to Manchukuo, Japan's "puppet" state. The Berrys were in Tientsin from 1933 to 1936 during the time of Japanese aggression. Notice how close Major Berry was to Vladivostok, where he had been in 1919–1920. Source: Map by Charles David, 2024.

a bus from Tai-Yuan-Fu to the village of Kai San. From there, bringing pack mules and saddle mules, they hiked for nine hours the first day and kept up the hike for two more days. On the third day of the hike, Berry sighted two deer and tried to bring them down but failed. He decided to return to their camp when he saw a leopard near a mountain slope. Quickly, he brought the leopard down. When they returned to Tientsin, Berry was considered the big winner of that expedition.[30] Today, the leopard skin from that hunt is in the possession of Berry's grandson Jim Berry.

Contributing to the sense that life in Tientsin was safe and pleasant, *The Sentinel* almost never reported on the Japanese aggression taking place in China, even though the Japanese were getting closer and closer to Tientsin in 1936. On the contrary, *The Sentinel* reported on new Japanese commanders arriving in Tientsin not as ominous signs of coming Japanese aggression but as social events. New Japanese arrivals were treated to welcoming dinners hosted by Commander Burt and then Commander Lynch of the 15th Infantry. Major Berry even gave Japanese dignitaries tours of the athletic department.

Disrupting this comfortable life, rumors arose that the United States was considering pulling out of Tientsin. A small article in *The Sentinel* in April 1936 denied the rumor that "American troops stationed in North China would be withdrawn."[31] As tensions grew between China and Japan, however, keeping a small US regiment with their families in Tientsin became harder and harder to justify. By July 1936, *The Sentinel* reported that many "old-timers" were going home on the USS *Grant*. Major Berry was one of them. He and his family left China in August 1936, setting sail for San Francisco and then on to their eventual destination, Texas, where Berry would be assigned to prepare soldiers for the coming world war.[32]

It was not until the next July that the entire 15th Infantry was pulled out of China. Desmond Power wrote, "The colorful 'Can Do' Regiment, the 15th US Infantry after twenty-six years in Tientsin, was pulling up stakes and going home."[33] The Japanese had taken possession of most of China's ports, the majority of its major cities, the larger part of the railways, and, by July 1937, occupied Peking and Tientsin.

CHAPTER 5

War is on the Horizon

OFFICER INSTRUCTOR BERRY

As Japan was planning to dominate in the Pacific region, Germany began to shake off its weak democratic government structure established by the Versailles Treaty after Germany's defeat in World War I. Its emerging leader, Adolph Hitler, was making plans to dominate Germany and then Europe. Steadily, Hitler's Nazi Party won more and more seats in the German Reichstag. By 1932, the Nazi Party was the largest political party in the Reichstag. And, in 1933, Hitler became the chancellor of Germany, ensuring the demise of democracy.[1]

President Franklin D. Roosevelt, aware of the increasing likelihood that the United States might be—however reluctantly—drawn into a future war, changed the status of the National Guard in 1933, making it a reserve component of the army in peace time as well as war. In 1936, after Japan and Germany signed a pact wherein they agreed to share intelligence and stop the growth of communism, there was a heightened sense of urgency in the United States to get more instructors in place to teach military tactics and strategy for the growing National Guard. When Major Berry returned to Texas from China in 1936, he was assigned to the 141st Infantry, the Alamo Regiment, which had become part of the 71st Brigade of the 36th Infantry Division National Guard. The division had been conducting military drills at home stations and annual training periods at Camp Hulen in Palacios, Texas, since Roosevelt had changed the guard's status.[2]

CHAPTER FIVE

141st Infantry Regiment Rifle Team, 1937. According to the August 19, 1937, edition of the *San Antonio Evening News*, the 141st Infantry Regiment, San Antonio unit of the Texas National Guard, won the Hulen Trophy at Camp Hulen in Palacios, Texas, for the second consecutive time. Maj. K. L. Berry, center front row, managed the team. Source: Author's private collection.

Berry's new position, officer instructor, was a position he was obviously good at since he had held similar positions several other times in his military career—at Camp Kearney, at Fort Benning, in Vermont, and in China. In addition to being an officer instructor, he managed the 141st Infantry Regiment Rifle Team, which won the Hulen Trophy as well as the National Guard Bureau's National Guard Pershing Trophy two years in a row under his leadership. Berry's magic touch as an instructor and manager, leading men to do their best, was evident in his work preparing his soldiers for the oncoming war. All the Texas National Guard units participated in the Third Army maneuvers in the vicinity of Camp Bullis, northwest of San Antonio, from August 6 to August 20, 1938. Officer Instructor Major Berry and his trainees were there. Federal inspection reports of the 36th Division units consistently showed "Satisfactory" and "Very Satisfactory" ratings.[3] It is more than likely that Major Berry's work as an officer instructor contributed to this success.

Roosevelt had been astute in 1933 to make the National Guard a reserve component of the army. With Germany and Japan moving closer to all-out war in 1939 in their spheres of aggression, the training of troops across the United States accelerated. Then, when European countries and Germany declared war on one another in 1939, Roosevelt secured the power from Congress to mobilize National Guard units. In September 1940, Congress went a step further, passing legislation to create a peacetime draft, making possible a huge expansion of military forces. In Texas, the 141st Regiment was inducted into federal service as part of the 36th Division on November 25, 1940.[4]

MILITARY BUILDUP

For his success at teaching infantry officers, Berry was promoted to lieutenant colonel on August 18, 1940. With this new rank, he became the executive officer at Camp Bowie, a military training camp in Fort Worth. But with war looming on the horizon, the Fort Worth site was considered to be too small. Accordingly, the War Department announced that Camp Bowie would relocate to Brownwood, Texas, where the department owned around 123,000 acres, and a new camp would be built with more room to train larger numbers of recruits. Berry moved his family to Brownwood so, as executive officer, he could oversee the teams building the new Camp Bowie, which would become the largest training center in Texas for the war. In November 1940, Lieutenant Colonel Berry commanded Camp Bowie for one month and then did so again from July 29, 1941, to October 25, 1941.[5]

Camp Bowie was the first major defense project in Texas, and 15,000 men were employed to build the camp. With the large acreage in Brownwood, Camp Bowie grew rapidly to include 8,000 acres for infantry and 28,000 acres for maneuvers on the ground. The artillery range was 23,000 acres. Because the camp was huge, eight divisions would eventually train there preparing for war. While the United States was building up its the military forces in 1940, Germany had invaded Belgium, France, and Holland. Great Britain was Hitler's next target. In Asia, Japan was taking more of China and moving toward French Indochina after France fell to the Germans.[6]

Germany began bombing London daily and was planning an assault on Britain from the sea. For months, British Prime Minister Winston Churchill had been imploring President Roosevelt to send Great Britain ships and weapons before the Germans invaded. Politically, President Roosevelt's hands were tied because of the Neutrality Act of 1937, which "prohibited US citizens from selling armaments to any nation at war."[7] But Roosevelt did come up with an idea to send some old destroyers to Britain if Britain leased the United States its military bases in Newfoundland and Bermuda. This was a huge risk for Roosevelt to take before the upcoming presidential election, because he did not know how the American public would view this agreement, which was circumventing the Neutrality Act. After this agreement was made public, Hitler postponed the invasion of Great Britain, perhaps because of Roosevelt's actions. At home, Roosevelt's opponents cried "foul" that the destroyers were given to Great Britain in this way, but the voters nevertheless reelected Roosevelt. With the momentum from his electoral victory, Roosevelt declared that the United States would become an "arsenal of democracy." He worked with Congress to pass the Lend Lease Act, which authorized the United States to lease war supplies to the nations under attack that were critical to America's safety. In the summer of 1941, the president took an additional step, freezing all German and Italian assets in the United States. After Japan expanded further into Asia and occupied French Indochina, Roosevelt froze all Japanese assets, a move that halted trade between Japan and the United States. Japan would have to look elsewhere for oil, namely the Dutch East Indies.[8] Because Roosevelt knew Japan would move southward for oil, he created a new command structure in the Philippines, the US Army Forces in the Far East (USAFFE). In so doing, Roosevelt hoped the US Army, naval, and air bases would appear as a threat to the Japanese in the Pacific. He then recalled Gen. Douglas MacArthur from retirement in July 1941, promoted him to lieutenant general, and placed him in command of USAFFE.[9]

NEW ASSIGNMENT

In his new role, General MacArthur needed experienced American officers to train the Philippine Army in the face of increasing Japanese aggression in the Pacific. In late October 1941, Berry, then commander of Camp Bowie, received orders sending him to the Philippines to get the Philippine Army ready for war. The Philippine scouts were already well-trained—an elite organization numbering twelve thousand men whose military prowess could be enhanced with some up-to-date training. It was the larger group of untrained, conscripted soldiers who would need all the training they could get from the American officers. MacArthur was certain the officers heading to the Philippines with Berry would have six months to get the Philippine Army in shape.[10]

Before he set sail, Berry went to pay his respects to Lt. Gen. Walter Krueger, a longtime military man who had served in the Spanish-American War and in the Mexican Punitive Expedition with Pershing, and who in 1941 commanded the Third Army and the Southern Defense Command at Fort Sam Houston.

Berry knew him from his various assignments at Fort Sam Houston and Camp Bowie. Krueger had been in the Philippines early in his career, and he also knew General MacArthur quite well. (Krueger would later serve with MacArthur during the invasion of the Philippines in 1944–45.) He wished Berry good luck in his new assignment.[11]

At the same time Berry went to say farewell to General Krueger, he reconnected with Dwight Eisenhower, his old army football friend. Eisenhower was serving as Krueger's chief of staff. Eisenhower enjoyed his position with Krueger because Krueger was not only a superb military man, but he was also likable. Eisenhower's previous service under Gen. Douglas MacArthur in Washington, DC, and the Philippines had been quite the reverse. Even though Eisenhower lauded MacArthur's "soldierly qualities," he did not appreciate the general's arrogance. In his later years, Eisenhower recounted his long association with MacArthur sarcastically, saying, "I studied dramatics under him for five years in Washington and four years in the Philippines."[12]

Eisenhower and Berry chatted for a while about Berry's new assignment with MacArthur to prep the Philippine soldiers for war. Because Eisenhower disagreed with MacArthur about the importance of the work in the Philippines as opposed to the work to prepare American soldiers for a war in Europe, Eisenhower encouraged Berry to try to get his orders changed, saying, "The big show was going to be in the other direction."[13] It is unclear if Berry could have changed his orders. At any rate, all his previous overseas duties had been in the Pacific arena, so going to the Philippines may have seemed like a good idea to Berry. But his military life and even his life more broadly certainly would have taken a different turn if he had tried to change his orders.

USS *COOLIDGE*

After saying goodbye to Krueger and Eisenhower at Fort Sam Houston in October 1941, Berry traveled to San Francisco where he met others bound for the Philippines. He wrote Alice a series of letters from October 30 to November 30, 1941, documenting his journey to the Philippines and arrival in Manila. He began, "I met a Lt Col Corkill, FA [field artillery] and Lt Col Mitchell Inf [infantry] who were headed same as I."[14] Corkill and Berry hit it off right away. They would become best friends in the wartime ahead and would remain very close friends when they were back in Texas after the war.

In addition, Berry saw old friends in San Francisco who were also headed to the Philippines—Brad Chynoweth, Charlie Steel, Phil Fry, Ross Smith, Adlai Young, and Ed Keltner. He jokingly wrote Alice that they all "look as old as h____."¹⁵ She must have known these men, too, and might have smiled at his humor. These old friends would become prisoners of war (POWs) with Berry after the surrender of the Philippines in April and May 1942.

In another letter to Alice, Berry wrote that he was sailing over on the USS *Coolidge*. His new friend Corkill had been made adjutant of the "boat," and Brad Chynoweth had been made executive officer. Also, in this letter, he asked Alice to write him letters. Always eager to hear about football, Berry asked Alice to send him football clippings and then expressed the hope that she could join him in Manila as she had done twenty-one years before. He was already missing her, but there would be no way she would join him this time because of the specter of war. While in San Francisco awaiting transport, he had not received a letter from home like other officers had, and he wondered why.¹⁶ Sadly, Berry would be asking this same question and wondering the same "why" for the next four years.

Despite not getting any mail from Alice, Berry continued to write to her. He set sail on November 1, he wrote, on the USS *Coolidge*, a "650 foot 33,000 ton palatial liner with swimming pool, gym, movie, etc." He shared a stateroom with Lt. Col. Loren Wetherby and Ed Corkill. He mentioned that "Brad [Chynoweth] is on the boat and as good looking as ever. Brad is a full colonel now as is Charlie Steele." He wrote about other friends: "Brig Young and Shorty McDonald are on the boat, but they were both passed over. Ed Keltner, Frank Brokaw, Hugh Dumas, Ben Hur Chastaine and many other Lt. Cols. are on the boat." Presumably Alice knew all these friends from previous assignments. About his new assignment, he heard that he might be commanding some Philippine regiments and be promoted to colonel, "but that must be taken with a grain of salt."¹⁷

In Honolulu, the USS *Coolidge* picked up escort ships and was put on blackout because the crew feared Japanese planes might spot them at night. The ship was not allowed to put any garbage in the water because the Japanese could trace the ship's location from garbage. The ship sailed only about 410 miles a day because its companion ship, the USS *Scott*, was a slower sailing ship by three to four miles per hour. Protecting them along the way was the USS *Louisville*, "a mighty ship for 10,000 tons. She carries 9 big 8' or 10' guns, several 5' ones and her super structure bristling with anti-aircraft guns and pointing shipward. It is a satisfying feeling to know she is along as it looks awfully competent in her battle grey paint," Berry wrote.¹⁸

He started recording the miles that they had traveled and the miles yet to come. One day, he wrote they were "1165 miles from Honolulu and 3673 miles from Manila." He thought they might reach Guam about the 15th or 16th and then reach Manila by the 20th. He and his roommates paid a visit to Gen. Maxon

Lough and chatted about possible assignments once they got to Manila. From what Berry could infer, he thought Lough and Chynoweth would be division commanders. For "entertainment" on the ship, Berry prepared to give a talk about his days in China for the officers. He told Alice, dryly, that he was never a good talker because he "never got much practice at home."[19]

By the 12th, he wrote, the USS *Coolidge* was "1996 miles from Honolulu and 2842 from Manila."[20] Berry and his fellow soldiers learned that when they got to Guam they would not be able to leave the ship; they were stopping only for the USS *Scott* to get fresh water. On the 14th, a plane flew overhead, and they wondered if it was a Japanese plane spying on them. When they stopped in Guam, they learned that American planes would be flying overhead to protect them. Even though war was supposed to be six months away, it seemed very close to those on the USS *Coolidge*.

Berry found sailing life was "very boresome," so to ease the boredom he started swimming daily in the swimming pool of the beautiful ocean liner, a routine that he enjoyed. He looked forward to the visits by the *Clipper*, a mail ship, but he failed to receive any mail from Alice. Others did receive mail, which peeved him. Expressing his boredom, he wrote to Alice, "Same dreary expanse of water and heat you can almost cut with a knife." By the 17th of November the ship was 1,138 miles from Manila, and passengers and crew learned that they had to stop at Corregidor, the island fortress in Manila Bay, to take on a boarding party bringing their duty assignments before proceeding to Manila.[21]

ARRIVAL IN THE PHILIPPINES

The USS *Coolidge* arrived at Corregidor Island in the early morning of November 20, 1941. The officers on board were supposed to get their duty orders there before heading to Manila. But there seemed to be confusion; some would go to the Regular Army Division and some to the Philippine Army. Others would join the USAFFE staff. A friend of Berry, Lt. Col. Ovid "Zero" Wilson (who had already been working in the Philippines for Gen. George Parker) told Berry he was probably going to the Southern Luzon Forces, because General Parker, commanding general of Southern Luzon Forces, had asked for him, Shorty MacDonald, and Ross Smith by name. Later, Ben Hur Chastaine told Berry that he was going to the 57th Infantry, which was part of the Regular Army Division. Then Berry heard that he would not know where he would be going until the next week. It must have been frustrating to encounter so much uncertainty after enduring a long voyage at sea.[22]

By November 22, Berry found a place to stay at Fort McKinley in Manila with Col. Harrison Browne, whom he called Brownie, rather than with the other officers at the hostess house. Lt. Col. Edmund Lilly roomed with Browne, too. At the time, Browne was chief of staff of the Philippine Division. At Fort McKinley, Berry met old military friends again: Ed Aldridge, Paul Stivers, Ovid (Zero) Wilson, mentioned above, Herb and Mary Harries, Arnold Funk, Tom Tarpley, and Marshall Quesenberry, all people Alice would have known from his previous postings.[23]

As the officers waited for orders, they experienced "blackouts" over the area, which Berry said "was kinda weird" for a war that was supposed to be six months off. "I believe it is nearer than people at home think," he wrote Alice.[24] In another couple of letters in late November, he told Alice he was angry with her for not writing. He wrote her his return address and told her that every other officer had gotten mail from home but him.

After another week, Berry still did not know what his assignment would be, but he had misgivings about what might lie ahead. He wrote, "All the officers who came with me were given a talk by the Southern Luzon Force Commander [General Parker] and I did not like the sound of it."[25] It seemed that Berry and others were going to be "glorified instructors," and some of them would be placed with units commanded by a Filipino commander. He hoped that a few Americans might get commands. "Don't like it over here worth a damn anyway but no can help," he wrote Alice.[26]

By the 30th, Berry had his assignment: Southern Luzon Force as a senior instructor with the 1st Philippine Constabulary Regiment, in Quezon City. Historian John Whitman explained that the term instructor was a "misnomer" in that the American high command deemed the instructors responsible for a regiment's success or failure even though instructors did not have the command of the regiment. Instructors were supposed to use their "leadership skills and great tact" to dissuade a Philippine commander from making a mistake.[27]

Presumably, being assigned to serve as an instructor was what Berry "did not like the sound of" in the talk given by the Southern Luzon Commander Parker. He wanted a command of his own. However, there was a potential benefit to being an instructor. Senior instructors with the Philippine forces generally held the rank of full colonel and were regiment commanders, and in another month, Lieutenant Colonel Berry would be both.

DECEMBER 8, 1941

It is worth noting that Berry got his assignment on November 30, ten days after he arrived, and only eight days before his war began. During those few days at the end of November and beginning of December, everyone was getting settled into their new assignments, getting supplies, and preparing to receive the Philippine soldiers, who needed at least six months of hard training from the American instructors to be battle-ready. Unfortunately, they barely had their uniforms, equipment, and assignments before the Philippine Islands were attacked. Many young cadets who would shortly become Philippine officers had not yet graduated from the Philippine Military Academy.

There had been some warnings of the fight to come. Commanders had been told that negotiations with the Japanese were not going well. Japanese troop movements in the Pacific sphere indicated that an attack on the Philippines could happen at any time, but the Washington military commanders wanted to be able to say that Japan attacked first.[28]

The attack came quickly. Colonel Lilly wrote, "At 6:30 am Dec 8, 1941 (Manila time) the phone in Quarters 39, Fort McKinley—which I occupied with Col. Harrison C. Browne, Chief of Staff, Philippine Division—rang. It was "Bish" [Hueston R.] Wynkoop asking to speak to Lt Col Kearie L. Berry. I told him that "K.L." was not in and, noting a tone of excitement in his voice, asked him if anything was wrong. It was then that he told me that Pearl Harbor had been bombed by the Japanese early that morning."[29] After that early morning phone call, the Philippines were attacked, and General MacArthur's war planes at Clark Field were decimated. The United States and Britain declared war on Japan that day. The Chinese declared war on Japan on December 9. Berry's war had begun.

PART III

War

"In every battle there comes a time when one group of warriors must be sacrificed for the benefit of the whole."

—*President Franklin Roosevelt, fireside chat, March 1942*

Map 3. Map of the Philippines. This map shows the island of Luzon and where Manila and the Bataan peninsula are in relationship to each other on Luzon. Source: National Museum of the United States Air Force.

CHAPTER 6

Lieutenant Colonel Berry in the Battle of Bataan

Part 1

A DISORGANIZED BEGINNING: DECEMBER 1941

The Philippines military was not prepared for war with Japan despite having some military forces there maintained by the United States for many years. The country was scheduled for independence from the United States by 1946, so the US had been slowly transferring defense to the government of the Philippines. The US Joint Army and Navy Basic War Plan Orange, which had been written in 1921 with future Gen. Jonathan Wainwright as one of its authors, limited the defense of the islands to the Manila area on Luzon and the island of Corregidor.[1]

When War Plan Orange was updated in April 1941, in case of an attack on the Philippines, the military was supposed to withdraw to the Bataan Peninsula and hold back the advancing enemy troops (map 3). A "prolonged defense" of Manila Bay was expected. However, even though the update still included the phrase that the United States would "relieve the Philippine garrison within six months after the outbreak of war," all who had written the update agreed privately that the Philippines would not be relieved in case of war.[2]

Thus, there was no plan to provide for reinforcements to come to the aid of the Philippines. Sadly, the commanders in the Philippines did not know this decision in the run-up to the war. In the spring of 1941, Maj. Gen. George Grunert, commander of the Philippine Department, had pleaded for more men, weapons, ammunition, and vehicles because his garrison was short of "everything needed to wage war against a modern, well-equipped enemy."[3] The reply from Washington was abrupt, stating that an adequate amount of war supplies would be available in 1942. However, the navy had already estimated that it would take at least two years for the Pacific fleet to "fight its way across the Pacific" to reinforce the Philippines if war broke out.[4]

In any case, other priorities had already been established. If the United States joined the Allies in the war, aid would go to Great Britain first. The reasoning behind this decision was offered by Gen. George C. Marshall, chief of staff of the US Army, in his July 1, 1941, to June 1943 biennial report to the Secretary of War: "Adequate reinforcements for the Philippines at this time would have left the United States in a position of great peril should there be a break in the defense of Great Britain."[5]

Despite the existing war plans described above, when President Roosevelt recalled Gen. Douglas MacArthur to active duty on July 26, 1941, MacArthur sought to discard War Plan Orange and instead build up the Philippine forces and to defend all the islands, not just Manila Bay. He proposed four major tactical command centers—North Luzon, South Luzon, a Reserve force on the Bataan Peninsula, and Visayan-Mindanao (map 4). MacArthur told General Marshall that he needed six months to put his plan in place instead of relying on Plan Orange. Surprisingly, Marshall and the Joint Board approved MacArthur's plan on November 21—two weeks before the Japanese attack. In MacArthur's plan there would be "no withdrawal from beach positions" to Bataan.[6] As soon as MacArthur's plan was approved in Washington, he moved his troops into the sectors he had outlined so they could defend all the beaches of the Philippines.

Originally, Lieutenant Colonel Berry was assigned to Gen. George Parker's South Luzon Force (map 4). Gen. Jonathan Wainwright's troops were sent to North Luzon, and Gen. William Sharp's troops were sent to the southern islands, of which Mindanao is the largest. Everybody had a false sense of security that war was six months away because MacArthur's plan had been approved, and he had told Washington he needed six months to make his plan successful. Consequently, when the Japanese attacked Pearl Harbor and the Philippines in quick succession two weeks later, US Army and Navy leaders were caught by surprise. Leaders had not considered that the Japanese were capable of attacking in more than one place at a time. But the Japanese forces showed the Allies what could be done in just three short weeks. Between December 7 and December 27, the Japanese destroyed the US Battle Force fleet at Pearl Harbor; destroyed MacArthur's air power at Clark Field on Luzon; invaded the Philippines at several locations; seized Guam; invaded Burma, British Borneo, and British Hong Kong; took Wake Island; and bombed Manila. The losses at Pearl Harbor and Clark Field spelled doom for the Philippines.

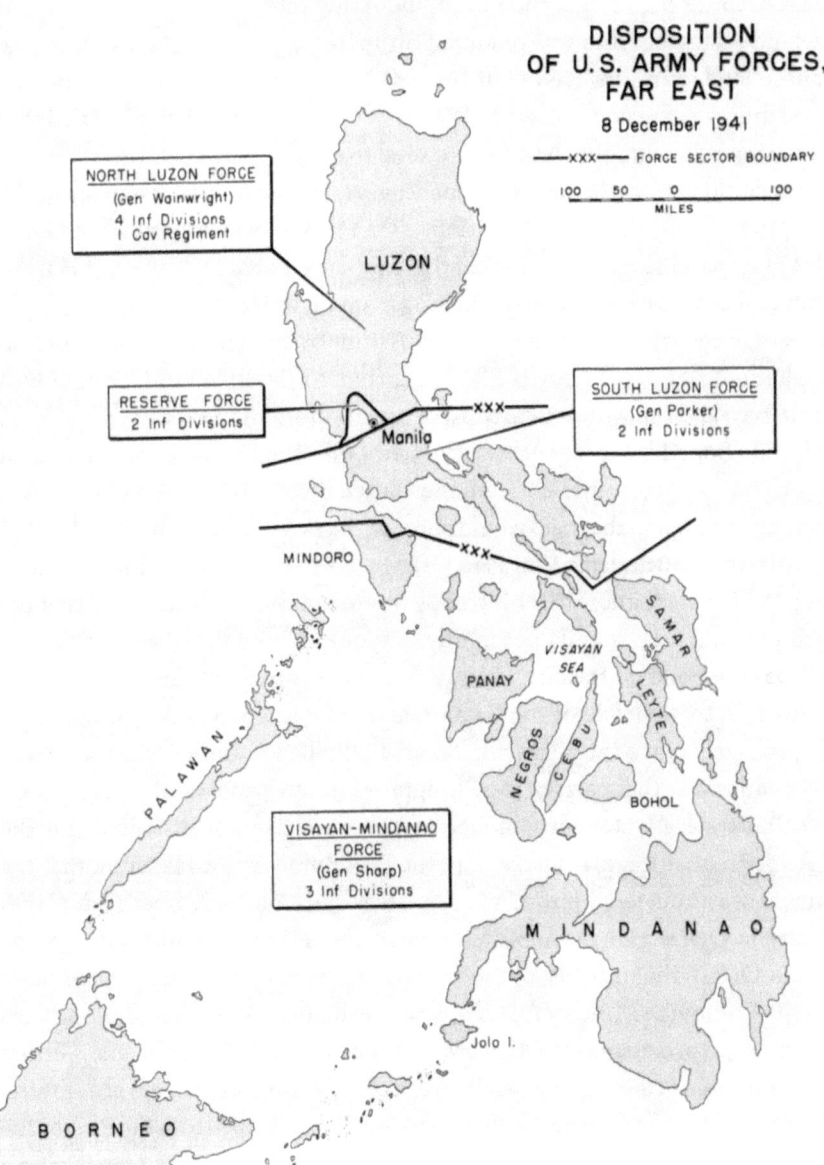

Map 4. Disposition of US Army Troops, Far East. This map shows MacArthur's plan to have four sectors defending the Philippines: North Luzon Force, Reserve Force, South Luzon Force, and Visayan-Mindanao Force. Manila is on Luzon Island. Source: *The War in the Pacific: The Fall of the Philippines*, by Louis Morton.

MacArthur's plan to get the Philippine Army into shape by June 1942 and to defend all the beaches was rendered futile by the Japanese attack. After the Japanese landed in Lingayen Gulf (map 3) in North Luzon on December 22, MacArthur was forced to revert to War Plan Orange. He had also learned that the assistance Washington had tried to send from Australia after the December 8 attack could not get through the Japanese "aerial and naval supremacy in the Philippines."[7] Lines of communication to the Philippines were cut when Guam and Wake Island fell, and the loss of the US Battle Force fleet at Pearl Harbor eliminated any possibility of help coming from Hawaii to the Philippines.

Out of necessity, on December 23 MacArthur ordered the evacuation of Manila, the removal of his USAFFE headquarters to the island of Corregidor in Manila Bay, and the withdrawal of all of his troops to the Bataan Peninsula on Luzon. He hoped his forces gathered together on the Bataan Peninsula could hold off the Japanese until help from the United States arrived. MacArthur was under the impression that help would come within six months as had been stated in the original and updated War Plan Orange. He declared Manila an open city, letting the Japanese know that his troops would not defend the city and that the Japanese could take Manila peacefully without hurting the local population. The Japanese bombed the city anyway. For the troops who were spread out in the three sectors outside of the Bataan Peninsula where MacArthur had sent them, even getting to the peninsula became a challenge as the Japanese moved into Manila and other parts of the Philippines (maps 3 and 4).[8]

At that time, Lieutenant Colonel Berry was the senior instructor for the 3rd Regiment, 1st Regular Division of the Philippine Army. He wrote that the regiment was "mustered into United States service on December 17 at Camp Murphy, in Quezon City near Manila. It was under the command of Philippine Captain Orias. The majority of the officers were young lieutenants graduated from the Philippine Military Academy the day before, December 16, along with a sprinkling of professors from that academy and other Reserve officers."[9] Eliseo Rio, twenty-two years old, who would work closely with Berry in the near future, received his commission as a 3rd lieutenant on December 16 from the Philippine Military Academy. One day later he was assigned to the 1st Regular Division of the Philippine Army 3rd Regiment, where Berry was the senior instructor.[10]

The 3rd Regiment, following War Plan Orange, was part of the desperate withdrawal to the Bataan Peninsula. First, Berry and Orias had to move their regiment north out of the Manila vicinity to the junction at San Fernando. At that point, the regiment would then turn southwest onto the Bataan Peninsula. They began their move north on December 26 and set up lines of defense along strategic roads with the infantry's three battalions.[11]

Young Lieutenant Rio, who became second in command of Berry's 3rd Battalion, wrote, "In view of the 'open city' declaration and perhaps the developments in the big battle raging in the north over the last four days, our entire battalion was hastily transported after evening mess on the 26th to a point about forty miles northwest of the city [Manila] to be a part of a defensive line being set up there."[12] Rio did not know at that time that he was a part of War Plan Orange, withdrawing to the Bataan Peninsula for a last stand against the enemy.

The 3rd Infantry connected with the 57th US Infantry in the San Fernando area of Luzon on December 28. Berry's friend Ed Lilly, who was with the 57th Infantry, visited him at his 3rd Regiment headquarters. In just a few short days, Berry's responsibilities had changed. Lilly learned that Berry was now the advisor for Gen. Fidel Segundo, commander of the 1st Regular Philippine Division. Interestingly, an advisor to a division commander should hold the rank of colonel, and in fact Berry had been promoted to full colonel (T—temporary) on December 24, 1941, by the War Department. However, because no one had told Berry, he was unaware of his new rank.

Segundo, a West Point graduate and former superintendent of the Philippine Military Academy, was appointed commander of the 1st Regular Philippine Division on December 19, when the division became a part of USAFFE.[13] When Segundo took over the division, it consisted of untrained young Filipino men who were "recent recruits, trainees, ROTC cadets, and civilian volunteers."[14]

On December 29, after forming the defensive line young Rio had mentioned, the 3rd Regiment received orders to assemble the battalions and prepare them to withdraw to the Bataan Peninsula. The peninsula is about twenty-five miles long and fifteen miles wide at its center. There were few roads in Bataan, only the north-south roads on each coast of the peninsula and one road bisecting the peninsula east-west. The only way into Bataan from Manila by land was the East Road along Manila Bay. As Plan Orange went into effect, the thirty-four thousand troops retreating north out of Manila had to use this road. The retreat was slow, crowded, and confusing. Another fifty thousand troops were coming from North Luzon and the southern islands of Mindanao and Viyasan. Adding to that number, twenty-six thousand civilians, trying to find safety, would inhabit Bataan after the retreat, bringing the total number of soldiers and citizens in Bataan to over one hundred thousand.[15]

The next day, two of the regiment's battalions were trucked to the position of the Moron-Bagac-Pilar roads to join the rest of the division, which had already arrived from Manila. Berry wrote, "During the night the remainder of the 3rd Infantry moved to Mauban Point [south of Moron] by marching and arrived the morning of January 1, 1942 [map 5]."[16] These soldiers marched down the east coast road to Balanga, then took the only road bisecting Bataan over to Bagac on the west coast. They were assigned to the west coast road from Moron to Bagac (map 5), protecting Bataan from a Japanese invasion that could come from Subic Bay, which was north of Moron.

Young Rio recognized his precarious position—his battalion was out on the front line, defending the rest of the regiment, which was behind him. "As days passed I began to see more clearly the role of our battalion in its current deployment. We were part of forces which capped the entrance to the peninsula of Bataan while the whole area to the rear was being organized into a bastion of defense."[17] Rio realized that the headquarters of the 3rd Regiment was being established a few miles behind where he was, and his battalion was at the front of the line defending the northwest side of Bataan from the Japanese who were coming from Subic Bay. For a newly minted 3rd lieutenant with no battle training, this realization was frightening.

JANUARY 1–15, 1942

As MacArthur's troops were retreating to Bataan, the Japanese were moving aggressively forward. By January 2, they had captured Manila and the naval base at Cavite. And they were planning attacks from the north into Bataan. The Japanese advance seemed unstoppable; consequently, on January 2, 1942, leaders in Tokyo, expecting Bataan and Corregidor to fall by early February, reassigned Japanese Gen. Masaharu Homma's 48th Division, sending it to invade the Dutch East Indies. Homma protested that he needed that division for the Bataan campaign, but his protests did no good. On January 7, Homma was ordered to begin the final assault on Bataan without the 48th Division. Tokyo expected a quick victory.[18]

On January 7, MacArthur organized the troops on Bataan into two corps. Colonel Berry's 3rd Regiment, which was part of the 1st Regular Philippine Division, became part of I Corps, commanded by Gen. Jonathan Wainwright. I Corps consisted initially of the 26th Cavalry (a Philippine Scout unit), the 1st Regular Philippine Division, the 31st Division, the 91st Division, and the 71st Division. Wainwright's assignment from MacArthur was "to defend Bataan's western half and prevent hostile landings along the west coast."[19] Gen. George Parker commanded II Corps, and his assignment was to defend the eastern part of the Bataan Peninsula and the East Road, the only road to Manila. The main line of resistance (MLR) ran across Bataan from Mauban on the west coast to Mabatang on the east coast, with Mount Silanganan/Natib separating the MLR line into two sectors (map 5).[20]

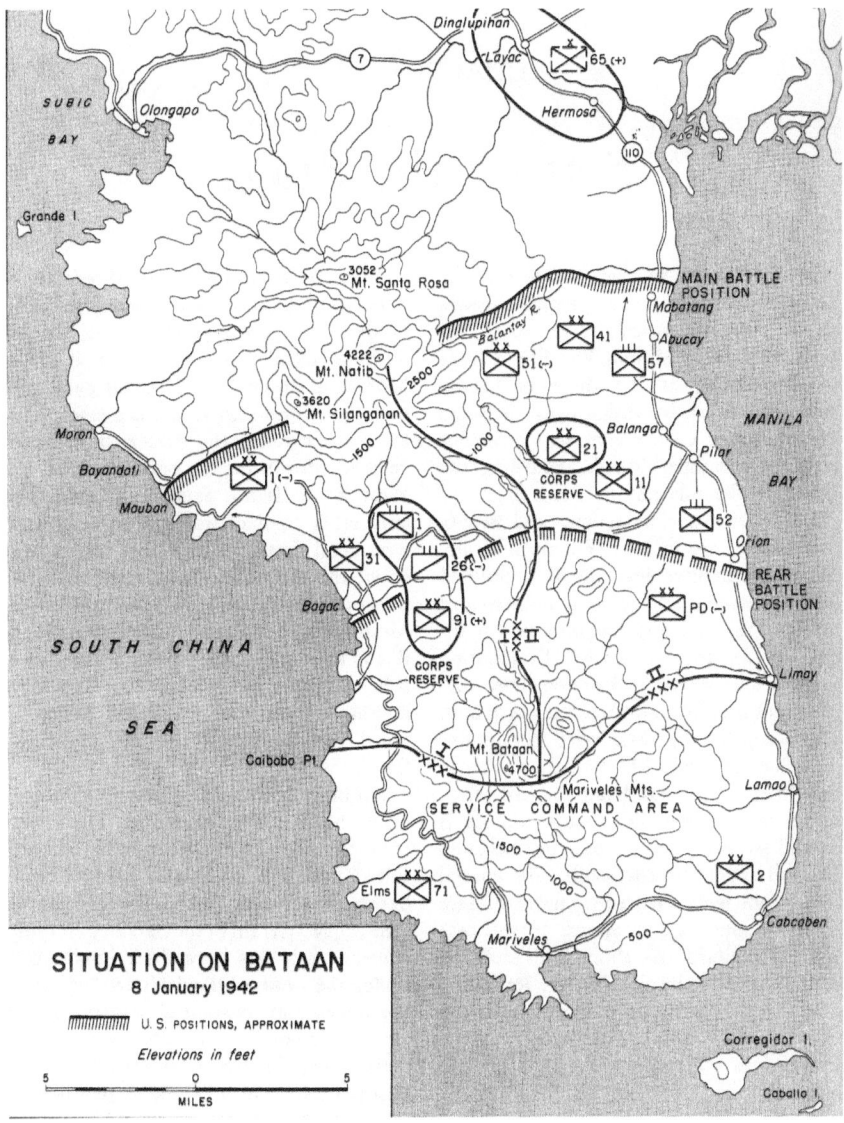

Map 5. Situation on Bataan. Two main lines of resistance (MLR) are shown on this map. The MLR from Mauban across the Mounts Silanganan/Natib over to Mabatang was the first MLR that I and II Corps established. It is clear to see how Mt. Silanganan/Natib divided the first MLR. General Parker's troops were on the line stretching from Mt. Natib to Manila Bay and General Wainwright's troops stretched from Mt. Natib to the coast of the South China Sea. The dark line drawn on the map from Mt. Natib to the Service Command Center shows the separation of the areas of responsibility for I and II Corps. After the Japanese penetrated the first MLR in January, both I and II Corps withdrew to the rear line of resistance from Bagac across Bataan to Pilar/Orion. Source: *The War in the Pacific: The Fall of the Philippines*, by Louis Morton.

Meanwhile, a January 3 inventory of food revealed that there was only a thirty-day supply of rations for the approximately eighty thousand American and Philippine soldiers and twenty-six thousand civilians on Bataan. Gen. Charles Drake called the inventory "heartbreaking," but when he went to Corregidor to inform Gen. Richard Sutherland, MacArthur's chief of staff, Sutherland did not believe that there were over one hundred thousand people to feed. Prewar plans had posited that forty-three thousand men would be on Bataan to fight for six months. According to General Drake, Sutherland believed that the quartermasters on Bataan were padding their numbers.[21]

Consequently, on January 5, believing the numbers were padded, MacArthur ordered half-rations (two thousand calories a day) for the more than one hundred thousand troops and civilians on Bataan.[22] These were the troops who were ordered to fight for six months and hold Bataan until help arrived. General Wainwright said, "It was hardly enough to hold body and soul together."[23] The lack of equipment, supplies, clothing, and especially food had a tremendously negative effect on the outcome on Bataan. Col. Richard Mallonee remembered, "Each day's combat, each day's output of physical energy took its toll of the human body—a toll that could not be repaired."[24]

Also on January 5, Colonel Berry was made commander of the 3rd Regiment (replacing Captain Orias). He had wanted a command of his own, and now he had one. Under his command were the engineer battalion of the 1st Regular Division; two engineer battalions (labor); one battalion of the 295th Artillery; one battalion of the 3rd Artillery; one battery of self-propelled artillery; and one two-gun battery of the 155th Coast Defense. With his regiment composed of untrained soldiers, Berry had to train and lead his men at the same time. Drawing on his engineering knowledge, Berry taught his young soldiers how to strengthen a battle line with fortifications, and with his tactical knowledge he taught them where to place artillery and how to use it. During the battles ahead, Berry would grow proud of his troops' performance. He knew how to train them to be soldiers.[25]

In I Corps, Wainwright placed Colonel Berry's 3rd Infantry Regiment, which had yet to see action, along the South China Sea coast along with the 1st Infantry Regiment. According to Berry, around January 8, the 1st Infantry was moved from Mauban up the coast to Moron about five miles north of the front to hold the road that ran north to Subic Bay (map 5). The Japanese had landed in the northern part of Luzon and were moving south toward Olongapo. The 3rd Infantry under Berry's command stayed on the MLR position from the coastline at Mauban to partway up Mount Silanganan/Natib to the east and continued to strengthen the line, building barbed wire fences, mine fields, and trenches. The decision not to extend the line up to the top of Mount Natib became an issue between General Wainwright and General MacArthur.[26]

Lieutenant Rio now understood the battle lines and the assignment of his division. He wrote, "The first line of defense ... actually ran across the entire width of the Bataan Peninsula and was divided into two Corps sectors of almost equal widths: with our Corps, I Corps, occupying the western side, and II Corps, the eastern."[27] He noted the separation point of the two corps was Mount Natib, an extinct volcano whose peak was 4,500 feet above sea level. He realized that the 1st Regular Division alone was assigned to protect and defend the entire line of defense of the I Corps sector, which was about nine and a half miles wide.

As Japanese Gen. Akira Nara started moving his troops against II Corps on the eastern side of Bataan on January 9, General MacArthur came to Bataan and visited his troops that same day for the first and only time. He gave them an optimistic talk, telling them that new planes were coming from the United States and bombers would attack the Japanese in Bataan. The speech was inspiring, but MacArthur knew it was not true. He then met with Generals Wainwright and Parker and told them he wanted the two corps on either side of Bataan to make contact with each other on the MLR at Mount Silanganan/Natib. The gap between I and II Corps troops was about five miles (map 5). Wainwright stalled on this order because he did not think it was necessary to close the gap in the line. The vegetation on the mountain was very dense and the terrain so rugged that Wainwright was sure the Japanese would not be able to get through the gap.[28]

However, when he returned to Corregidor the next day, MacArthur persisted in his order to close the five-mile gap. Through General Sutherland, MacArthur ordered again that the two corps make "actual and physical" contact with one another, so that there were no gaps in the line of defense for the Japanese to get through. He ordered both corps to "shift troops into the unoccupied areas on Mounts Silanganan/Natib."[29]

Prodded again to act on MacArthur's order, General Wainwright then ordered General Segundo to get his 1st Division troops to unite with II Corps at Mount Silanganan/Natib. General Segundo sent the K Company of the 1st Infantry to make contact with II Corps. But the young troops could not get the job done. The terrain was extremely rough, dense, and seemingly impenetrable, and they came back to the main line saying they could not penetrate the jungle to make contact with II Corps. According to Rio, "The defensive line up to Mount Natib was situated on high and severely rugged mountains covered with virgin tropical rainforests and undergrowth so thick they seemed almost solid. Numerous deep ravines crisscrossed a wide area up to the slopes of Mount Natib."[30] Then on the 12th, for a third time, General Wainwright was ordered to fill the gap. Lieutenant Rio, who had become an aide to General Segundo during this time recounted, "The several reconnaissance parties ordered to hack their way through the area and contact the troops of II Corps on the other side of the demarcation point were never able to substantially penetrate that particular section of the defensive line."[31]

In II Corps, Gen. Albert Jones, commander of the 51st Division, was ordered to ensure the two corps touched because his division was located on the western half of the II Corps line up to Mount Natib-Silanganan. His troops did not have enough time to prepare fortifications, nor did they have regular tools—they used mess kits and spoons—to build barbed wire barriers up the jungle slope of the Natib-Silanganan. Their fortifications were understandably inadequate. Major Maury of the Scout Artillery in II Corps, however, was not worried about the Japanese coming through the gap; rather, he was amused to think anyone could believe that the Japanese could penetrate the jungle and deep ravines. He told Colonel Mallonee, "The enemy would have to have a brigade of human flies" to cross the area and gain access to the rear of the troops.[32] Like Segundo's K Company, the troops of the 51st Division in II Corps did not make contact with the 1st Division in I Corps. Nor did I and II Corps leaders think it was necessary or possible.[33]

In I Corps headquarters, Rio, now Segundo's aide, listened as, "the Division Staff, with the concurrence of I Corps Headquarters [Wainwright's headquarters], concluded that the area was completely inaccessible and impassable to man or any of his war machines, and, therefore, did not require any kind of manned defense establishment."[34] In any case, Wainwright and Segundo lacked the manpower to provide coverage all the way to the point of contact with II Corps.[35]

Consequently, there was no physical contact between the two corps on the first main line of resistance. The gap on Mount Silanganan/Natib was ostensibly impossible to penetrate. For their part, the Japanese troops who were moving into Bataan from the north saw the impenetrable barrier of Mount Silanganan/Natib as an opportunity to break through the MLR established by I and II Corps. And they did.

JANUARY 15–26, 1942

As Colonel Berry and his colleagues in I Corps prepared to defend the Bataan Peninsula, the Japanese continued to make huge gains elsewhere in the Pacific. In the second half of January, they had advanced into Burma; taken North Borneo; taken Rabaul on New Britain in the Solomon Islands; invaded Bougainville; and had begun the siege of Singapore. At the same, the assault of Bataan was underway.

On January 15, Japanese General Nara's 141st Infantry Regiment attacked General Parker's II Corps and "found a seam between the 41st and 51st Divisions . . . and began to peel back the 51st Division" along the MLR (map 5).[36] General Jones's 51st Division began to collapse. On the same day, General Wainwright's I Corps intelligence reports indicated that a Japanese column was on its way from Olongapo south to Moron. As noted earlier, General Segundo had most of the 1st Infantry in place at Moron north of Mauban and Colonel Berry's 3rd Infantry in place to defend the main line position at Mauban. Only the part of the line held by the 3rd Infantry was reinforced with a double apron of barbed wire. The rest of the line up the mountainside was not reinforced except by the jungled terrain. There had been no time to build barbed wire fences up the mountain in I Corps just as there had been no time in II Corps to build them on its side of the mountain by the 51st Division.[37]

Among the units that had been brought up to the defensive position at Moron was a new rifle company commanded by Rio's military academy friend 3rd Lt. Daniel Ledda. They had graduated together in December, just the month before. On January 16, the Japanese crossed the Batalan River, attacked Ledda's I Company first, and then pushed through to Moron and toward the waiting 1st Infantry. Fighting broke out, casualties increased, and the Japanese became stalled at that point. One of the first officers to be wounded was Rio's friend Ledda, who was struck by a fragmentary hand grenade in his torso and head. He was conscious as General Segundo and Colonel Berry congratulated him on the good performance of his company in helping to stall the Japanese at Moron.[38]

Also wounded in the fight was Maj. O. S. McCollum, commander of the 1st Infantry. He was evacuated to the rear, and Segundo immediately placed Berry in command of the 1st Infantry as well as the 3rd Infantry. Berry did not have many trained American officers with him to assist in leading the two infantry battalions. Most of the officers were young, inexperienced Philippine officers. Berry remembered, "I visited the front and conferred with the American officers there. On my way up I was fired on by a sniper and along with another officer personally returned the fire and silenced the sniper."³⁹ General Wainwright, always eager to give moral support and assistance at the front line, visited Segundo and Berry and "ordered part of the 26th Cavalry to reinforce the 1st Infantry."⁴⁰

With the reinforcement of the 26th Cavalry, the 1st Infantry commanded by Berry counterattacked and drove the Japanese back across the Batalan River and out of Moron. Segundo is reported to have said, "Say, K. L., your boys were supposed to drive the damned Japs only across the river, not all the way back to Tokyo."⁴¹ By midafternoon, the counterattack was deemed successful, and most of the infantry was pulled back. But success was short-lived. The Japanese reinforced their troops and resumed their attack, taking Moron the next day.⁴²

At the same time Berry and the 1st Infantry were fighting to force the Japanese back out of Moron, another report came in about a large Japanese contingent further inland to the east of Moron and Mauban moving toward Mount Silanganan, the unreinforced part of the I Corps line. At Segundo's division headquarters, a Philippine Army captain from the 1st Battalion turned up at 3:00 a.m. telling Segundo that he had barely escaped enemy capture. His battalion had been in charge of patrolling the middle section of the MLR. He reported that "a large enemy contingent sneaked up on his command post and attacked while he was sleeping."⁴³ He believed the Japanese came through the right flank, which had been thought to be impenetrable. The young captain also thought they might already be behind the MLR.

Indeed, the Japanese were behind the MLR. On the 16th and 17th, a Japanese contingent penetrated the infamous gap that Wainwright and Parker had been told to close. Wainwright woefully recounted, "On January 16 my I Corps was hit. Fresh Japanese troops had landed at Port Binanga, on the northwestern coast of Bataan, and now, after a march southward through the jungle, they hit my 1st Philippine Division."⁴⁴ Those troops were Berry's 1st and 3rd infantries. Berry had discovered that Homma had greatly increased his forces against the I Corps MLR. Then Homma split his forces and kept the larger force fighting Berry's 3rd Infantry, which was placed at the MLR facing north. Berry wrote, "The 3rd Infantry was along the beach and the left two kilometers of the MLR facing towards Moron. Continual efforts were made to overcome this force, but all efforts failed. Snipers from this group scattered thruout [sic] the area and were a great source of worry and danger. I was active along the front during this period directing efforts to expel the enemy group and to hunt down the snipers."⁴⁵

While Berry's 3rd Infantry was fighting the Japanese at the MLR near Moron, Homma sent a company of his soldiers to infiltrate the line along the jungled, mountainous slope. Following Wainwright's order, Segundo had previously sent K Company from the 1st Infantry up the slopes to extend the right flank up Mount Natib-Silanganan, but the troops, not well trained, stopped and retreated when the terrain became too thick. They needed an American officer to lead them. It was reported that, "Colonel Berry gave them hell and sent them forward again, but with the same results."[46] When Berry was with the young rifle company trying to urge them forward to clean out the enemy penetrating through the jungle, he heard heavy fire breaking out on the MLR near Moron. Berry wrote, "I hastily went to the section of the line during the firing and found about ½ a company of infantry and a platoon of machine guns out of their fox holes and in front of the MLR firing to the rear into the hill I was coming over to visit the MLR."[47] In other words, they were firing in Berry's direction. He learned that "they had been fired on from the rear and instead of searching for the sniper, all men left their positions and faced to the rear and opened a heavy fire back into their own lines and into the hill on which our Ops [outposts] were located."[48] The enemy had gotten behind Berry's front line by coming through the jungle on Mount Silanganan, and Homma, with the majority of his troops, kept up the pressure on Berry's 3rd Infantry, who were facing the enemy in front of them at the MLR (map 6).

At this point, Wainwright questioned whether General Segundo did, in fact, have any troops on the slopes of the mountain. He sent Maj. Houston Houser to take a patrol out to see if Segundo's soldiers in the jungle were still in place. Houser had not gone one thousand yards before he was fired upon by the enemy. On the 20th, after more scouting in different places up the slopes, Houser did find pockets of Filipino soldiers, but he also encountered more of the enemy. The jungle was so thick that the enemy could slip through unnoticed past the Philippine troops that had stayed in place. Houser reported that the enemy had not only penetrated the line but were now in positions to the rear of Segundo's front line and were setting up a roadblock on the road where supplies and food were brought into his troops (map 6).[49]

Apparently, toward the end of the day on the 20th, Segundo had ordered Company B to try to stop the enemy from fully establishing the roadblock, but they were unable to move them. Maj. Alva Fitch sarcastically explained why Company B was not successful: "On the morning of the twentieth, Segundo promptly took command of the situation and tied hell out of it. He took our reserve battalion [commanded by Captain Laird] of infantry, deployed them across the road and started beating the jungle back towards the rear echelon."[50] Because of the great amount of time that was spent beating back the jungle, the

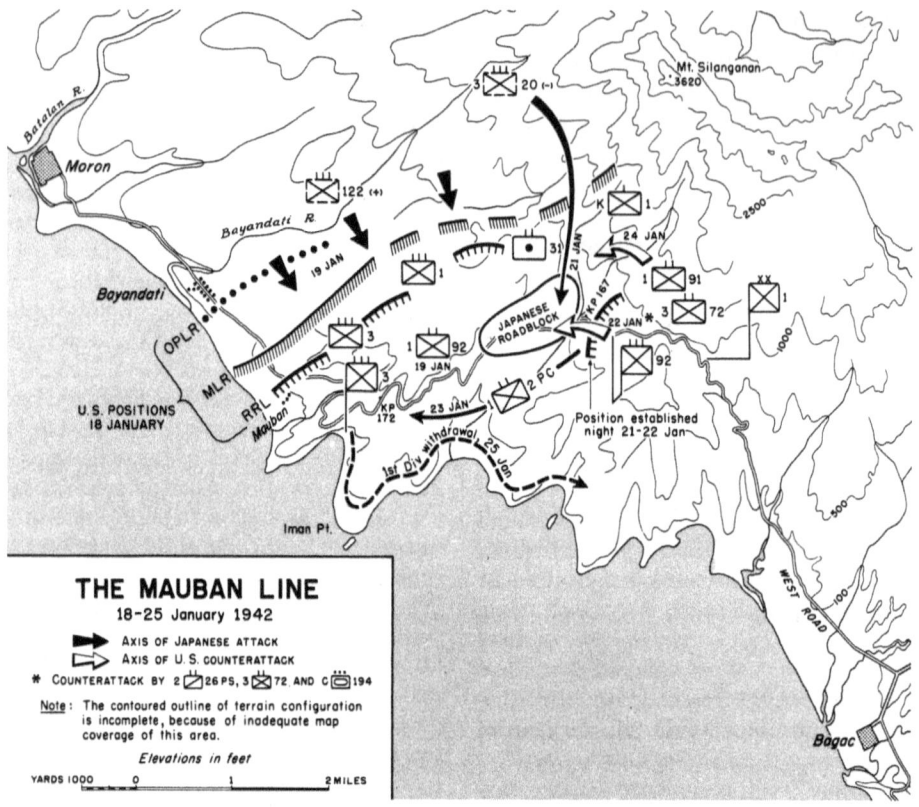

Map 6. The Mauban Line Positions. The enemy had gotten behind Berry's front line by coming through the jungle on Mt. Silanganan and had set up a roadblock. Homma, with the majority of his troops, kept up the pressure on Berry's 3rd Infantry, which was facing the enemy in front of it at the main line of resistance (MLR). When Berry's division was surrounded, he took his troops and withdrew down the cliffs to the beaches (the dotted line on the map labeled "1st Div Withdrawal") and most of them made it to Bagac. Source: *The War in the Pacific: The Fall of the Philippines*, Louis Morton.

Japanese had ample time to organize their position, according to Fitch. They laid mines and set up anti-tank barriers and obstacles, and more of their soldiers made it through the Philippines' front line. As a result, when Segundo ordered the company to engage the enemy at the roadblock at the end of the day, the company could not dislodge the enemy because of their good fortifications. At that point, Segundo decided to escape on foot to return to his 1st Division headquarters to wait for further orders from Wainwright.[51]

The Japanese roadblock on the Moron-Bagac Road cut off Segundo's 1st Division troops—all now under Berry's command—from the rear (map 6). No food or supplies could get through the enemy roadblock. Meanwhile, Berry and a few other American officers were with their young Filipino officers and troops fighting the Japanese in front of them, behind them, and to the east of them. This was the first battle for these young troops.

General Wainwright and General Bluemel visited Segundo at the 1st Division headquarters on the 21st to ascertain what could be done to unblock the road. Wainwright decided to direct a counterattack with elements of the 26th Cavalry and 92nd Infantry. He too was unsuccessful in dislodging the Japanese, but he did not give up. On January 22, he ordered Col. John Rodman, commander of the 92nd Infantry, to send his tank platoon to break through the roadblock and establish contact with the stranded 1st Division troops. (On map 6, Rodman's tank platoon is number ninety-two.) However, as Major Fitch had noted, the Japanese had had ample time to construct anti-tank obstacles and emplace land mines, thereby repulsing Rodman's attack. Rodman then sent other motorized troops to fight, and they were able to open a hole for some Filipinos to reach a ridge near the roadblock.[52]

In the end, though, all efforts failed to dislodge the roadblock. Rodman had enough troops, but they were all "understrength, tired, poorly fed . . . and had no automatic weapons at all."[53] Word reached Berry that rescue attempts had failed. He and his 1st Division troops were stranded. Berry wrote, "The situation was desperate and rapidly getting worse. Only a small amount of food was on hand when the Japs cut the road and local supply was almost nonexistent. Ammunition was short. The division was now being attacked from three sides."[54]

Then communication lines to Berry's command post were cut when Japanese mortar fire knocked out his radio. So, Segundo, Wainwright, and Berry could not communicate with each other as Berry was deciding what to do. Captain Laird's troops, who were with Berry, had not gained any ground against the Japanese at the roadblock, and Major Fitch's efforts to turn his limited artillery against them had also failed. "Snipers picked at the defensive positions; mortars inflicted casualties and the artillerymen with supporting infantry remained cut off," wrote Fitch. "We had received no word from General Wainwright and had no communication whatsoever. Our position was extremely precarious." Fitch recalled how Berry had decided that he and the other officers would gather their troops "during the night and fight their way back to the defensive lines."[55] It would have been a bloodbath had Berry and his colleagues followed through with the plan. But other developments led him to hold off.

On the other side of the roadblock, Wainwright believed he could still help Berry's 1st Division troops by getting them food and supplies by boat. With radio communication knocked out, Wainwright ordered Segundo to get that message to Berry by messenger. Rio was chosen for this task.[56] Just as Berry was getting ready to have his troops fight back to defensive lines, Rio arrived with a message from Wainwright: "Hold your positions. Plenty of help on the way. Food will reach you tomorrow."[57] As a result, Berry did not issue the order to attack. Instead, he tried to consolidate the ground they held. As fortune would have it, the navy did not send any food because of the dangerous situation in the China Sea with enemy ships in the area.[58]

That night, the Japanese slipped into the area and started a firefight with Captain Laird's battalion, which was protecting the rear of the 1st and 3rd infantries near the roadblock. The Filipino battalion collapsed and slipped away during the night, leaving the men under the command of Berry, Col. Halstead Fowler, and Major Fitch extremely vulnerable. Berry learned of the battalion's collapse only when he observed Japanese soldiers arriving near his command post in buses that belonged to the collapsed battalion.[59]

When a radio was found at the Field Brigade Artillery headquarters, Berry was finally able to send a radio message to Wainwright. He sought permission to withdraw down the cliffs to the beaches (map 6). To do so safely, he wanted Wainwright to order a counterattack to distract the Japanese. He received a coded message back from Wainwright that, unfortunately, his signal men could not decode because it was a 26th Cavalry code. Ultimately, Berry made the decision to withdraw without authority from Wainwright.[60]

Military historian Morton wrote in his summary of this situation that there was little Berry could have done but withdraw his troops down the rocks to the beaches. He had to abandon the main line of resistance. Supplies were gone, and the men were exhausted and hungry. Colonel Fowler had no ammunition for his artillery unit. Because Berry did not want equipment and guns to fall into enemy hands, he ordered all of it destroyed before he withdrew his troops down the rocky cliffs to the beaches to head south to Bagac.[61]

The trek on the cliffs to Bagac took two days and two nights. It was extremely perilous because in some places the ledges were so narrow that some men fell down the cliffs, injuring themselves. They cried for help, but others had to leave them behind because they could not reach them. The rocks were sharp and cut right through men's shoes. All were barefoot when they made it to the beach near Bagac. Then they had to climb up the jagged cliffs to get to the coast road. Colonel Berry stopped his men before they went into Bagac because he wanted to go before them to see what the situation was. He was surprised to see Wainwright's Packard coming down the road toward him. Berry worried about what the general would say to him about retreating without orders.[62]

Berry later wrote to Alice, "I feel that if I had waited another half day before withdrawing that the great bulk of the division would have been cut to pieces and cease to exist. The Corps Comdr. [Wainwright] did all he could to get help to us and when he saw me on our withdrawal and I told him that I got most of the personnel without too big a loss. He seemed very pleased. He told me that he would see that I was mentioned in orders. What he meant I don't know as I have heard nothing further about the matter."[63]

Berry's decision to withdraw the troops was the right decision, a decision for which he would be recommended for the Distinguished Service Cross. When Berry made the decision to withdraw, he did not know that General Parker's II Corps had also been pushed back from the MLR by the Japanese and was in retreat. Also unknown to Berry, General Sutherland had already ordered Wainwright's I Corps to fall back and establish a new line of resistance along the Pilar-Bagac Road to be in line with II Corps. Wainwright was, in fact, very glad to see that most of the 1st Division men made it south along the beaches. He needed the men badly, and he put them right to work on the Pilar-Bagac rear line of resistance (map 5).

CHAPTER 7

Colonel Berry in the Battle of Bataan
Part 2

MACARTHUR'S DIFFICULT ORDERS AND THE RESILIENCE OF THE TROOPS

The withdrawal to the rear line of resistance from Bagac to Pilar/Orion went well, and Wainwright was actually relieved to be on the new line and away from the jungled, mountainous area of Mount Natib-Silanganan, despite the fact that vehicles and artillery had to be abandoned. The Japanese 65th Brigade had lost over 1,400 men, and their commander reported that the remaining men were exhausted from the fight. The young, inexperienced Filipino troops had performed well, thwarting the Japanese advance.[1]

Weak, hungry, and worn out from the fight at Mauban and the treacherous retreat down the cliffs to the beaches, Berry's troops had no time to rest as they were put to work right away on the new line of resistance on January 27. They were positioned in the center sector of the 1st Philippine Corps, constructing trenches and wire entanglements without proper tools. They dug holes with their mess kits and bayonets.[2]

The line from Bagac to Orion was fourteen miles long and supposedly easier to defend because there was no tall Mount Natib to separate the two corps. But the Mariveles mountain range, while not as tall as Mount Natib, did have some difficult terrain—hilly and densely covered with thick undergrowth and trees. In the battle that would come, this dense terrain would prove to be difficult for the soldiers to see the enemy through the foliage, and many weapons proved useless in the undergrowth. The new front line of the I Corps was six-and-a-half miles long. Three divisions, the 1st, the 91st and the 11th, as well as the 2nd Philippine Constabulary and the 26th Cavalry, were to cover the I Corps' front line. However, Wainwright abruptly received word from MacArthur to defend the entire western coast region, from the new line all the way to the southern points of Bataan. His I Corps would have to fight a two-front war.[3]

Two days earlier, MacArthur had had a meeting with his leadership at his headquarters on Corregidor. He let it be known that he thought Bataan would fall soon, and Corregidor would be the last hope for the Philippines. But to be the last hope, he would need more food and men. As a result of this thinking, he ordered the transfer of food from Bataan to Corregidor. He also wanted to bring the Philippine division to Corregidor before Bataan fell, a transfer that never happened. Generals Wainwright and Parker already were suffering with meager supplies and food, so when MacArthur decided to take even more away from Bataan, the two generals understood that MacArthur had written off Bataan. Wainwright worried more about the lack of food than he did the Japanese. The men's health was deteriorating rapidly because of lack of food. They had been on half rations since January 6. They needed protein and vegetables. After one month on Bataan, men were losing considerable weight, and health problems such as malnutrition, beriberi, scurvy, avitaminosis, and amoebic dysentery increased every day. Men who needed to be alert became listless.[4]

Not wanting to give away what food he had—and knowing full well what the answer would be—Wainwright requested more food for Bataan. MacArthur turned him down, saying that the men on Corregidor needed the food and supplies so they could hold out after Bataan fell, a response that verified what Wainwright and Parker had inferred. MacArthur argued that the troops would have to fight on with less food, declaring, "If they run they will be doomed but if they fight they will save themselves."[5] In his biography, Wainwright wrote that had there been food in their bellies and any hope of being rescued by the United States, the men might have held out longer. As it was, there was little food; not enough quinine to treat malaria, which was on the increase; hot ninety-five degrees days in the shade; and the knowledge that the Japanese were coming.[6] Historian Schultz wrote, "By the end of January there was an eleven-day supply of meat and fish, a six-day supply of flour, and only four days' worth of vegetables."[7]

Regardless of their poor health and starving conditions, the troops had no choice but to fight. On January 29, the Japanese attacked Wainwright's new area near the southern part of the peninsula. The Japanese landed at Longoskawayan Point, Quinauan Point, and Anyason Point (map 7). In addition, they landed near the road south to Mariveles, where supplies went back and forth

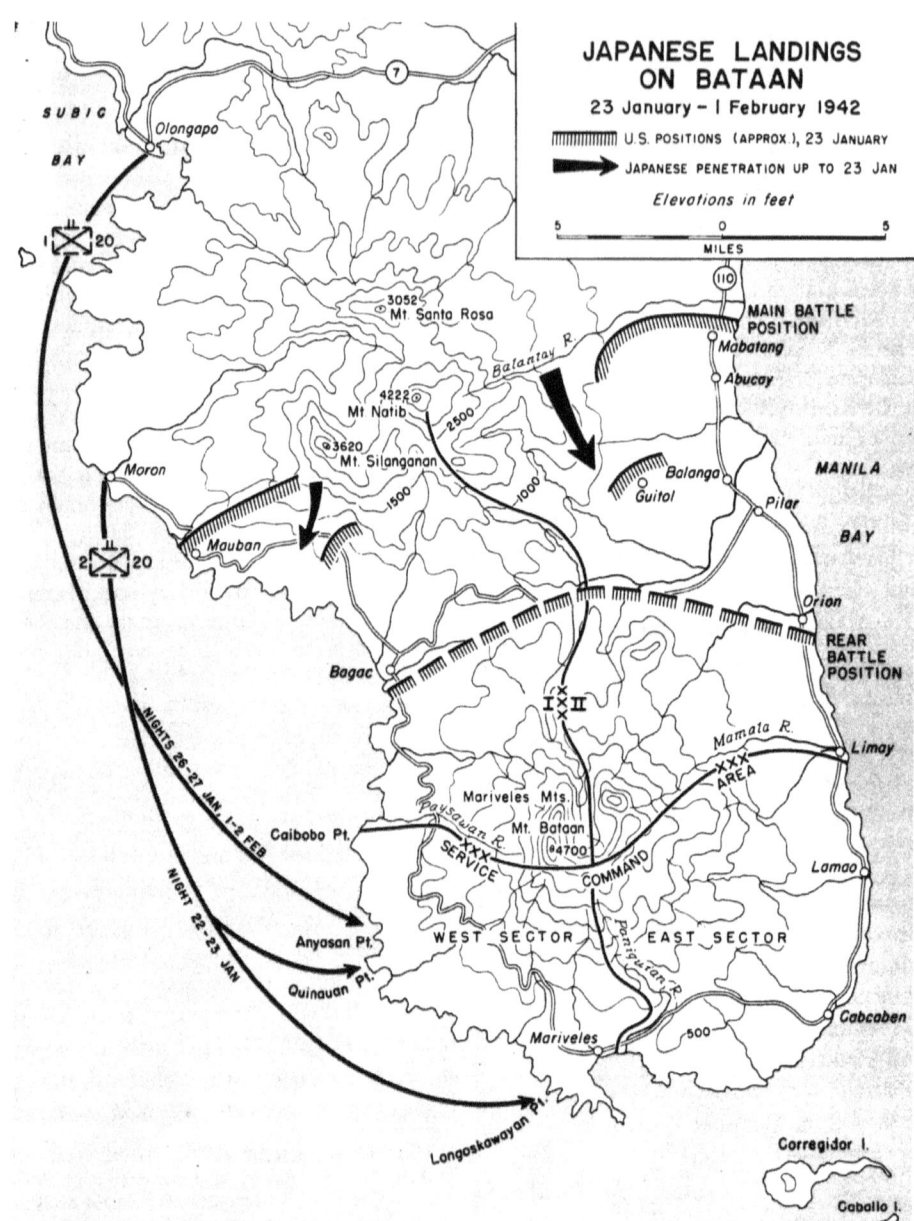

Map 7. Japanese Landings on Bataan. The map shows the new main line of resistance (MLR) running from Bagac to Orion. The Mariveles Mountains sit in the middle of the new area to defend. Corregidor Island can be seen in Manila Bay. Source: *The War in the Pacific: The Fall of the Philippines*, by Louis Morton.

to Corregidor. Berry and his 1st Division troops were engaged in battles at the new main line of resistance near Bagac and not in the southern region near the points.

This engagement, in which the Japanese tried to land on the southern coast, came to be known as the Battle of the Points. The battle lasted three weeks into February, with many different segments of the hungry, sick, and exhausted I Corps troops pushing the enemy back to the China Sea. The Japanese were shocked to have been beaten back; Bataan should have been theirs by now. It was a courageous, stunning, and all too brief victory for the defenders of Bataan.[8]

THE BATTLE OF THE POCKETS

In January and February, the Japanese continued to advance and win battles in the Pacific, invading Java, Singapore, and Sumatra. In Bataan, at the same time as the Japanese started the offensive at the points, they started a second offensive against the new MLR of I Corps, where Berry's troops were. Establishing the new line was not complete because the Japanese landings at the points had drawn off Wainwright's reserves. He was waiting for reinforcements, namely the 45th Infantry and the 11th Division. The 11th Division reinforcements arrived missing an entire regiment. Worse, as soon as the 45th Infantry arrived, MacArthur ordered it to move to his headquarters on Corregidor.[9]

Wainwright was dismayed. He had counted on the 45th Infantry to be in the center of his front line. Instead, he was forced to use the shambolic 1st Division that had reassembled just two days before the new attack. Berry's troops had not recovered from their withdrawal down the beaches. Weak from hunger and exhaustion, they had not finished setting up fortifications along the new line—using, it should be remembered, only mess kits and bayonets to dig foxholes and clear undergrowth. Even more significant, this division lacked essential artillery and automatic weapons.[10]

After first trying and failing to gain ground against the 91st Division troops on the front line, the Japanese found the weak spot where the 1st Division was still building entrenchments and attacked them on January 29, making it through the front line during the night. Rio recalled, "The firefight continued unabated through the night and well into the following day. The exceedingly thunderous din of battle gave evidence to what was, to me, the fiercest encounter so far in the war."[11] Communications to troops in the rear were cut off once again.[12]

Berry recalled that the Japanese came at the line hard and penetrated 1,500 meters through one of his company's outpost sectors, setting up fortified areas

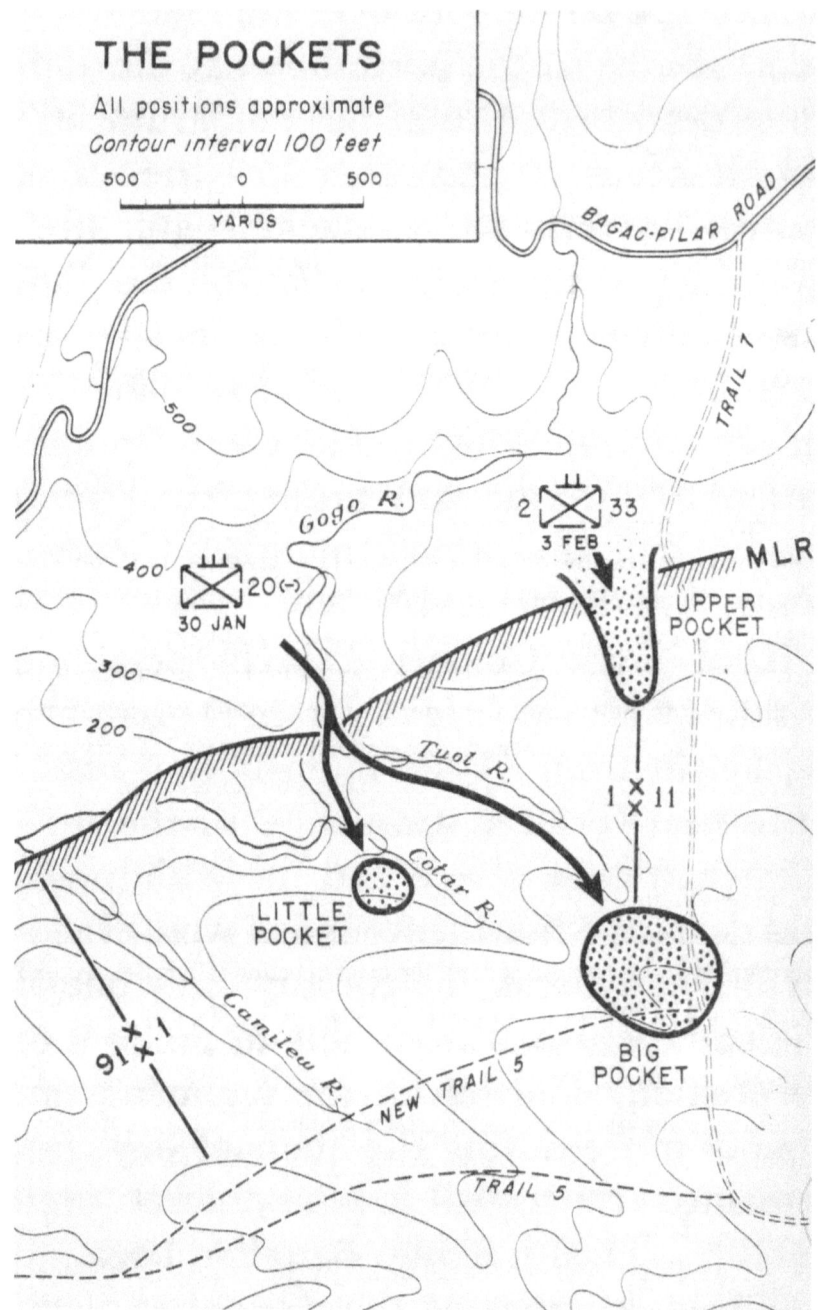

Map 8. Battle of the Pockets near the Tuol River. The numeral I designates the location of Berry and the 1st Division, and the numeral II designates the location of Brougher and the 11th Division. Source: *The War in the Pacific: The Fall of the Philippines*, by Louis Morton.

that bulged into I Corps' territory but surrounded by I Corps troops, mainly the ill-equipped 1st Division. The fortified areas became known as "pockets."

Once through the line, the Japanese divided themselves into two groups—one small force was on top of a hill four hundred yards behind the line. It would come to be called the "little pocket." Another, larger force stopped a mile behind the front line to the east, "close to the junction of two trails that US forces had been using to move troops and supplies."[13] That location came to be called the "big pocket." Days of brutal fighting followed. Very little progress was made by Segundo's worn-out 1st Division troops toward wiping out the enemy in either pocket (map 8).

In a letter dated February 2, a downhearted Berry wrote the following to Alice from the battlefield:

> Moved down here and am now in front lines again and have been fighting for 3 days. Japs have penetrated the lines and we have been trying and are still trying to eject them. Have had little success so far. We are fighting in the jungles where trees are thick and tall and at places the underbrush so thick you can hardy [sic] thru. Snipers get thru over lines and worry us no end and kill and wound men at times. A shot has just gone off near me and I have sent a patrol to investigate. Guns are roaring about 700 yards away and have been going on now for 3 days and nights. It is not pleasant and you can tell the world that what Sherman said was true.[14]

Berry told Alice that there were only five American officers, including himself, in both the 3rd and the 1st infantries that were under his command. Because American officers had the military training that the young Philippine officers lacked, the fact that only five were available meant that they were spread too thin to provide the leadership necessary. He explained the difficulties inherent in the Filipinos' lack of military training:

> I have a nice bunch of Filipino officers on my staff but are green and have a lot to learn. My regiment was organized Dec. 19 so you see it has had no training and that makes it very hard to get them to hold a position or recapture one that is lost.[15]

In closing his letter, Berry wrote that he was feeling very much alone and was preparing for the worst in the battles ahead. Adding to his depression, he had received no mail from Alice since he left Texas in late October 1941. Understandably, he worried something was amiss back home. Most troubling, he did not know if he was loved. The fact that he wrote to Alice from the battlefield showed how much he loved and missed her. The fact that Alice had apparently not written to him from the comfort of her home made him feel disheartened, forgotten. He wrote:

> Have never heard from you, Sweet, since I left home. Trust and hope everything is OK with you and the kids. Anxious about Tom's appointment. Hope you got my radios as I sent you two. Hope you are still writing me as I will get them someday I hope and will enjoy them so much when they do come.
>
> I am making my will which appears next below.[16]

He bequeathed all his worldly possessions to Alice and had three Filipino officers of his regimental staff witness his signature, "Lt. Col. K. L. Berry," reflecting the fact that he did not know he had been promoted to colonel, temporary status, on December 24, 1941.

In the meantime, no tactic, not even using tanks, had been successful in dislodging the Japanese from either pocket. General Segundo had sent in his troops to the little pocket but could not get the small force of Japanese dislodged because his young soldiers were unsure in battle and would not stay in place. It might have been that the young troops needed Segundo, their mentor, to lead the charge. Wainwright sent Gen. William Brougher's 11th Division to the big pocket, but tanks had immense problems with visibility getting through the thick, dense terrain. In many cases, the 11th Division Philippine Igorot soldiers jumped on the tanks, and, using a rifle or a club, pounded on the right or left side of the tank to guide it in the correct direction. Igorots were tribesmen from mountains of northern Luzon who joined the 11th Division. Before war broke out, the Igorots' "headhunting activities had been suppressed and now they found a new and lawful release to their pent-up spirit as they jumped off tanks and shot down into the Japanese foxholes and went after the Japanese with knives and grenades," as told to historian John W. Whitman by Zenon Bardowski. Wainwright called these soldiers the "greatest guides" he ever saw.[17]

According to Lt. Col. Edwin Aldridge, Wainwright was frustrated by the "repeated halfhearted attempts of the 11th Inf [infantry] because they failed to make any headway against the large group of Japs."[18] (Aldridge was General Jones's chief of staff and privy to private information.) Consequently, Wainwright went to Segundo's command post on February 5 to discuss changes in leadership. He wanted one leader to command all the attacks on the pockets. Generals Segundo, Brougher, and Albert Jones, commander of the 51st Division, were called to the meeting. Wainwright chose General Jones to command all the troops of the divisions currently fighting the pockets.[19]

Jones developed a plan to isolate each pocket and move in troops to surround each pocket. His plan was that the little pocket would be cleared of the enemy first. His forces would then attack the big pocket. Wainwright approved the plan, and February 7 was set as the date for executing it. Results of the first day were disappointing. Troops of Segundo's 1st Division surrounded the little pocket, but the Filipino weaponry on hand—rifles, a few machine guns, and a few automatic rifles—were ineffective in such jungled close quarters. Segundo's troops resorted to hand-to-hand combat with grenades and bayonets. These young soldiers had to make their own grenades out of dynamite sticks attached to bamboo stems. If the fuse was too short, it exploded before it was thrown. If it was too long, the Japanese just threw it back at the Filipinos. The Filipino soldiers were inexperienced and "timid" in offensive movements and needed to be led into combat by experienced men.[20]

Exasperated by Segundo's lack of success, Jones removed him from commanding the assault on the little pocket and put Berry in charge of encircling and taking the little pocket. Jones also wanted Berry to put American officers, rather than Filipino officers, in charge of offensive units, whenever possible. As noted earlier, Berry had only five American officers with him, but Jones wanted those officers to lead their young soldiers in the fight. Berry agreed. Jones and Berry discussed the plan in detail that night. One machine gun unit would be in place to scare the Japanese while other units surrounded the pocket, attacked, and pinched the encirclement tighter. Berry recalled, "The men would not advance it seemed unless some of the American officers were present and forcing them to do so but I believe taking street charge of the firing line exerted such an influence on their morale that we had no trouble in keeping the men in their positions during the night and were able to hold such ground as we had gained."[21]

The fighting went well on February 8, and the Japanese were close to being wiped out. Aldridge recounted, "Berry eliminated the small group first and after considerable difficulty repaired all wire defenses and later on changed the fence and put up additional bands."[22] Wainwright took note of Berry's aggressiveness in battle, and in the middle of the Battle of the Pockets, he appointed Berry to be commander of the 1st Division, replacing Segundo. Aldridge remembered, "The change in the 1st Div was immediate. Berry was everywhere and saw that things were done. I made one trip to his CP [command post] after he was given command. He had moved his CP forward where he could control things."[23]

Berry's letter to Jones confirms Aldridge's assessment, "On the final day of the Little Pocket fight, I personally visited the rifle company and machine gun platoon which forced the enemy out of the pocket and drove them into the open where during the next three day we killed at least 153 of them."[24] Even as he assumed command of the 1st Division, Berry continued to believe that he was still a lieutenant colonel.

On February 9, as Jones was preparing to execute the second part of plan, the encirclement of the big pocket, the Japanese attacked and pushed more troops through the line to assist their soldiers trapped in the big pocket. This incursion created another pocket of the enemy behind the lines, and it became known as the "upper pocket." The Japanese, having lost too many men, had decided to withdraw, and these new troops came to assist in the withdrawal of troops from the big pocket. To respond to this development—and without permission—Brougher took some of Jones's encircling troops away to stop the Japanese in the upper pocket from connecting with the troops in the big pocket. Brougher's "unauthorized transfer of troops"[25] forced Jones to delay his plan until he could secure additional other troops from elsewhere. On the other hand, Brougher's action prevented the Japanese from reaching their comrades in the big pocket.

To assist Jones with more troops, Berry brought his troops from the little pocket. Berry's troops had fought hard and learned that their jungle tactics worked. Berry placed his men in a position to prevent the two sets of enemy troops from joining forces at the upper pocket. By holding off Japanese reinforcements, Berry's soldiers made it possible for other units to continue their pincher movement inward around the big pocket. Fighting went on for three days. By the 12th, the fight for the big pocket was over, as the last of the Japanese slipped away, defeated. It had been a hard battle with many losses on both sides. Berry was proud of his young troops. They had fought hard to stop reinforcements from coming to the enemy. They were proud of themselves, as well.[26]

In the meantime, the battle for the upper pocket continued. Brougher, now in command because Jones had fallen ill with dysentery, took on the challenge of pushing the Japanese out of the upper pocket. On the 12th, units of Berry's 1st Division moved up to the west boundary of the upper pocket. The Japanese fought hard, but some began slipping away—they no longer wanted this ground because their comrades, survivors from the big pocket, had reached safety. In any case, the battle was not going well for them. To end the battle, the Filipinos laid down "a barrage of incredible intensity . . . turning the battlefield into an inferno,"[27] but still the Japanese fought on. Hand-to-hand combat with bayonets and grenades ensued. The Filipinos advanced, and the Japanese withdrew. By the 16th, the enemy was gone. Berry's troops had done their job well, as he knew they would. They had gained fighting experience the hard way. And Berry, who had written his will on February 2, lived to fight another day.[28]

Despite the brutal reality of war, the Battle of the Pockets was not without its poignant moments of humanity. Two Japanese came out of the pockets alive, and one "with a broken thigh was taken to Berry's command post." Apparently, Berry gave the almost-naked Japanese soldier something to wear and a can of milk mixed with sugar to drink. The soldier tried to stand and bow to Berry in gratitude, but the pain overtook him and he vomited up the sugared milk.[29]

FEBRUARY 16–28, 1942

Beyond the intense fighting in the Philippines, the broader war was a mixture of some successes and great losses for the Allies. One bright spot for the Allies was the attack on the Japanese at Wake Island by the USS *Enterprise* on February 24. However, demoralizing losses for the Allies persisted. The Japanese struck violently against Darwin, Australia, and invaded Bali. They even attacked an oil refinery near Santa Barbara, California, using submarine shelling. Even more troubling, the Japanese sank the USS *Langley*, an aircraft carrier, on February 26 and then went on to sink the largest US warship in the Far East, the USS *Houston*, in the Battle of the Java Sea.

Meanwhile, the fact that Japanese General Homma had not yet won in the Philippines angered the leadership in Tokyo. The delayed Japanese victory in the Philippines was consequential for the overall war effort in the Pacific. The ferocious battles in the Philippines kept precious Japanese forces tied up there. Moreover, Homma had to request more troops and more supplies after his defeats at the Battle of the Points and Battle of the Pockets. The Japanese were winning almost everywhere, but not in Bataan. Homma's position with Tokyo was uncertain.[30]

The Battle of the Points and Battle of the Pockets had given the Filipino soldiers a new spirit, a sense of accomplishment and pride. They had defeated and pushed the mighty Japanese back. They would be forever proud of their accomplishments in these battles. In better spirits than in his previous letter, Berry wrote home about his soldiers, "My division is rapidly gaining in experience and staying qualities and I have hopes that when the Japs next push comes that we will hang a lot of them on this wire [barbed wire defenses] and deal them a bloody repulse."[31]

The second half of February was comparatively calm on Bataan. Homma withdrew his troops to the north and waited for additional troops, food, and supplies. Additional troops were a necessity. Over the course of the battles on Bataan, the Japanese 16th Division of fourteen thousand soldiers had been reduced to seven hundred duty-worthy men. MacArthur and his commanders knew that the lull in the action was because Japan was taking time to reinforce its troops for a final assault on Bataan. MacArthur had no reinforcements, supplies, or food for his troops, and he knew none were coming. Depressed, MacArthur told two war correspondents "the end here will be brutal and bloody."[32]

In spite of the fact that the Americans and Filipinos had fewer men and supplies and less food, their commanders worked the troops hard preparing for the next battle they knew would come. Both corps prepared defenses, clearing ground and laying down a vast array of mines in all locations they believed the Japanese would use to reach them. The Philippine Scouts in I Corps even took this time to build a new road to connect with II Corps.[33]

Lack of food and medicine continued to be major problems. Col. Glen Townsend, 11th Infantry, wrote in his diary that their sixteen ounces of food a day had been cut to four ounces a day. There was no flour or vegetables, and the quinine was gone. But because the troops still held out hope that reinforcements and supplies would come, they soldiered on.[34]

During the brief lull between battles, Berry wrote Alice from the battlefield that he was now a colonel. He learned about his promotion in a conversation he had with Wainwright in late February, as they were supervising their troops building defenses. Wainwright told him that on February 6, he had been "made a jawbone Colonel." "Jawbone" meant that his rank of full colonel was no longer temporary but permanent. Berry was confused because he did not even know he was a temporary full colonel; no one had bothered to tell Berry he had been promoted twice in the space of six weeks. Wainwright informed him that the War Department had promoted him to full colonel (Temporary) on December 24. When Wainwright made Berry a division commander on February 8, he knew that Berry had been made a "jawbone Colonel" on February 6. Division commanders were usually generals, but Berry was wearing his "colonel's eagle." He wrote Alice, "Perhaps, if my division does well for a sufficient length of time I will get a star."[35]

Feeling hopeful after the success of the recent battles and learning he had been promoted, Berry continued his newsy letter to Alice. "My health is pretty good but feel I don't weigh over 170 at the outside.... Have pretty good supply of clothes, toilet articles and blankets now thanks to my many friends."[36] One of those good friends was Ed Aldridge. Aldridge had supplied Berry with "a lot of stuff—clothes and toilet articles" after Berry came off the beaches with nothing in late January. As the division operations officer for General Jones, Aldridge had not yet been promoted to full colonel. Berry explained, "Men on command jobs are promoted first and the staff comes second."[37] He continued, "Haven't seen H. C. Browne (Hal) since old Ft. McKinley days. Don't know where he is or what he is doing. I am pulling for him and can't understand why he has not been made."[38] Berry thought that Colonel Browne should have been "made" a general by then. Colonel Browne thought so, too.[39] Berry continued writing Alice that many of their friends that she would know were there, but "I am kept so damn busy that I never get to see them and don't know where they are even."[40]

He then expressed his longing for her by adding, "Sweet, please write me a long newsy letter and charter a phantom plane and send it to me. You might visit me a few days and live in my jungle home for a while. I'd build a special dugout for you and give you all the choice rice, fish, hash, etc., you could eat but you would have to go easy on the coffee."[41] He certainly longed to see her as his imagination indicated. The question was whether Alice longed to see him.

In February, Washington military leaders and the president knew that Bataan would fall soon because aid could not reach them. In a fireside chat on February 22, President Roosevelt made no mention of sending aid to the Philippines. His chat was all about sending help in all other directions. It was widely inferred by commanders and American soldiers on Bataan that the United States had forsaken them. Privately, Roosevelt did not want his top general in the Pacific to be killed or taken prisoner; consequently, he decided to pull MacArthur out. On February 23, he ordered MacArthur to leave. MacArthur was surprised by the order and made no move to leave at first. He knew what his troops might think of him if he left. He wanted assurances that people would know he was ordered to leave. He was given those assurances. After a few weeks, MacArthur decided on March 12 for his escape.[42]

MARCH 1–15, 1942

MacArthur's decision to leave was kept a complete secret from just about everyone, including General Wainwright, who was chosen to command thousands of starving, sick men in both I and II Corps to fight to the bitter end. When the word got out that MacArthur had escaped during the night of March 12, a crestfallen Rio wrote, "Our initial reaction was one of utter dejection; it was as if the final nail had been driven into our coffins."[43]

Rio's reaction to the news of MacArthur's departure from the Philippines was widely shared among the troops. The morale of enlisted men and junior officers turned bitter. Senior officers kept their dismay to themselves at this time, trying to keep morale up among the troops. Some officers, like General Brougher, vented their bitterness later when they had become prisoners of war.[44]

The decline in morale following MacArthur's departure was not owing to his popularity among his troops. On the contrary, before his departure, MacArthur was already called "Dugout Doug" by the troops because he never visited the front lines and only came to Bataan one time, on January 10. After MacArthur's departure, his reputation only worsened. Hostile poems and songs were created about him and his leaving them to fight and die. Here is the first stanza of the song "Dugout Doug," which was sung to the tune of "Battle Hymn of the Republic":

> Dugout Doug MacArthur lies ashaking on the Rock
> Safe from all the bombers and from any sudden shock
> Dugout Doug is eating of the best food on Bataan
> And his troops go starving on.[45]

To keep the troops' spirits positive after MacArthur's sudden departure, commanding officers suppressed their feelings about MacArthur and touted the leadership of General Wainwright. The troops loved Wainwright, who was known to visit the front lines frequently, sometimes at great peril. This was his way to boost morale—the only thing he could give his soldiers, he said. Later when Berry was a POW, he wrote about Wainwright, describing him as "a soldier's general—where the fighting is you find 'Skinny;' he takes no thought of his own safety but gets up where the bullets are flying; generally carries a rifle so he can have some of the fun (?) himself; I have been mighty proud and fortunate to have been under his command and hope to serve under him again."[46]

Washington, DC, put Wainwright in total command of the US Forces in the Philippines (USFIP) and promoted him to lieutenant general. This move upset MacArthur because he had planned to direct all operations in the Philippines from Australia, but he had failed to communicate this idea to Washington, DC, in a timely fashion. As it turned out, his poor timing did not matter. When Gen. George Marshall in Washington heard of this plan, he overruled MacArthur, noting to the president that directing a battle from four thousand miles away was not logical.[47]

Wainwright turned to Maj. Gen. Edward King to serve as commander of Luzon Forces Corps I and II on Bataan. Gen. Albert Jones became I Corps commander. The decision to make Jones the I Corps commander angered General Brougher, according to Aldridge, who wrote, "Was Brougher sore and believed he should have had the job. Has often said that he was the one recommended but I happen to know different."⁴⁸ General Parker remained the II Corps commander. Wainwright said bravely to his aides as he took command, "Lee marched on Gettysburg with less men than I have here. We're not licked by a damn sight."⁴⁹

MARCH 15–31, 1942

Before he left for Australia, MacArthur ordered the rations for the men on Bataan cut yet again to "three-eighths of normal."⁵⁰ Food and supplies dwindled to almost nothing as the Japanese kept a tight circle of ships blockading the Philippines. After considering all that he could do to help his starving men, Wainwright made the decision to order the horses of the 26th cavalry be shot and eaten. A cavalryman at heart, Wainwright found his decision unusually painful. It was made no less painful by the fact that there was no more food to feed the horses. To demonstrate to those near him that he meant all the horses, Wainwright ordered that the first horse to be shot had to be his horse, Joseph Conrad.⁵¹

Assuming his new role as commander of USFIP, Wainwright moved into his new headquarters on Corregidor on March 21. One of his first decisions was to send food and supplies from "the rock" to the troops on Bataan for as long as he could. He had inherited a command of two corps who were fading daily in their capacity to fight. Dengue, malaria, and starvation had cut their combat readiness to a shockingly low level. By mid-March, getting a little extra food to Bataan from Corregidor was almost too late to help the troops. Generals King, Jones, and Parker knew they had only meager supplies and food to stretch until April 15. They reckoned they would be totally without food by that date. After MacArthur left, their troops now knew they had been written off.⁵²

In the meantime, the Japanese had begun a propaganda campaign, dropping leaflets addressed to the Filipino soldiers, trying to entice them to desert the Americans. The leaflets invited the soldiers to accept a "safe conduct pass" and pictured pretty girls eating good meals at Manila restaurants. Loudspeakers near the front lines played Filipino songs sung by soft female voices at night. But these propaganda efforts failed; most soldiers made fun of these leaflets and remained in the Philippine Army.⁵³

Sporadic but intense fighting continued during this time as the Japanese probed the front lines of both I and II Corps, deciding where they would make their big push when a date was set. Berry noted that from mid-February until the start of the April Japanese offensive, things were fairly quiet along the division front except for small fights along the outpost, patrol skirmishes, and enemy shelling and bombing. Berry had to keep his troops alert because he knew thousands of fresh Japanese soldiers were on their way for the final push.[54]

APRIL 1–10, 1942, JAPANESE FINAL ASSAULT ON BATAAN

In late March, in his new role as commander in the Philippines, Wainwright reached out to Gen. George Marshall in Washington to secure food for his starving troops. When Wainwright explained his requirements for food sufficient for more than one hundred thousand men on Bataan and Corregidor, Marshall was astonished. He asked again for the count. He could not believe that there were that many men on Bataan and Corregidor. Marshall's skepticism was not surprising. Earlier in the year, General MacArthur and General Sutherland had rejected the Bataan quartermaster's requests for food for over one hundred thousand men, believing that the request was overblown. Subsequently, they had sent Marshall an update memo reporting that there were seven thousand white combat troops and thirty thousand Filipinos on January 3. This number was less than half the actual number of troops in the Philippines.[55]

Wainwright told Marshall, "To be utterly frank, if additional supplies are not received for Bataan by April 15, the troops there will be starved into submission."[56] Rations were down to 1,500 calories a day in February, and by March the calorie intake was 1,000. Soldiers were seriously losing muscle. Marshall asked for MacArthur's reaction to this news. Still angry that he was not in charge of conducting the Philippines from Australia, MacArthur took a gratuitous swipe at Wainwright. He disputed Wainwright's claim that food would be gone by April 15 and hinted to Marshall that perhaps Wainwright had not been rigorous enough with conserving food.[57]

In any case, Wainwright's request came too late for anything to be done about food. The Japanese had begun their offensive, choosing Mount Samat, in Parker's II Corps sector, as their main target (map 9). Wainwright had predicted that the Japanese would choose the II Corps line because the East Road, within II Corps' sector, was the only road that led to Manila. He had warned Parker accordingly. Unfortunately, the troops in I and II Corps, starving and sick from malaria and dysentery, had no fight left in them. They were at their weakest point after a three-month-long starvation diet. Soldiers were sneaking off to towns to try to find food. Thus, when the final intense push in II Corps began on April 3, Good Friday, many of the units just fell apart as young soldiers ran away.[58]

In one of the coincidences of war, on April 3 the troops of Japanese Gen. Akira Nara faced the troops of Philippine Division Cmdr. Gen. Vicente Lim. As

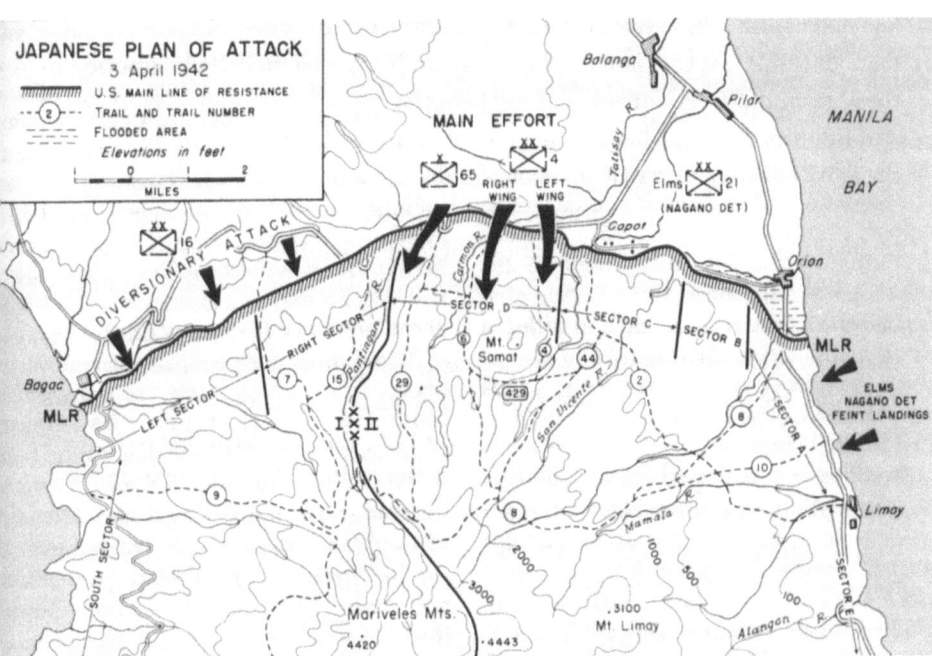

Map 9. Japanese Plan of Attack, April 3, 1942. The map clearly shows how the Japanese planned to take Mt. Samat and use most of their forces against II Corps. Source: *The War in the Pacific: The Fall of the Philippines*, by Louis Morton.

it happened, Lim and Nara had been classmates at the US Army's Fort Benning in 1926–27. Nara is reported to have said that he liked Lim. It would be interesting to know whether they were aware that they were facing each other in battle. Their past connection extended to Berry, who as a young captain had also taken classes in military tactics at Fort Benning in 1926–27, at the same time as Nara and Lim. It is likely that Berry knew both men. In the intervening years, much had changed. Berry now commanded a division, and Nara and Lim, no longer classmates, were generals on opposing sides of war.[59]

On the fateful morning of April 3, 1942, American and Filipino observers could see and count the batteries of artillery and mortars being aimed at them—nineteen batteries of artillery and ten batteries of mortars (map 9). A little to the east, 196 big guns were arrayed against them. In addition, Japanese bombers took off from former US air bases and dropped sixty tons of bombs that day. Philippine troops were strafed continuously from planes above. The assault focused narrowly on Gen. Maxon Lough's sector in an effort to blow a hole in the line and reach Mount Samat. Philippine and American troops were outgunned and outmanned. After five hours of bombardment from guns and planes, Commander Lim's 41st Division was blown apart. Nara then turned his attention to the 42nd and 43rd infantries, tearing them apart as well. Col. Loren Wetherby's 41st Infantry retreated because he did not receive the orders to stay and protect the flank of Brougher's 11th Division. Chaos reigned.[60]

For the next few days, fighting was extremely fierce. The Filipinos lost ground in the II Corps sector, and the Japanese succeeded in reaching their goal, Mount Samat. Wainwright came over to Bataan from Corregidor to see for himself how bad the situation was. He had ordered a counterattack, but what he found was that most of the infantry units were starving, disorganized, and depleted of men and guns. He was able to get some food to Bataan, but the effort to distribute the rations widely to the troops in the middle of the fighting failed. More soldiers were deserting to find food. In addition, as the relatively healthy men starved, over eight thousand men lay in the two hospitals on Bataan suffering from battle wounds, dysentery, dengue, and malaria. One colonel went to Hospital No. 1 and told the sick men that volunteers were needed desperately, and after a period of silence, about seventy-five men very courageously got out of their hospital beds and joined the colonel. Just as they were leaving, Japanese airplane bombs hit the hospital, killing seventy-three men and wounding 117 more.[61]

On April 6, after a day of trying to counterattack, the American and Philippine troops in II Corps found themselves instead ordered to form a new defensive line. The 41st and 21st divisions and one regiment of the 51st Division were gone. General King ordered Col. Edmund Lilly, with remnants of the 51st Division, and Gen. Clifford Bluemel, with what was left of the 31st Division, to form a defensive line together. The effort failed. In the fog of battle, Parker's II Corps headquarters did not even know their troops had withdrawn to form a new line. As the Japanese advanced with tanks and artillery, Lilly and Bluemel's men panicked and ran away.[62]

Bluemel, commander of the 31st Division, which was part of Parker's II Corps, was more than angry at Parker's headquarters, and by April 8, when II Corps headquarters called Bluemel and told him to form and hold a new line, he refused to talk to anyone but General Parker. When Parker got on the phone, Bluemel blasted him for the lack of help from his headquarters with reconnaissance, food, ammunition, and even Parker's headquarters' officers help on the line. Bluemel thought Parker should have sent his headquarters' staff to help on the battle line. Bluemel then spoke with General King and finally with General Wainwright, who told him to use his own judgment to maintain a line.[63]

Meanwhile, seeking to gain the initiative, Wainwright ordered General King to launch an attack with Brougher's 11th Division in I Corps. Instead, King sent a member of his staff, Gen. Arnold Funk, to Corregidor to tell Wainwright that the situation was critical and surrender imminent. Funk told Wainwright about everything he saw when he had surveyed the front himself: II Corps had been destroyed, and I Corps was in no shape to fight. But Wainwright had orders from MacArthur not to surrender, so he told Funk that he was ordering General King and General Jones to attack with I Corps.[64]

To be clear, Wainwright ordered General Jones, I Corps commander, to take his forces north and east and counterattack. Jones made it known that his soldiers, who were already falling back to the Binuangan River to form a new line, had nothing left in them to attack the Japanese to the east. It would have taken a couple of days to fight their way into that position, and Jones knew his starving troops could not do it. Food rations were essentially depleted; they did not have enough to even make it to April 15, as first thought.[65]

General King checked with Wainwright to make sure he was still in charge of all forces on Bataan because the order to Jones had come from Wainwright. Wainwright, realizing he had skipped the chain of command, agreed that King was still in charge of I and II Corps. With that knowledge, King told Jones to continue retreating with his troops. Afterward, King ordered "all depot commanders and warehouse personnel to prepare to destroy their installations that night."[66] Without telling Wainwright, he had already made up his mind to surrender I and II Corps on Bataan. King made the surrender decision himself on April 8 without conferring with his officers, presumably so they would not get any blame. More importantly, knowing that Wainwright had orders not to surrender, King did not inform Wainwright of his decision. With these actions, King took full responsibility, believing he would be "court-marshalled for surrendering the largest force the United States had ever lost."[67]

CHAPTER 8

Surrender and the Bataan Death March

THE CONFUSION OF SURRENDER

Berry heard about the surrender late in the afternoon of April 9. His officers then tried to pass the information to all the division units, which were spread over broad territory—a battalion of four rifle companies of the 1st Infantry was on the first delay position at the front line, covering the main body; another part of the division was near the Pantingan River; and another was near the division compound. As the word of the surrender was being delivered to Berry's soldiers, the Japanese continued to attack his four rifle companies—it's not clear if it was willfully or unknowingly—even after his troops had stacked their arms and were flying white flags north of the main line.[1]

Sadly, Berry's small battalion was wiped out, and he wrote, "This small battalion of the 1st Regular Infantry (PA) put up a glorious fight against overwhelming odds and was destroyed. The two American officers with this battalion were lost as were many of the Filipino officers and men for few survivors were ever found."[2] Mourning the loss of these troops, Berry blamed General Brougher, commander of the 11th Division, and Colonel Townsend, commander of the 11th Infantry, for not being in the right place to have helped the small battalion. Berry wrote, "As it [the small battalion] was pushed west on Trail 8 into the sector of the 11th Philippine Division it was found that this division, in violation of orders of the 1st Philippine Corps, had gone to the rear and abandoned its position."[3] In his private notebook, Berry wrote that General Brougher "pulled out with his entire division on Apr 9–10 leaving rt. flank of 1st Div. exposed and probably caused 1st Div. to lose one entire infantry bn. [battalion] including two American officers."[4]

Because Berry expressed such animosity toward Brougher and Townsend, it is important to know why the 11th Division was not where Berry thought it should have been. According to Brougher, General Jones's orders were to: "Assemble the men in bivouacs, stack arms, put up white flags, and wait for the Japanese to come in and receive surrender." On the evening of April 9, Brougher did as he was ordered to do. His troops stacked their weapons, flew white flags, and lit bonfires to show the Japanese the way to come in. Ignoring these signs of surrender during the night, "The Japanese came in with machine guns blazing and shot into our soldiers disarmed and huddled in their bivouacs," according to Brougher. His terrified men "took off down Trail 7 toward Mariveles." Brougher and his officers "debated pro and con the merits of sticking it out or following the men to the rear." The terms of the surrender were unclear to Brougher, but "with machine gun bullets whistling all around," he wrote, "we got the impression the Japanese were giving no quarter and that we would all be slaughtered if we remained where we were. Finally, we decided to follow our men to the rear and took off."[5] Brougher found his men at the rear on April 10 and surrendered his division.

Meanwhile, according to Aldridge, General Jones "directed [Gen. Luther] Stevens to go up trail and see if he could straighten out the 11th Div [Brougher's] which had abandoned its position, and were coming down en masse, throwing away unit arms, etc., along the trail."[6] Stevens reported back to Jones that earlier in the night "a bunch of Japs came up trail 8, and when near CP 11 Div, near Brougher's command post, and fearing for their lives, Brougher and the rest of the 11 Div just pulled out. They did not notify Murphy, Berry, McDonald or our CP [command post]. They were just too darned scared to stop. It was a perfect example of what might be expected of Brougher," according to Aldridge.[7]

Apparently, there was no love lost between General Jones and General Brougher. Brougher had angered Jones in the Battle of the Pockets when, without authority, he took Jones's soldiers away just as Jones was going to attack the big pocket. The incident of Brougher's men running down the trail and then Brougher himself leaving his post was a second time Brougher did not follow Jones's orders. Jones had ordered him to stack arms, fly white flags, and await the Japanese. And, a third time was when, contrary to Jones's order, Brougher allowed two of his regimental commanders and Major Volckmann to escape to join the guerilla fighters on Luzon. Evidently, General Brougher did not think he had to follow General Jones's orders.[8]

The 11th Division's retreat soured Berry's opinion of Brougher. He believed that Brougher had run away like a coward instead of being there for Berry's battalion. Berry's assessment of Brougher was also influenced by Aldridge's account of Brougher's actions that night as reported by General Stevens. With the 11th Division having retreated, Berry placed an engineer battalion and an infantry battalion astride Trail 9 to guard against any further Japanese attack. By the next morning, April 10, the Japanese knew about the surrender and there was no more shooting.[9]

However, in his autobiography after the war, Wainwright extolled Brougher's virtues, much to General Jones's dismay. Brougher received medals for his leadership, and his division received presidential unit citations. It may be that the chaos and lack of coherent communication as to where this or that company or infantry was supposed to be, together with the lack of knowledge about the terms of surrender, meant that the loss of the small battalion was no one's fault but that of the Japanese, but Berry, Aldridge, and Jones had not thought so.[10]

COLONEL BERRY SURRENDERS THE 1ST DIVISION

When General King surrendered, he believed his surrender applied to all the troops on Bataan, but that was not the case. Under Japanese order, each American and Philippine unit had to find a Japanese unit and unconditionally surrender to that commanding officer.[11] Lieutenant Rio, who was at 1st Division headquarters with Colonel Berry when the news of the surrender came in, said that Berry was ordered to "contact the nearest Japanese Commander and formally surrender the Division."[12] Rio could not detect "any particular emotional reaction" from Berry when the surrender news came to him, even though he was about to place his fate "in the hands of a ruthless enemy." Rio recounted that Berry "went about his last tasks as Division Commander—to prepare to surrender his command in as orderly a fashion as possible."[13]

On the morning of April 11, Berry prepared to surrender the 1st Division by drawing a map for his driver and Rio showing them where he thought the Japanese might be. Rio's account of the surrender follows. He began, "The latest reports mentioned of the enemy having penetrated far to the rear of the II Corps lines and it was toward that area the Division Commander decided to look for them."[14] Berry asked Rio to dress in his finest officer's uniform and told him not to bring a gun. Rio was fearful about the risk they were taking because he was not sure that all the Japanese had heard of the surrender. The trip could be dangerous. To mitigate the danger, Rio made a large white flag out of a bedsheet, sat in the front seat with the driver and held the flag out the window. Berry sat in the back seat. Rio wrote, "We had hardly gone five miles on the trail in the southeast direction when a group of Japanese soldiers stopped us and surrounded the car with guns ready. One of them, with grunts and hand motions, ordered us out of the car. My heart leaped within me and my mouth started to dry up fast, but I held on to the white flag as we were lined up against the car and searched for any weapons."[15]

The Japanese soldiers proceeded to take anything of value from them, including their watches. When a young Japanese officer appeared, Rio wanted to complain about the loss of their watches, but Berry cut him off, saying "Never mind that, Lieutenant. Just try and tell him instead what our purpose is."[16] The officer did not know English or any Philippine dialect, forcing Rio to use charades and sign language to make the point, "My colonel wishes to talk with your superior officer about surrender."[17] After some minutes, the officer nodded and motioned them back into the car where he sat in the back with Berry. Another soldier was up front with Rio and the driver. Nervous, Rio continued holding the white flag out the window as they drove for fifteen minutes and then stopped at a large encampment of Japanese soldiers. Berry, Rio, and the driver stayed in the car while one of the soldiers ran to get an officer of the camp to come to the car. Rio wrote, "As they were approaching us, the soldier in the back seat got out of the car; Colonel Berry and I did the same."[18] The officer saluted Berry and then began speaking Japanese-accented English to Berry. With a bit of humor, Rio wrote that he had to jump in and act as an interpreter because the two men could not understand each other even though they were both speaking English. Berry spoke with a "full-blooded Texan accent" and the other with an accent that was "obstinately Japanese." Rio wrote, "I repeated in my best 'orientalized' English whatever Colonel Berry said to the Japanese officer and restated the utterances of the Japanese in something more understandable to the American."[19]

A meeting was arranged with the Japanese colonel who commanded the regiment. In his tent, they sat awkwardly six feet across from each other on the dirt floor, where the Japanese commander started the conversation through his interpreter. Rio wrote, "He started the conference on a rather cordial note by drawing Colonel Berry into conversation about their respective homes and families, as well as their favorite sports. Now and then he would crack a joke which made the officers with him roar with laughter but which did not come out as funny in the English of the Japanese translator. Colonel Berry and I laughed anyway, more out of politeness than in mirth."[20]

The business of the surrender took place after a few minutes, and the Japanese commander instructed Berry, "to have the personnel of his Division assembled in three separate groups: all the Americans, the Filipino officers, and the rest of the Filipino soldiers."[21] When Berry and Rio returned to their car, several Japanese soldiers had commandeered it, so Berry decided that they would walk back to where the division was waiting instead of trying to argue with the Japanese soldiers. Apparently, Rio told the interpreter that they needed the car. After hearing about the situation, the Japanese commander approved of Berry taking the car.

On the morning of April 12, the Japanese regiment came early, and its commander sent for Berry. The handover of American and Filipino soldiers took place without incident. Rio had planned to stay with Berry, but the Japanese ordered the Americans separated from the Filipinos. At that point, Berry told Rio to join the rest of the Filipino officers. Rio remembered their farewell, "'Goodbye and good luck, sir,' was all I could say as I saluted him for the last time. 'Goodbye, Lieutenant, and thank you very much' he said, shaking my hand."[22]

As it turned out, Rio did not join the other Filipino officers. After seeing Filipino enlisted men driving trucks for the Japanese, Rio got the idea to shed his officers' insignia so he could blend in with the enlisted men. He hoped he could get a job that would also afford him the opportunity to escape. He did just that. By April 25, he was on his way to a new life. He eventually became a guerrilla fighter against the Japanese for the rest of the war. After the war, Berry made sure Rio received the Bronze Star Medal for his actions in the Battle of Bataan.[23]

Hundreds of other Philippine officers and noncommissioned officers (NCOs) in the Bagac area from the 91st Division were not so lucky. They were separated from their American officers, just as Berry's troops had been, and then ordered to move east along the road from Bagac to Balanga on their own. Japanese soldiers who were camped along the Pantingan River stopped the prisoners and divided them into two groups—officers and NCOs in one group and privates in the other. The Japanese then ordered the privates to keep on marching down the road to Balanga. The officers and NCOs (about four hundred of them) were bound, hands behind their backs, and tied to one another. At that point, the Philippine prisoners were marched to the edge of a ravine, where the Japanese soldiers shot them in the back. As fate would have it, Berry's young Filipino officers were far more fortunate than those with the 91st Division. The treatment of the prisoners depended entirely on which group of Japanese soldiers held them captive.[24]

The surrendering Philippine and American troops thought history would be unkind to them even though they fought their best under the most trying conditions. General King assumed he would be court-martialed for surrendering the largest contingent of soldiers in war. Despite what some may have thought was a failure, the battle on Bataan was actually successful: The prolonged battle for the Philippines had tied down a large force of Japanese soldiers for four months. Moreover, Japan could not complete the conquest of the Philippines without committing additional troops to the battle, troops that would otherwise have been sent to fight elsewhere. As a consequence of the heroic efforts of the Philippine and American soldiers, the Japanese timetable for invading Australia was delayed, giving Americans and Australians additional time to gather their strength to fight back.

THE BATAAN DEATH MARCH

After American and Philippine forces surrendered on Bataan, the Japanese had to figure out what to do with their seventy-eight thousand prisoners. They were woefully unprepared for the number of prisoners they received from the surrender. They had estimated that the battle would continue until at least the end of April, giving them time to make preparations for prisoners to be trucked to the San Fernando train station from Bagac or Mariveles. Their estimate was based on the assumption that the American and Philippine troops would be capable of carrying on the fight throughout April. This assumption was wrong. The troops on Bataan had been on extremely slim rations for weeks and would run out of food completely by early April. Nor did the Japanese expect that a high percentage of the prisoners would be suffering from malaria, dysentery, and dengue. As a result, thousands of sick and starving prisoners were going to have to walk to San Fernando, sixty to sixty-five miles away. American soldier Lester Tenney, who had been an enlisted man with the 192nd Tank Battalion, lamented, "If only we had heeded Gen. King's message to save some vehicles for moving our forces to another location, if we had not destroyed all of our trucks, maybe would have been able to ride to prison camp."[25] In fact, King had saved some trucks, but they were confiscated by the Japanese, who were moving equipment toward Mariveles for the attack on Corregidor.[26]

A lucky few were spared the long march that lay ahead for most of the prisoners. A group of 125 prisoners from the Luzon Force headquarters and the Tank Group headquarters were trucked north, arriving at the first prisoner of war camp, Camp O'Donnell, on the 11th. A few days later, another 250 arrived in fourteen trucks. They were supposed to get the camp ready for all the prisoners who were coming. While these 375 officers and soldiers prepared the camp, the vast numbers of those who surrendered on Bataan were marched to San Fernando, the train depot, in the heat, with almost no food or water, over

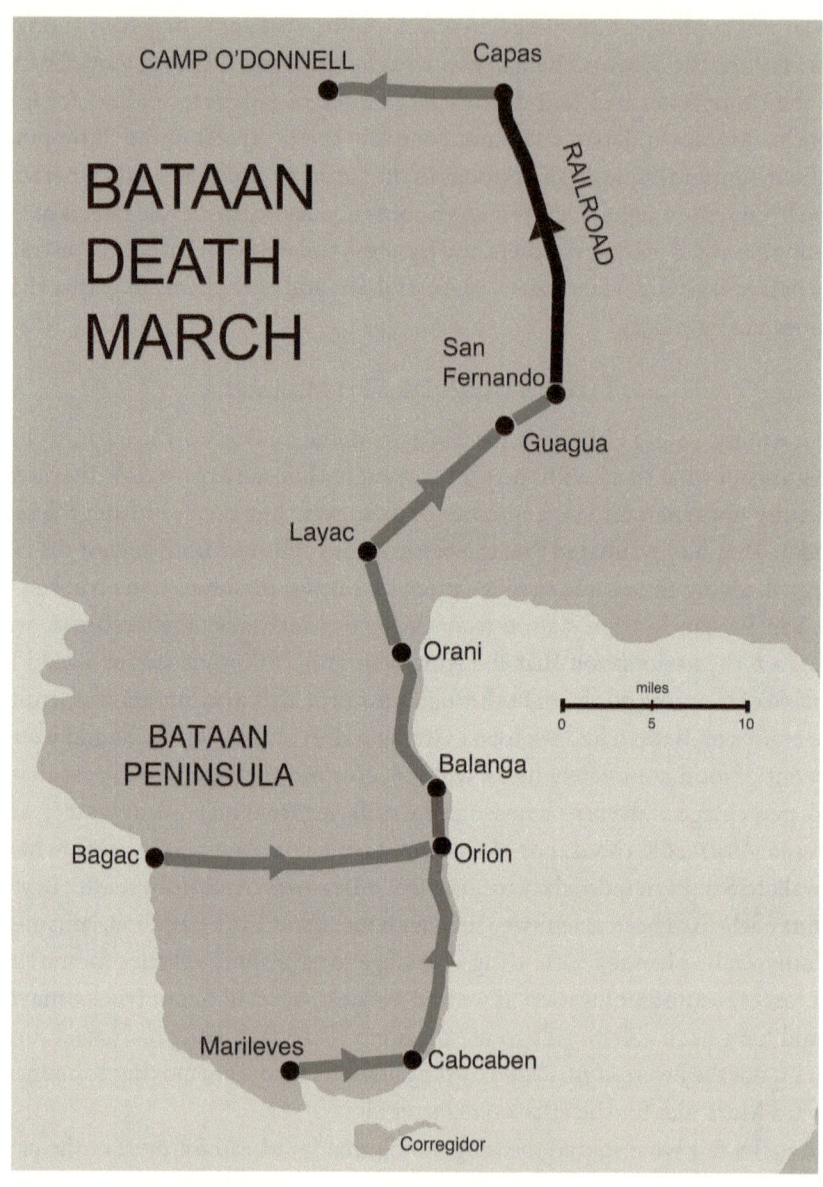

Map 10. Death March Map. Colonel Berry started the Bataan Death March at Bagac and hiked to Balanga, where he connected with Colonel Aldridge for a while. From Balanga, Berry hiked to Orani, then Lubao (not on this map), and then to San Fernando, where he got on a train bound for Capas. From Capas, he hiked to Camp O'Donnell. Source: "Bataan Death March," National Museum of the United State Air Force.

a span of three weeks. This prisoner march north became known as the Bataan Death March. But it was not just one march; there were several marches of men coming from different areas of Bataan over a three-week period, all walking to San Fernando.[27]

The Death March had already begun on the eastern side of Bataan with II Corps troops when Berry and his troops on the front line in I Corps were surrendering. Most of II Corps and a portion of I Corps in the rear started the march from Mariveles on the 10th and 11th. But those with Berry at the front of the I Corps' line started the march from Bagac. After Berry handed everything over to the Japanese on April 12, he was held in Bagac for two nights and one day with only one ball of rice to eat before setting off on the grueling march.[28]

Depending on where a given prisoner started, the march to San Fernando was between sixty and sixty-five miles—all in the unbearable heat of April, the hottest month of the year in the Philippines. Filipinos called this time of year "the days of dryness, the season of drought."[29] The prisoners had no relief from the hot sun beating down on their bodies and, especially, their heads. Some men luckily had their helmets, others had caps, and some had rags for protection, but many more had nothing at all. A desperate Capt. Mark Wohlfeld from the 27th Bombardment Group saw a dead Filipino soldier with a helmet by the side of the road. He asked the Japanese soldier guarding him through words and pantomime if he could get the helmet. He was not sure at all if the guard would allow him to get it or if he would be shot trying to get it, but he finally said, "Fuck it, I'll try it."[30] He succeeded in getting the helmet, which probably helped him survive the tortuous "sun treatments" when the guards made them sit out in the open all day long under the hot sun with no water. One wonders whether he was able to keep the helmet. Many of the Japanese took helmets away from the Americans so they would suffer from the sun treatment.

To the Japanese soldier, who was trained to fight to the death, surrender was seen as an act of cowardice and not to be pitied. "The warrior philosophy associated with the traditional Bushido code was reawakened," said Lester Tenney, who had served with Company B of the 192nd Tank Battalion on Bataan.[31] Consequently, the surrender of thousands of Filipino and American troops was an abomination to their captors. Barbarous treatment of the "cowardly" prisoners ensued. A Japanese interpreter told Tenney and others with him, "You are lower than dogs. You will eat only when we choose to feed you; you will rest only when we want you to rest; we will beat you any time a guard feels the need to teach you a lesson."[32] The Japanese were hardest on the prisoners who were weak and sick or appeared weak. Tenney wrote, "I vowed to walk with determination, my head high, shoulders back, and chest out. This posture would make me feel righteous, and the guards did not harass or belittle men who looked healthy and in control of themselves."³³

Another who held his head high was Berry. While still in Bagac, he had had a skirmish of sorts with a Japanese soldier who hit him because he wanted Berry's pen. After Berry gave him the pen (which he said was broken anyhow) they left him alone most of the way on the eighteen-mile "hike," as Berry called it, from Bagac to Balanga, permitting him to lead his officers without guards around them (map 10). If Berry's pen had been made in Japan, he would have suffered a different fate. William Dyess, a lieutenant and then captain with the 21st Pursuit Squadron, saw Japanese soldiers drive "crushing blows" with the butts of their rifles into the faces of Americans found with Japanese-made items. Others were beheaded.³⁴

Because of the battles, the roads were in horrible condition for walking. Tenney remembered, "The entire road was now nothing more than potholes, soft sand, rocks, and loose gravel. Walking for any long distance or for any extended period of time was going to be a painful and difficult experience."³⁵ Hundreds of men developed blisters so large and painful they could not walk and fell behind the columns of prisoners marching north. Some sat in the road or off to the side where Japanese soldiers found them and either prodded them with their bayonets to get up and keep going, or, in many cases, just killed them with the bayonet. Even Berry was prodded along by a bayonet at one point in his hike. He and one of his officers, Lt. Col. Victor Gomez, were carrying a young American major from the 91st Division named Judson Crow until the Japanese forced them at bayonet point to leave Crow on the roadside and hike on without him. In hopes that Crow could care for himself, Berry left Crow his meager medical supplies. After the war, Berry learned that Crow died on the march.³⁶

If physical torment or outright murder were not enough, the Japanese would not allow the prisoners to obtain fresh water from artesian wells by the side of the road. Dyess remembered, "Artesian wells lined the road. It seemed to me I could smell water. But we all knew a bullet or a bayonet awaited the man who might try to reach the wells."³⁷ Their bodies, craving water, went into deep dehydration and "they stopped sweating and urinating. Their saliva turned adhesive and their tongues stuck to their palate and teeth. Their throats started to swell, and their sinus cavities, dry and raw from the dust and heat, pounded with a headache that blurred their vision. Some men got earaches and lost their hearing."³⁸ Sometimes the Japanese guards allowed them to drink from dirty, feces-laden streams with dead bodies in them and then laughed at them. Of course, this water made the men sick, and if they became too sick to continue walking, they were beaten, shot, bayoneted, or beheaded. If one prisoner tried to help another, he could suffer the same fate. Lt. Kermit Lay, from Berry's I Philippine Corps, tried to help a Captain Miller, who could barely walk because of the huge blisters on his feet. As he and another prisoner held Miller up as they walked with him, a Japanese guard appeared and "rammed his bayonet right through Captain Miller."[39] Dyess remembered seeing murdered bodies of those who needed help or who had lagged behind all along the march. He wrote, "We remained keenly aware, however, that these murders might well be precursors of our own if we should falter or lag."[40]

By August 11, the first prisoners from II Corps began stumbling into Balanga (map 10) after completing a nineteen-mile grueling march from Mariveles. Balanga, the capital of Bataan, was the site where two days earlier, on April 9, General King had surrendered to the Japanese in the courtyard of an elementary school. Berry and others from the front lines of I Corps made it to Balanga on the afternoon on the 14th. Col. Ed Aldridge had been trucked north from Mariveles with some of the generals, including General Stevens. His truck journey was not without incident, as he was taken out of the truck at one point, severely beaten, and left tied up for the night before someone relented and allowed him to proceed again by truck. In his diary, Aldridge remembered seeing Berry: "Next morning, April 14th, we were sent with an escort to Balanga.... Our old Chev-4 wheeler did some heroic work and it was in poor shape. Got to Balanga and reported to Jap HQ. Could not take anything out of our truck so we were left with what we had on our backs. Were then marched to concentration camp where we mixed with the rest of our companions. While on the truck, we passed Berry and MacDonald with their units footing it. They arrived that afternoon" (map 10).[41]

Chaos reigned in Balanga because the town was the junction point for I Corps and II Corps before the prisoners headed north toward San Fernando to the train depot. Groups of prisoners initially made up of one hundred to two hundred people grew to more than ten thousand in a few hours. No one seemed in charge of the guards or the thousands of prisoners. As a result of the mass confusion, the Japanese guards grew more impatient and more brutal. John Toland wrote, "Rank made little difference in the treatment at Balanga. While Colonel Harrison Browne, chief of staff of the Philippine Division, was being looted, he heard a groan. Turning, he saw his commander, Brigadier General Maxon Lough, on his knees. A Japanese lieutenant was viciously swinging a four-foot hardwood club at the general's head" for no reason at all.[42] Incredibly, Lough lived through the ordeal.

There was an attempt to feed the prisoners at Balanga, but because of the confusion, many prisoners got nothing. Berry was able to get "a small meal of rice upon leaving." Because he had somehow hung on to his canteen, he most likely had a chance to fill it from the few spigots of dripping water at Balanga. During the next miles of the march, the atrocities committed by the Japanese captors worsened. On the way to Orani, the next stopping point, some of the prisoners were forced to march in "double time" for a distance until they were stopped and made to witness a fellow American being beheaded by a guard with a sword just for show. Lester Tenney wrote, "I have relived this scene hundreds of times since that day; I will never be able to get the scene out of my head."[43]

From Balanga, Berry's hike to Orani took one day. Once in Orani, the prisoners were either put in a "rice paddy enclosed by barbed wire" or into a crowded enclosure with human feces all over the ground. In this enclosure, men were packed in so close together they had to draw up their knees to make room for others. The stench from the open latrine was overpowering and sickening. Berry wrote that he spent one night at Orani and received no food before setting off on the one-day hike to Lubao. Along the way, he saw several Americans shot for no reason. And in Lubao he saw, "two Filipinos killed in front of prisoners there—one with 2 butt strokes and the bayonet and the other with a pick handle and another or two taken to the bushes and not returned."[44] Berry told author John Toland after the war that there was just one water spigot for several thousands of men who were jammed into an old 150-by-70-foot mill. Desperate soldiers crowded in a long line to get a few drops of water. The guards watched for soldiers fighting over water, and if they did, the Japanese hurt or killed them.[45]

Aldridge and Berry had found each other at Balanga but had a hard time staying together. Berry mentioned hiking with other colonels, Virgil Cordero and Loren Wetherby, to Lubao. In Lubao, Berry stayed "4 nights and 3 days" and was given "1 small meal rice each morning." The final leg of the march was from Lubao to San Fernando, about fifteen miles. It took Berry one day to hike to San Fernando, where he stayed two nights and one day with four meals of rice.[46]

Prisoners stumbled into San Fernando sick, thirsty, and starving. Toland interviewed Aldridge after the war and learned where men were "housed" in San Fernando, including in pottery sheds, the Blue Moon dance hall, empty lots, old factories, school buildings, and a large cockpit area near the railroad station. Men who arrived at the cockpit, where Aldridge was, were on death's door and all of them had dysentery.[47]

At San Fernando, conditions were similar to those at Orani: Prisoners were put in a barbed wire compound for sun treatment as they waited for their turns to board a train. Some men waited there for days, all the while being brutally treated. When it was Berry's turn to board, he was jammed into a railroad boxcar that was six feet wide and eighteen feet long—one hundred prisoners to a car. Men in the middle of the cars had a very hard time breathing because of the heat, crowding, and lack of air. Many suffocated to death. Other deaths aboard the trains were attributed to disease, starvation, or lack of water. Dysentery and diarrhea were rampant. There was no help for these prisoners who were "compelled to live in their own filth."[48] If prisoners were lucky enough to have their boxcar doors open, the Filipino people who lived near the train tracks threw food into the boxcars for them. Lester Tenney remembered, "Little did they know at the time that their actions and generosity saved many of us from starvation."[49]

After a twenty-five mile train ride that took three to four hours, they stopped in Capas (map 10) where prisoners detrained. When Berry got off the train, he hiked another seven to eight miles to Camp O'Donnell, his first prison camp. The fates of men who made it to Camp O'Donnell differed sharply depending on whether they were trucked or forced to march to the camp. The men who were trucked north never endured the Bataan Death March—they never suffered the dehydration, the starvation, or the atrocities on the march. Conceivably, they were stronger and survived the harsh reality of O'Donnell better than others. Out of the more than seventy thousand men who marched from Bagac or Mariveles, only fifty-four thousand made it to O'Donnell. Approximately seven to ten thousand died on the march, 2,330 of whom were Americans. Berry's courage and resolve, coupled with his athletic strength, fortified him in his quest to survive the Death March.[50]

PART IV

Prisoner of War

"Prisoners of War shall be treated with a spirit of goodwill and shall never be subjected to cruelties or humiliation."

—*1904 Japanese Army Instruction No. 22, Chapter 1, Article 2, from* My Hitch in Hell: The Bataan Death March.

CHAPTER 9

Luzon POW Camps and Hell Ships

POW CAMP O'DONNELL

Colonel Berry's Time at This Camp: April 24 to May 10, 1942

On April 24, 1942, after twelve grueling days on the Bataan Death March, a dehydrated and half-starved Colonel Berry entered POW Camp O'Donnell. Prisoners were beaten and pushed into a compound where they were ordered to stand at attention. Berry, like all the other prisoners, was forced to strip in the hot sun and lay out all possessions he had with him, which were nothing more than his canteen and the clothes on his back. When his friend Col. William E. Corkill found Berry at O'Donnell, Corkill gave him some clothes and a bedroll. Corkill had been one of the fortunate prisoners to be trucked most of the way and, as a result, was able to bring clothing and other items with him.[1] Meanwhile, as the guards inspected the prisoners' things, they would beat or kill the prisoner if they found anything Japanese. The word to get rid of those things spread quickly through the ranks. Lester Tenney wrote, "Any of us who had anything like this chewed and swallowed it or pushed it up his anus as far as it would go."[2]

Berry was forced to stand at attention and listen to the camp commander, Capt. Yoshio Tsuneyoshi, give a speech for two hours. Thousands of prisoners who started the Bataan Death March on April 10 or 11 from Mariveles had already heard this speech. The commander gave his speech to three thousand prisoners at a time. In part of his speech, he made a proclamation, which had come from the commander in chief of the Imperial Japanese Forces on April 11, 1942:

Any of those captives who commit the following acts shall be shot to death:
1. Those who escape or attempt to escape.
2. Those who attempt to escape disguised as civilians.
3. Those who inflict injury upon inhabitants or those who loot or set fire.[3]

In the eyes of the Japanese, the prisoners were not prisoners of war (POWs). Japan had never ratified the articles of the 1929 Geneva Convention, which described how prisoners of war were to be treated, so they were not bound by the articles. Their treatment of prisoners at O'Donnell was thus brutal and unrestrained by international law.

Camp O'Donnell had been a prewar training camp for the American-Philippine army but had been abandoned because the water supply was not adequate for eight thousand men to be trained there. When it was abandoned by the army, the water supply was in part destroyed so the Japanese could not use the camp for themselves when they invaded. Unfortunately, the Japanese chose this abandoned camp as the first POW camp. O'Donnell had only two water pumps still partially working, which could not provide enough water for the approximately sixty thousand prisoners who made it to O'Donnell.[4]

The lack of water was devastating. Pvt. John Falconer recounted, "Mostly, you stood with your canteen in line, waiting to get water. You stood for hours. I remember seeing men who had died waiting in line. I don't claim to be a mathematician, but I knew if a certain number of men died each day, you knew sooner or later your own number was going to come up. I'd sit by the fence and watch the bodies being carried out in blankets slung over poles. Endless movement. We sat and counted. Then there were so many we lost count."[5]

Surrounding the nipa-roofed (thatched) barracks at the camp were a high barbed wire fence and wooden gun towers. Enlisted men claimed spaces on the ground in the compound as their sleeping areas. In contrast, colonels and generals had sleeping areas inside buildings. The generals had better quarters as they did in most of the camps. General Stevens had a small hut with a wooden floor, and, according to Aldridge, he invited Aldridge and Colonels Harrison Browne, K. L. Berry, Ed Keltner, Marshall Quesenberry, Robert Hoffman, and Stuart MacDonald to sleep on his floor. Aldridge wrote, "Eventually all got a blanket and put straw on the floor. This was a good group."[6] Other colonels shared a double-deck bamboo barracks. Sixty were in that building. After some days, all the colonels were moved into an old regimental recreation hall and chapel.[7]

Death was everywhere in the camp. Berry recalled the long line of corpses being carried to their graves each day: "I remember distinctly counting 299 in one day as they passed me."[8] Soldiers died from dehydration, starvation, dysentery, malaria, and from just giving up. One of the barracks became the "hospital," but the prisoner doctors and their assistants had only the supplies they had managed to carry with them on the march. Capt. Loyd Mills reflected, "There was no medicine, no quinine. If you got something—that was it.... I know now that if I stopped to realize the odds, I never would have made it."[9] Soldiers who were put in the hospital were put there to die.

Enlisted men were assigned various duties by the American officers in the camp. Some of the duties were burial duty, cutting wood, and hauling in water for boiling rice. Not many men were fit for these duties because of their severely weakened conditions, but those who forced themselves to eat were able to work. They ate *lugao*—watery rice soup that, unfortunately, had worms in it. "Tried to eat it in the dark, if you could, so you wouldn't see the worms," wrote Pvt. Daniel Stoudt.[10] However, many men just stopped eating. They were too sick, or they had stopped trying to get food down. Others suffered from malaria, which made keeping food down a struggle. Those who were blessed with a strong will to live kept forcing themselves to eat. Those who were not so blessed simply died. Pvt. Robert Brown lamented about his friends who had stopped eating, "I buried so many of my friends and I couldn't understand why they were there. I'd see them stacked up like cordwood at O'Donnell. Guys out of my squadron. People I'd known. The shock of it was hell."[11]

Approximately nine thousand Americans passed through O'Donnell, and around 1,200 died there. Deaths were attributed to the lack of water, malaria, and dysentery. Fortunately for Berry, the generals and colonels were moved out of O'Donnell on May 10 to POW Camp Tarlac. Other American soldiers were moved to POW Camp Cabanatuan in June. Both Tarlac and Cabanatuan were, like O'Donnell, on Luzon. Thousands of Filipino soldiers were left at O'Donnell, and many sources say that approximately twenty thousand of them died from the horrendous conditions there.

Col. Richard Mallonee wrote, "O'Donnell was our lowest ebb. The many rough experiences of the years ahead could not approach the despair of O'Donnell. I have known men decorated for less courage than that required to endure the daily life there."[12]

POW CAMP TARLAC

Colonel Berry's Time at This Camp: May 10 to August 11, 1942

Berry, along with the other colonels and the generals, traveled by truck convoy to a second POW camp, Tarlac. In the town of Tarlac, the prisoners noted some damage from the recent battles, but the town had not changed much. As they

"Our Happy Tarlac Home". The Life of a J.P.O.W. Tarlac, P.I. May 10, 1942

The old cadre barracks of the Philippine Army at Tarlac, P.I., was the first home of Senior Officer (Generals and Colonels) group.

POW Camp Tarlac. "Our Happy Tarlac Home. The old cadre barracks of the Philippine Army at Tarlac, P.I. was the first home of Senior Officer (Generals and Colonels) group." Source: *The Life of a P.O.W. Under the Japanese in Caricature as Sketched by Col. Malcolm Vaughn Fortier*, by Malcolm Fortier.

drove through the marketplace, the POWs could see vendors selling mangos, eggs, canned goods, and bananas—all things they craved. Camp Tarlac was another former Philippine Army camp that had been partially completed at the start of the war in December 1941. The Tarlac compound was two hundred yards by one hundred yards and had a high barbed wire fence around it, with one guarded entrance. There were four buildings within the compound: "a two-story wooden barracks, a small latrine-bathhouse behind the barracks, a nipa bamboo hut where batmen (orderlies) were quartered, and a guardhouse."[13] Col. Malcolm Fortier drew Camp Tarlac in one of his caricatures. Fortier chronicled daily life in the POW camps with his caricatures from 1942 to 1945. Because he did not portray the Japanese in a bad light, they allowed him to keep his caricatures. He posted his caricatures on the barracks' walls in every camp, and his fellow POWs looked forward to seeing them. He illustrated what others wrote about in their diaries; however, he captured daily events and POW activities with a twist of sarcasm and humor.

Our modern plumbing was so old and the water supply so uncertain that we depended on the rains for a bath.

Showering in the Rain. "Our Ultra Modern Bath House—We bathe when it rains. Our modern plumbing was so old and the water supply so uncertain that we depended on rains for a bath." Source: *The Life of a P.O.W. Under the Japanese in Caricature as Sketched by Col. Malcolm Vaughn Fortier*, by Malcolm Fortier.

Berry, the other colonels, and the generals (along with their aides and enlisted men) from POW Camp O'Donnell were the very first prisoners at POW Camp Tarlac. All seventy-seven officers and fifty-two enlisted men had fought on Bataan.[14] When they arrived at Tarlac on May 10, the POWs found the barracks to be filthy, and there was nothing with which to clean. They did what they could. Some of the highest-ranking officers got single beds, but most of the officers were on "double-decker iron beds with wooden mattresses." Berry and other colonels received "one blanket, one sheet, one very small dollsize pillow and a mosquito bar [from which to hang a protective net around a bed], all hard used but better than nothing."[15]

The enlisted men built six latrines for all to use, and each person was allowed ten squares of toilet paper a day, which were handed out by a POW officer in charge of quarters. Every prisoner, regardless of rank, did his own laundry outside using a small bowl of water. Eventually the POWs configured a larger laundry facility with a bigger wash space with a pipe leading water to it. The plumbing was old, and the water supply unreliable, so whenever it rained, the prisoners went outside and bathed.[16]

Japanese Colonel Ito, the commandant of all POW camps on Luzon, visited Tarlac a few times during the two months that Berry was held prisoner there. General Brougher remembered Colonel Ito as "the only Japanese camp commander who was always kind and humane in his dealings with our group of prisoners."[17] A junior officer, Lieutenant Ugi, was nominally in charge at Tarlac, followed by Lieutenant Ura. Ugi rarely was seen in the camp, so Corporal Nishiyama, who supervised the guards, was the one really in charge of the camp at the beginning.[18]

By June 3, the number of POWs at Tarlac increased as POWs who had surrendered on Corregidor on May 6 began arriving from Manila. These prisoners from Corregidor had not suffered the desperate, horrific conditions of the Bataan Death March or Camp O'Donnell. After they were captured on Corregidor, they were allowed to carry their personal belongings with them as they traveled by boat to Manila, where they were imprisoned in various places around the city for several weeks. Life was not good for them, but it was not a Death March or Camp O'Donnell. After several weeks in Manila, they were trucked north to POW Camp Tarlac where they joined the Bataan Death March survivors, who now called themselves the "Battling Bastards of Bataan."[19]

Their arrival meant there were more people to fit into the barracks, use the water, and eat the food. Col. William Braly (Berry's future roommate) was in the June 3 group. A few more came from O'Donnell on June 5. More from Corregidor arrived on June 9, June 28, and July 11, bringing the number of American POWs up to 178. General Wainwright was in the June 9 group. After Wainwright's arrival, General King handed over the POW leadership reins to him.[20]

The quality and quantity of food were substantially better at Tarlac than at O'Donnell. However, assessment of its quality and amount depended on each prisoner's previous experience. General Wainwright, for example, thought the food was awful, but he had not experienced life at Camp O'Donnell; rather, he had previously been held prisoner at the University Club in Manila, where the food, while not great, was better than at Tarlac. Colonel Berry, on the other hand, later wrote when he was a prisoner at POW Camp Karenko, "plenty of rice issued with fair amount of vegetables and meat about every 5 or 6 days; permitted to purchase candy and mangos every day or two for individual and mess permitted to purchase some eggs, beans, bananas and coconuts; issued some flour and milk, sugar, and fat—Tarlac a fine place to be compared to O'Donnell and Karenko."[21]

Braly also wrote about the food: "A couple of times each month we would have meat of some kind such as a young carabao or a pig, but fresh vegetables were a rarity. Sometimes a few camotes or a basket of eggplant would be brought in and several times we had 'Hong Kong' greens." These greens were mung beans, which Braly and others called mongo beans. They must have had a powerful effect, because Braly said the latrines were of great importance after the men ate "Hong Kong" greens.[22]

The scarcity of even small amenities is highlighted in the following humorous anecdote from Colonel Mallonee about the rare arrival of two pounds of coffee at the camp for the senior officers, the generals:

> I know it will be most difficult for you to believe me—in fact, it seems almost impossible—but that two pounds of coffee, eked out by charred rice brewed with it, had a strange effect upon those grown men, senior officers. There may have been just enough caffeine to produce the effect, or it may have been purely psychological, but in their weakened condition, about half the group got as high as if they had drunk several Scotch and sodas. Old army songs were rendered in barbershop harmony. Impromptu jigs and square dances shook the building. The Interpreter came running, fearing a riot. He sat there for over an hour, incredulous but highly amused. After it wore off there were, incredibly enough, many hangovers.[23]

Given the scarcity of food, and especially of fresh vegetables, prisoners soon realized that they needed to grow their own food. According to Colonel Braly and Colonel Fortier, General Brougher was credited with starting gardening. It is likely that Berry, who later wrote about his own garden, would have been among the early gardeners at Camp Tarlac, too. Initially, gardens were voluntary; later the Japanese would make farming mandatory.

During his two months at Tarlac, Berry wrote that he was the personnel adjutant, an administrative position where he would have kept track of personnel—numbers, names, ranks, and so forth. Keeping this data assisted him when he put his private notebook together. In this notebook, which he kept for his eyes only, Berry wrote specific details about his fellow officers—height, weight, hair color, eye color, hometown, marital status, birthdate, rank, health, and his personal opinion of each officer.

General Brougher initiates voluntary gardening; what a headache that turned out to be, as well as a backache.

Gardening at Tarlac. "Brougher's G Leaners—I.W.W., PWA. Or What Have You? General Brougher initiates voluntary gardening: what a headache that turned out to be, as well as a backache." Source: *The Life of a P.O.W. Under the Japanese in Caricature as Sketched by Col. Malcolm Vaughn Fortier*, by Malcolm Fortier.

For pleasure at Tarlac, Berry walked "a great deal" trying to keep himself strong and fit. Many people walked to have something to do, as can be seen in Fortier's compound caricature. Gen. Lewis Beebe wrote that "four laps up and down the area make about one mile."[24]

Several other POWs kept diaries describing daily life at Tarlac. Reveille was held at 6:30 a.m., followed by calisthenics. Berry told family members after the war that he often led the morning calisthenics. Breakfast was around 8:30 a.m. Officers purchased wash bins from the modest post exchange and did their own laundry if the weather was nice and the water was turned on. Among other things requested by the POWs for purchase at the post exchange were thread and needles so they could repair their clothing. The guards fulfilled the POW requests from town if they could—and for a price. American officers were put in charge of dispensing and selling the food or personal items that came in. There were not many items to sell at Tarlac, but the prisoners looked forward to their time at the post exchange, nevertheless.[25]

Reading material at Tarlac was sparse, but those who surrendered on Corregidor brought some books with them. The prisoners could also read the *Japan Times* and *Japan Advertiser*, newspapers printed in English, when the guards brought them into camp. Two of the POWs, Col. Bob Hoffman and Col. Stuart Wood, could read and speak Japanese, so they helped translate other newspapers printed in Japanese. They also became interpreters for both sides. Interestingly, Hoffman and Berry had served in Siberia at the same time in 1919. Hoffman had been the aide for Commanding General Graves with the American Expeditionary Forces in Siberia.[26]

In May 1942, Japanese newspapers were filled with propaganda about their many recent victories, including the capture of Burma, Tulagi in the Solomon Islands, and Corregidor and Mindanao in the Philippines. In addition, papers reported that Japanese forces had reached India. Unsurprisingly, the Japanese newspapers did not mention losing the epic Battle of the Coral Sea. Because they were provided only selected information about the progress of the war, the guards at Tarlac rejoiced, convinced that Japan was invincible.

Prisoners played cards to relieve the boredom of daily life. Other activities included "Indian wrestling," set up by Corporal Nishiyama. He would choose the pair of men for each round. Colonel Ito brought in a volleyball set for exercise and fun. Colonels Virgil Cordero, Allen Stowell, Chester Elmes, and James Monihan took up embroidery to pass the time. And after the guards took away their razors, many men began growing beards. Beards and long hair became ongoing topics of conversation throughout imprisonment at all POW camps. At future camps, Berry, too, grew his hair and sported a mustache.

Music was also an important diversion for the prisoners at Tarlac. Colonel Braly, who had managed to keep his violin with him since he was captured on Corregidor, became part of a musician's group at Tarlac. This group continued to play together throughout their imprisonment at all the other camps. Other prisoners formed a choir. And both groups would perform for birthdays, church services, and holidays. At Tarlac, where there was a mess hall, they were sometimes permitted to use the hall to sing hymns on Sunday mornings.[27]

The first death among the Tarlac POWs was Col. Edwin F. Barry, who died very suddenly from streptococcus on July 18, 1942. The Japanese guards told the Americans to build a coffin for Colonel Barry but gave them no materials for it. As a result, the enlisted men made a coffin for him out of boards from the side of the barracks, the only wood they could find. Colonel Ito arrived for the funeral procession; no service was allowed. Officers stood in rows and saluted the coffin as it was loaded into a truck. Only six officers (those who knew him best) and two enlisted men were allowed to ride with the coffin to the Tarlac Cemetery where Colonel Barry would be buried.[28]

In early August 1942, Lieutenant Ura told the prisoners that they would be moving to Japan or Taiwan in a few days. Colonel Ito arrived with a special truckload of food and offered a feast to the surprised prisoners before they left. On August 11, all Tarlac prisoners marched to the railroad station and boarded a train bound for Manila. Mallonee described the scene: "We were formed and marched through the streets of Tarlac to the railroad station.... No Filipino uttered a sound. However, in the middle of one large group, a boy of about fourteen bravely whistled *Auld Lang Syne* while our column passed."[29]

NAGARU MARU

Colonel Berry's Time on This Ship: August 11 to August 16, 1942

After leaving Tarlac, the prisoners were taken to Manila to await transport to a POW camp on the island of Taiwan. Little did they know that they would look back on Tarlac as one of the best camps in which they had been imprisoned. In Manila, 178 POWs boarded the "hell ship," the *Nagaru Mara*. Ships transporting prisoners earned the nickname "hell ships" because of the inhumane treatment prisoners received on board these ships. Common conditions included severely overcrowded spaces below the deck, very few port holes for air, lack of toilet facilities, and lack of food and water. Fortunately, Berry's time on the *Nagaru Maru* was relatively short and proved to be not as dangerous as future POW journeys on hell ships.[30]

At this point in the war, the sea journey for POWs was moderately safe because the Japanese controlled the South China Sea between Taiwan and the Philippines. Later in the war, other POWs would not be so lucky, as American submarines roamed the sea looking for Japanese warships. Because the Japanese did not mark ships that were transporting POWs, submarines and airplanes could not distinguish them from other targets. For example, in September 1944, the *Rakuyo Maru*, carrying 1,317 Australian and British prisoners of war from Singapore to Taiwan, was torpedoed by Americans, and 1,159 POWs died.[31]

Quarters on the *Nagara Maru* were considered fair for most of the generals, who were assigned cabins next to a group of Japanese officers. They had mats to sleep on and access to a relatively clean bathroom. They were permitted to go up on deck anytime they wanted. However, Berry and the rest of the POWs

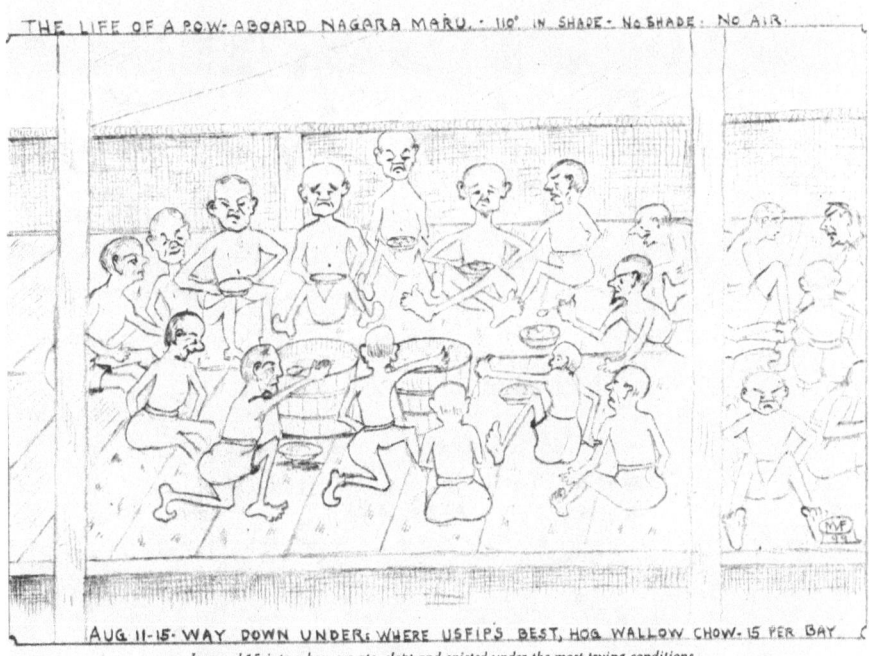

Jammed 15 into a bay, we ate, slept and existed under the most trying conditions.

Eating Aboard the Ship. "Aboard Nagara Maru—110 degrees in Shade—No shade—No air. August 11–15—Way Down Under: where USFIP's Best, Hog Wallow chow—15 per bay. Jammed 15 into a bay, we ate, slept and existed in the most trying conditions." Source: *The Life of a P.O.W. Under the Japanese in Caricature as Sketched by Col. Malcolm Vaughn Fortier*, by Malcolm Fortier.

were ordered down into the ship's hold, where an extra wide shelf had been built between decks so the ship could transport larger numbers of men. Fourteen men occupied a space on a rough wooden shelf in each of the thirteen-foot-deep berths. While the generals had access to the deck whenever they wanted to be outside, the POWs in the hold had to take turns to go up on deck for fresh air. However, they did have relatively free access to the outhouse-style toilets located on deck.[32]

For the men in the hold, there were no tables. Meals, served in a bucket for fifteen men, consisted of rice and a small fish for each prisoner. To eat, Berry and his fellow berth mates squatted around their bucket of food either in their berths or outside on the deck. Tea was the liquid served, and there was always a long line of men waiting to fill their cups.[33]

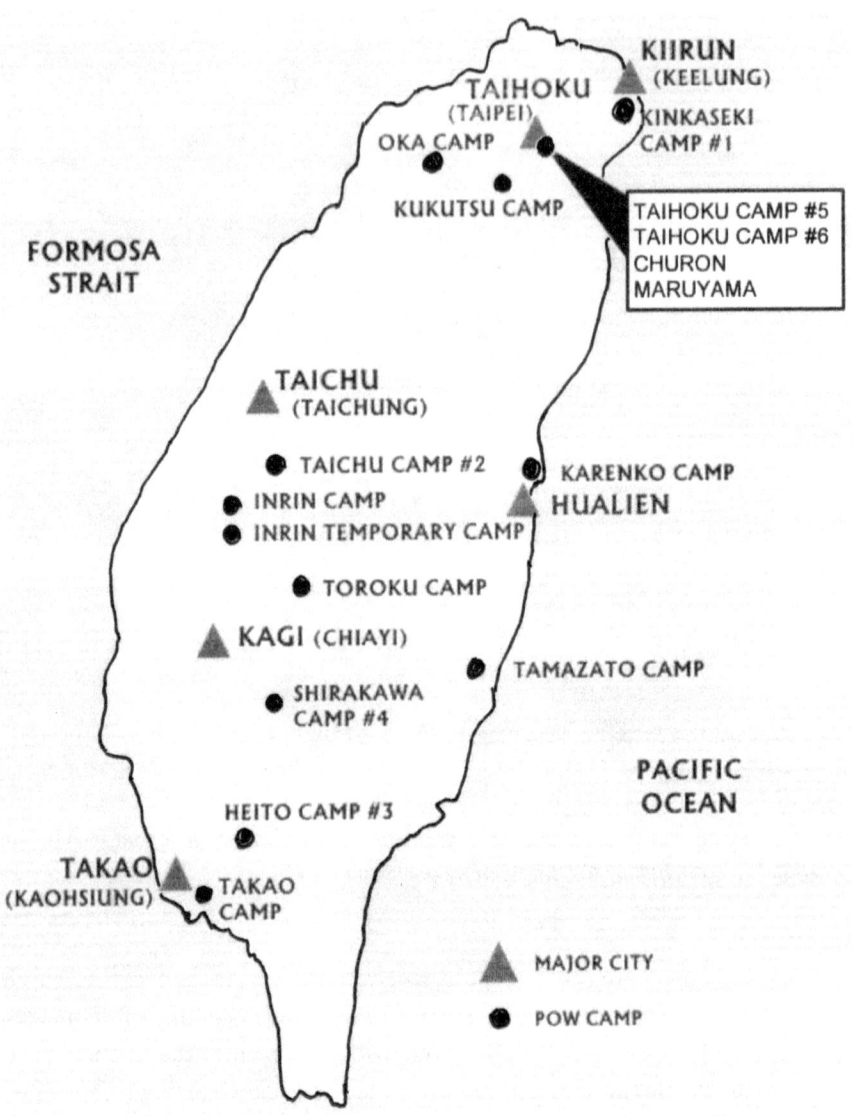

Map 11. Map of POW Camps on Taiwan. This map of Taiwan shows the city of Takao where the *Nagaru Maru* docked after five days sailing across the ocean. From Takao, the POWs boarded the *Suzuya Maru* and traveled around the southern tip of Taiwan to the port city Hualien. From Hualien, they walked for an hour to reach Camp Karenko. Source: Taiwan POW Camps Memorial Society.

SUZUYA MARU

Berry's Time on This Ship: August 16 to August 17, 1942

After five days, the *Nagaru Maru* arrived at Takao (Kaohsiung) Harbor, on the southwest coast of Taiwan (map 11). All the POWs on board were transferred to a smaller ship, the *Suzuya Maru* (sometimes called the *Otaru Maru*). The *Suzuya Maru* was in terrible shape. Mallonee wrote, "Its barnacled bottom should have been sent to its graveyard many years ago."[34] Conditions were far worse than on the *Nagaru Maru*. There was almost no ventilation inside the prisoners' quarters, and the guards allowed each man only thirty minutes on deck. Fortunately, the POWs aboard the *Suzuya Maru* did not suffer long. During the night, the ship traveled at a snail's pace around the southern tip of Taiwan and then dragged itself midway up the east coast of the island to the inlet of Karenko (Hualien) on the morning of August 17, 1942. From the port, the POWs marched inland five kilometers to POW Camp Karenko (map 11).[35]

CHAPTER 10

Taiwan POW Camp Karenko
Part 1

COLONEL BERRY'S TIME AT THIS CAMP: AUGUST 17, 1942, TO JUNE 6, 1943

Camp Karenko on Taiwan first opened when the POWs with Colonel Berry arrived from the Philippines on August 17, 1942. While life was not the horror it had been at O'Donnell, Berry's time there was difficult, occasionally brutal, and always humiliating. He and fellow prisoners endured ten months of hard labor, a critical lack of food, and the ever-present likelihood of suffering arbitrary abuse at the hands of the Japanese guards. But with help from one another, almost all the POWs survived their time at Karenko.

ARRIVAL

When Berry entered the compound, he saw a cookhouse, a two-story wooden barracks, a few smaller buildings the Japanese used, and the Japanese administrative building, which sat thirty feet up a hill on a terrace. The POWs referred to that building as "the hill."[1]

As Berry had done at the previous camps, he was forced to strip down to his underpants and lay out his meager possessions. If he had matches, a lighter, or a knife, they were taken away. Everyone went through the same procedure. But there was a new humiliation at this camp. The POWs were issued wooden clogs to wear instead of their regular shoes.[2]

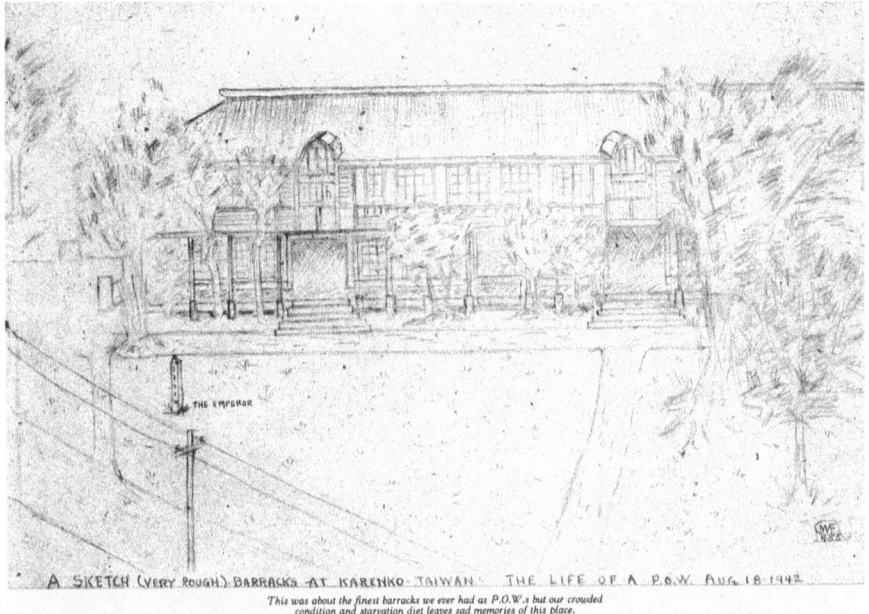

A SKETCH (VERY ROUGH) BARRACKS AT KARENKO-TAIWAN. THE LIFE OF A P.O.W. AUG 18 1942.
This was about the finest barracks we ever had as P.O.W.s but our crowded condition and starvation diet leaves sad memories of this place.

POW Camp Karenko. "A Sketch (very rough) Barracks at Karenko-Taiwan. This was about the finest barracks we ever had as P.O.W.s but our crowded condition and starvation diet leaves sad memories of this place." Source; *The Life of a P.O.W. Under the Japanese in Caricature as Sketched by Col. Malcolm Vaughn Fortier*, by Malcolm Fortier.

CAMP COMMANDERS AND RULES

With the move across the ocean to Taiwan came new POW rules, guards, and commanders. At Karenko there was no humane Colonel Ito who brought them a feast of food. Instead, Lieutenant Colonel Nakano commanded all the Taiwan POW camps, and because of his foul disposition the POWs called him "Old Sourpuss." Giving derisive nicknames to the Japanese commanding officers and guards was a way to assert some control over their situation by diminishing those in whose hands they were forced to place their lives. The commandant of the camp was Captain Imamura or "Little Snake Eyes." He greeted the POWs to Karenko by bragging about his country's victories in the Pacific region. Then Imamura issued the camp rules and told the prisoners that if they did not obey, their lives could not be guaranteed.[3]

Stripped and stood in Tropical sun for three hours, while our effects were searched for contraband.

Make Acquaintance with Clogs. "Another Shakedown (3 hours) Arrival Karenko—PM—Aug-17–1942. Stripped and stood in Tropical sun for three hours while our effects were searched for contraband." Source: *The Life of a P.O.W. Under the Japanese in Caricature as Sketched by Col. Malcolm Vaughn Fortier*, by Malcolm Fortier.

Another ranking Japanese officer was Second Lieutenant Nakashima, nicknamed "Boots." The POWs came to despise Boots. He was described as "mean and vindictive" because he would look for ways to humiliate the prisoners. Another guard was Second Lieutenant Wakasugi, nicknamed "Baggy Pants" because his pants sagged. Berry referred to Baggy Pants as BP when he began to write about him in his diary. BP became a semilikeable central figure for the POWs at Karenko. It was he who sometimes gave them information about the war and warnings about other angry guards. Some of the POWs thought BP was the "best Jap around."[4]

Berry, like everyone else, had to sign a document pledging that he would not escape. Those who refused to sign the pledge would be locked in a windowless, dirt-floored shed with practically nothing to eat or drink for a few days as a punishment. All the Americans signed the pledge. However, the Dutch and British, who arrived later in the summer, fall, and winter, found out about the windowless shed punishment the hard way when they refused to sign the document.[5]

In addition to other rules, Captain Imamura explained to the prisoners that paying respect to the Japanese emperor was required. Paying respect meant that the POWs had to bow to a wooden post, the emperor's stand-in, which sat in the middle of the compound. After their greeting from Captain Imamura on August 17, the POWs had to practice bowing several times to the post, so that Imamura could see if they were doing it correctly. From then on, prisoners were forced to bow to the effigy when they lined up for *bango*, or roll call, which occurred mornings and evenings every day. Naturally, the POWs found a way to undermine Imamura's attempt to humiliate them. Braly wrote, "It was fortunate perhaps that the Japanese officer could not hear the muttered imprecations of these vituperative veterans during this little formality." Col. Paul Bunker wrote, "Up as usual and adored the emperor, with Berry's calisthenics after wards."[6] Berry led morning exercises again as he had at Tarlac.

QUARTERS, ROOMMATES, AND QUARRELS

Colonel Berry was appointed squad leader of 2nd Squad by Captain Imamura. He served in this capacity from August 17 to December 4, 1942. Braly became his assistant squad leader. Each POW squad was alphabetically organized for the most part, and, as a squad, they ate together—there was no mess hall—worked together, went to morning and evening bango together, told each other the same old stories, helped one another, and quarreled with one another for ten months.

Berry's squad was assigned three rooms in the barracks. In an eight-man room were Colonels Ted Chase, Alex Campbell, Virgil Cordero, Pat Callahan, Nick Carter, Frank Brezina, Paul Bunker, and Vic Collier. Across the hall were Colonels Berry, Louis Bowler, and William Braly in one room and Colonels Roscoe Bonham, Harrison Browne, and Pembroke Brawner in the other three-person room. A few days after this configuration was established, Colonels Alfred Balsam and John Boatwright were moved into the eight-man room, making that a very crowded ten-man room.[7]

Berry's roommates, Braly and Bowler, had not been in the battles on Bataan, so it is not known whether he knew either one of them before he met them at POW Camp Tarlac in June 1942. Both men had worked in staff positions under Maj. Gen. George F. Moore, who had commanded the Coastal Artillery Command. Consequently, Braly and Bowler had been captured on Corregidor as opposed to Bataan and had not walked the Death March or experienced POW Camp O'Donnell. Instead, they had been held in camps around Manila before being trucked north to Tarlac. Because they had been able to carry some of their possessions with them, Braly brought his beloved violin. Braly wrote, "I found myself in Squad 2 and fortunately assigned to one of the smaller rooms with Colonels Louis Bowler, and K. L. Berry. With the later addition of one or more officer, who was changed a couple of times, this congenial family carried on for more than two years."[8]

Berry liked his roommates. About Braly he wrote, "violinist, stamp collector, studying Spanish, Christian Scientist, good card player, church worker, good mixer, educated, unselfish and a nice fellow in general; great friend of 'Louie' Bowler's—they are 'Alfonso and Gaston.'" (Berry was referring to a comic strip, *Alphonse and Gaston*, created by Frederick Burr Opper, that featured a pair of Frenchmen, Alphonse and Gaston, who were overly polite to one another.) About Bowler he wrote, "gentleman; conducts services for the 2nd Squad; well educated; gardener; a member of the 'Alfonso and Gaston' team; 30# or more underweight; not standing our captivity very well; needs someone to 'kinda' look after him which Bill Braly does; like him very much." As Braly mentioned, this trio stayed together as roommates for over two years, with the addition of Colonel Browne at Karenko and then Colonel Nicoll Galbraith at Shirakawa. At Karenko, Berry and Bowler gardened together to supplement their room's meager food rations.[9]

Several POWs kept diaries during the months at Karenko before Berry started his own, so some of Berry's daily life is known from their diaries. Colonel Bunker, one of Berry's squad mates, had been a classmate of General MacArthur at West Point. "Bunk" as he was often called, was twelve years older than Berry and had been a permanent colonel since 1935. During the Philippine battle, he was the seaward defense commander for the harbor defenses of Manila and Subic Bays. His office was moved to Corregidor when MacArthur moved headquarters to Corregidor in January 1942. Like Braly and Bowler, he was captured there after Wainwright surrendered on May 6.[10]

Berry described Bunker: "White [hair], blue [eyes], heavy and erect, active physically for his age, walked great deal at Tarlac and did some running. Spends time at Karenko braiding string and lining out squad positions on parade ground; former All-American footballer at USMA, I understand." Berry admired former football players, so this was a good connection between the two men. However, Berry added to his notes, "kinda crusty and inclined to be sarcastic at times; had to give him two call downs." As squad leader, Berry rebuked Bunker a couple of times over rooming details. Berry's "call downs" angered Bunker, causing him to write harshly about Berry several times in his diary. After several months together, Berry wrote, "Seems like a pretty nice old file at times and I like him better now; biggest old growler in camp but likeable at times." Bunker became quite ill from malnutrition in January 1943, and Berry visited him in the makeshift hospital, bringing Bunker fresh tomatoes from his personal garden.[11]

Quarreling among POWs was inevitable given their tight quarters. In the ten-man room, Bunker got into an argument with Boatwright about blankets. Berry stepped in and scolded Bunker. This was one of the "call downs" Berry mentioned in his private notebook. Bunker was annoyed that Berry took Boatwright's side because he thought Boatwright was a disagreeable type who clashed with several of his squad mates. In fact, Boatwright and Virgil Cordero came to blows at one point.[12]

Quarters became increasingly crowded as more Americans arrived in late August, September, and December 1942, and February 1943. In addition, defeated British, Australian, and Dutch officers who had been captured in Singapore, Hong Kong, and Java arrived in September, December, and February 1943. Karenko, thus, became jampacked with prisoners. Everyone was shifted around; space and small cupboards for POWs' precious possessions were coveted. Bunker and Berry butted heads about the cupboards—the second "call down" Berry mentioned in his notebook. Annoyed, Bunker wrote about the coveted cupboards in his diary: "Yesterday p.m. BP [Baggy Pants] was coaxed into relieving the congestion in our room by ordering 1 officer to move into each of the rooms across the hall. Berry doesn't like it and so the move has not been made." Then the next day Berry finally moved two officers, as BP had ordered the day before. According to Bunker, Berry "demanded 1/3 of our total locker space! This created a furor. Berry appointed Boatwright as room corporal. Glad I argued him out of appointing me." It is interesting that, despite their differences, Berry wanted to appoint Colonel Bunker as "room corporal," but Bunker declined the position. Apparently, Bunker wanted things to change, but did not want the responsibility for doing it. Somewhere in this period of shifting people around, Col. Harrison Browne, Berry's good friend, moved in with him, Braly, and Bowler.[13]

FOOD AND MEALTIME DUTIES

At mealtime, POWs received two-thirds of a bowl of a watery rice soup. They were starving and losing a lot of weight on this meager ration. They were getting less food than they had received at Tarlac. General Wainwright wrote, "Food became a mania with all of us. We woke up each morning with hunger paramount in our minds, ate our rice and hot water, and were as hungry after it was finished as before, yearned for food through the long day, went to sleep to dream of food."[14]

To the rice soup, a POW chef would add whatever came into the kitchen for flavor. Taro root and napa cabbage, which were easily obtainable, were added more often than scarce flour or curry. Very rarely, there were tiger shark pieces, beans, or a hint of duck or fish flavor. On the whole, however, rice was the meal. Unfortunately, it was not pure rice. Wainwright lamented: "The rice contained a white worm about an inch long, with a black head.... At first we carefully took them out of our rice with our chopsticks and put them to one side. But in final desperation we closed our eyes and simply ate, telling ourselves that there might be protein in a worm."[15]

Each squad had a duty schedule for retrieving food from the kitchen and then dishing out the food to squad members. Braly described this precarious duty: "Food carriers from each squad carried buckets from the kitchen to their quarters where squad food servers distributed the soup and rice into the respective bowls. And woe to the disher-outer who accidentally slighted some bowl a trifle!" Bunker described a confrontation over food distribution that involved Berry and Col. Frank Brezina. Apparently, Berry "accused Brezina of partiality in dishing out the soup and Brezina, as usual, acted like a sulky child, told him he was quitting his job." Bunker then took his turn complaining about Brezina. "Brezina is on chow detail now and he is lousy," he wrote. When Berry was the server, Bunker complained about him, too, saying, "We are all glad to see Berry's tour as soup server draw to a close." Understandably, Squad 2 was not alone in experiencing conflicts over food. Every squad, even the generals in Squad 1, argued about food. As General Wainwright observed, "If a man received a bean in his soup, and another did not, it made for hard feeling."[16]

To add much needed protein to their diets, the POWs at times went to extremes. Large snails were plentiful around Karenko in the fall and were considered to be a good source of protein. Consequently, many POWs hunted them. In spite of their occasional quarrels, Berry and Bunker hunted snails together for themselves and their roommates.

> Berry and I went snail hunting and collected a washbasin full including many big ones, 4–5 inches long. After lunch I got my shower while waiting for the snails to boil. Then Berry and I took them outside the kitchen and cleaned them—a messy, slimy job because they hadn't boiled long enough.... They tasted like chicken gizzards and are not my idea of a delicacy but they are edible and, I hope, serve to change one's metabolism—furnish a welcome change from the eternal rice. We ate several and took the remainder to our rooms, including the yellow broth.[17]

The Nips practically never gave us meat or fish; large snails being plentiful we made a run on them, until they were exterminated.

Eating Snails for Protein. "Doc Worthington Gets a Prize Beauty. The Demand for Meat Must Be answered. The Nips practically never gave us meat or fish; large snails being plentiful we made a run on them, until they were exterminated."
Source: *The Life of a P.O.W. Under the Japanese in Caricature as Sketched by Col. Malcolm Vaughn Fortier*, by Malcolm Fortier.

For his part, Colonel Braly was not keen on eating snails and recalled, "My first sample was from a batch prepared by Colonels Bunker and Berry. I can't say that I enjoyed it but many stayed with the snails as long as the supply lasted."[18]

The scarcity of food made inequities in food distribution from the kitchen to the squads another source of resentment among the POWs. The Japanese simply ignored the problem. Apparently, those who worked in the kitchen had opportunities to "taste" food ahead of time and apportion the buckets to favor some squads over others. Starving and angry, Bunker complained, "It is humiliating to see the pot bellies on our enlisted men who work in our kitchen, and know their fat is due to the food they steal from us! Gen. Sharp is OD [officer on duty] tonight and used Chastaine to help apportion the food at night meal. Yet the men [enlisted men] robbed the soup of Squads 1 and 2 gave most of the beans to other squads, specially 7 and 8, their own squads. Our squad was ripe for mutiny."[19]

INDIVIDUAL GARDENS

To supplement the starvation rations provided by their captors, many POWs started gardening again, as they had at Tarlac. Braly noted: "Several garden-minded officers and men persuaded Baggy Pants to assign them a few square yards of ground each for private gardens in order to supplement the ration. Among the first to get started, as I recollect, were Brig. Gen. Brougher and Cols Berry, Bowler, Richards and Steele."[20]

Berry and Bowler grew tomatoes, potatoes, carrots, garlic, and kohlrabi. Bowler made salads for their room from the produce they raised, supplementing the starvation diet provided by the Japanese. For Berry, gardening became more than a way to supplement the prisoners' diet; it was also a gratifying diversion from the rigors of camp life, as he noted: "Sure having fun out of my garden and lots of—it surely helps."[21]

The food raised in their garden became vitally important to Berry and his roommates. They had more food than other rooms that did not have gardeners. Even thus supplemented, however, their diets were inadequate. Berry and his roommates lost weight and became very weak at Karenko. Berry reflected on the perilous situation he and other POWs faced: "We are slowly starving to death and I am afraid that some are not going that way very slowly either. Only reason we in this room have fared better than others is our little gardens which gives us a vine raw leafy salad every meal. It is about played out now however and will be slim picking for a few weeks until we can grow more."[22]

NEW POSITIONS

From time to time Berry, who enjoyed gardening more than being a squad leader, asked Braly to attend squad leader meetings for him. Berry disdained interacting with his captors. In December, the camp commander changed many squad positions, and Braly and Berry exchanged roles. For both men the change was welcome. Berry could concentrate on gardening, and Braly, who welcomed the responsibility, could go to meetings with other squad leaders and be in the know. Braly retained a squad leader position in all future POW camps. He liked the job and was good at it.[23]

Most likely, Berry eschewed the squad leader position for a reason other than his disdain for the Japanese and his love of gardening. He did not hear well. Bunker observed, "Col. Berry tried to pass on the dope to us, but his deafness evidently caused him to miss some important points."²⁴ Bunker was most likely correct about the reason Berry got orders wrong for his squad. Explosions on the battlefield caused many infantry officers to become hard of hearing. Recognizing his own hearing difficulties brought on by the bombardments on Bataan and Corregidor, Wainwright gave the leadership of Squad 1, the generals' squad, over to General King. He feared his officers might suffer at the hands of the Japanese if he failed to hear the Japanese commanders correctly and misinformed his officers.²⁵ Berry may well have come to the same conclusion as Wainwright when he asked Braly to attend squad meetings. Braly could ensure that the Japanese orders would be transmitted correctly back to the squad.

PROMOTION WORRIES

On March 27, 1942, a week and a half before General King surrendered I and II Corps on Bataan, Wainwright had evidently promoted Berry to brigadier general. Because there are no accounts of this promotion from Berry at the time (i.e., in his diary or a letter to Alice), the battlefield promotion was unknown to family until Berry started writing about his promotion not materializing.

In November of 1942, the Japanese at Karenko allowed some of the POWs to send radiograms home. Berry was one of the lucky ones. His November 10 radiogram to his wife, Alice, was not broadcast in the United States until April 4, 1943, almost a year after his capture. His radiogram read:

> Am well. Have not been sick or wounded. Find out why my promotion did not go through. Ask friends to try correct injustice now. Write frequently. Address as Time Magazine or Army and Navy Register suggested. If permitted, send package concentrated foods including shelled nuts, sweets and Bull Durham. Beautiful scenery, excellent climate and comfortable quarters. Write frequently giving all family news since I left home and [illegible] letter might not arrive. Will write tomorrow November tenth. Kiss children for me. Hope Mister South appointed Tom. Have children write and send Christmas packages containing nuts and sweets. Love and Kisses "K.L."²⁶

Berry's radiogram makes clear that he knew his promotion did not make it through. Perhaps he learned this news at Tarlac when Wainwright and his staff arrived there after the surrender at Corregidor. Berry's concern about the injustice of not being promoted recurs frequently in later letters and entries in his diary. On November 16, for example, Berry wrote a two-page letter to Alice, making the case that he should have been promoted and directing her to "leave

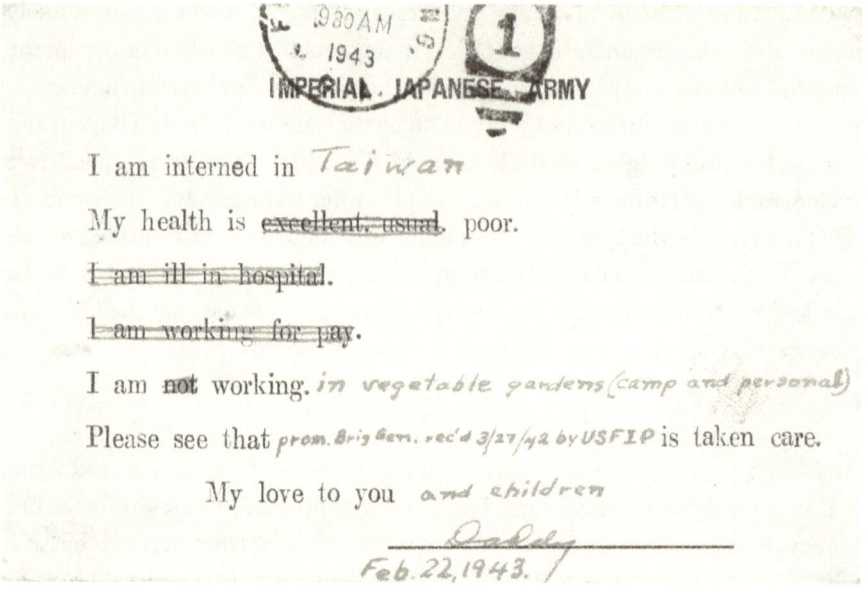

Postcard Home. Colonel Berry's postcard home, dated February 22, 1943. Source: Author's private collection.

no stone unturned" in her efforts to secure the promotion for him. The Japanese never sent his letter, but—tellingly—Berry saved it for his records.

About ten days after recording his radiogram, Berry wrote a letter to Gen. Albert Jones, who was at Karenko with him. Jones had been Berry's commanding officer for part of the Battle of Bataan and had firsthand knowledge of Berry's role in the Battle of the Pockets. Nevertheless, Berry's letter summarized his account of the battles that he had directly and indirectly commanded. Berry hoped that Jones, once armed with this summary, would push for his promotion when the war ended. It is very curious that Berry did not give such a summary to General Wainwright while at Karenko. After all, Wainwright had been the one who promoted him to general on the battlefield.[27]

Again, Berry wrote another letter to Alice, saying much the same on December 28. Like his letter to Alice written several weeks earlier, this letter was not sent. The Japanese read the POW letters but ultimately decided against allowing POWs to write long letters home. Instead, in February 1943, the POWs had the opportunity to send a fill-in-the-blank postcard home. Berry seized on this unlikely opportunity to press Alice once again to pursue his promotion.

The generals were taken out of Karenko in the spring of 1943. Thinking he might not see them again, Berry wrote both Wainwright and Jones notes of appreciation. In his diary, he summarized a subsequent meeting with Wainwright, and it is clear Berry was concerned that he would not get the recognition he thought he deserved.

> Gave Gen. Jones and Wainwright letter of appreciation and extract from my private notes. Both seemed greatly effected and Gen. Wainwright told me, 'I just love you for that letter.' He is a grand man and soldier. Told me I was a grand fighting man and that had recommended me Brig Gen[eral] but crash came too fast. Said he had a very fine recommendation for a decoration for me, which he would see went thru [sic] as soon as he were able. Said I deserved great credit for the Tuol River fight. Did not mention a DSM for Moron so I asked Beebe about it as he told me at Tarlac that I would get DSM. He denied saying it but he did. I then spoke to Gen Jones about it and he said I was recommended by Gen Wainwright, that he saw it and that I was only colonel so recommended.[28]

Interestingly, Berry's private notebook indicated that twenty-nine other officers were promoted between March 28 and May 4, 1942—one to major general, two to majors (aides of Wainwright and Moore), and the rest to full colonels. Apparently the "crash" that Wainwright mentioned to Berry did not affect those promotions. In fact, Wainwright managed to promote one officer just two days before the surrender on Corregidor. No wonder Berry felt like an injustice had been done to him. In the end, he did not get his star until after the war, and even then, it did not come easily.[29]

ACTIVITIES

When the POWs were not being forced to work, they created activities that added a touch of humanity to their lives. The activities made Berry feel better, and he was able to sustain his optimistic spirit. Playing cards became a constant pastime in the off hours. Berry and Brownie played cribbage together, and Berry played solitaire, which he called Sol. He would use playing solitaire as a way to predict when the war would be over. He would ask Sol a question about the war, and if he won, he would say Sol had predicted a positive answer. "Sol said that one of the Axis nations would crack by July 1st and that war be about over by Dec 1st. Hope Sol did not lie." If he lost the game, Sol's answer was a negative response to the question.[30]

There was a bare-bones post exchange (PX) at Karenko, which Col. Pat Callahan ran. POWs could buy some items with their POW pay, including salt, cigars, and cigarettes. Braly explained what the pay was for a colonel: "310.00 yen per month. Of this the Japs placed 271.00 yen in postal deposits, deducted 15.90 yen for food, 3.10 yen for enlisted men's pay and generously gave me the remaining 20.00 yen in cash."[31] All colonels received the same pay. Callahan repeatedly asked the Japanese to buy playing cards for the PX so he could sell them to the POWs, but they refused. So, the POWs started making cards out of cigarette boxes.

Church services were held each week, with different squads taking the lead and preparing the service. Berry mentioned roommates Bowler and Braly as two who contributed to church services. Braly played his violin at church services, and a choir of POWs sang hymns. Berry wrote on April 25, "Easter and what a beautiful day—2nd one in a row. Sun out all day and it was warm at last. Had church outside and 121 Am. Dutch and British attended and 2 Nips—Mort and BP. Australians ran services as today is also Anzac day. They put on a nice impressive service and I feel like it did me good to attend." On Memorial Day, Bowler gave a beautiful service. "The best little talk I ever heard and most appropriate," pronounced Berry.[32]

On one occasion, the guards allowed the POWs to put on a program of singing. Not a fan, Bunker referred to this activity sarcastically as a "Sing-Song." About the performance, Bunker wrote curmudgeonly, "Terrible to relate a British soldier recited 'Gunga Din' (also terribly) and our Cornell butchered 'Casey at the Bat.' I stuck it out until they started singing 'Old Black Joe' and then I quit. The Jap officers all attended, and the camp commander presented the performers with 5 bottles of sake!" Colonel Braly, on the other hand, loved the event: "Hit of the evening was the American Double Quartet, assembled and trained by Brigadier General Lewis Beebe. The singing of this group was especially fine and the octette became a standby from then on."[33]

Officers giving talks to one another on interesting topics became another form of entertainment. However, this entertainment was not sanctioned by the guards, so the audience was limited to a few visitors sneaking into someone else's room. Berry attended a talk given by Brig. Ivan Simson, a British chief engineer in Malaya before he was captured. Simson named his lecture "The Vale of Kashmir," and Berry enjoyed it very much because it was about hunting tigers in India.[34]

Reading was another pastime prisoners enjoyed; the POWs swapped books, especially after the prisoners from Corregidor, Singapore, and Hong Kong arrived with books in their luggage. Early on, a library was an informal entity within a squad, but when the new, more lenient camp commander, Lieutenant Kojima, arrived in March, a formal library was established. The British and Dutch had the most to contribute, bringing the total number of volumes up to three hundred, according to Braly.[35]

Berry mentioned two books that he was reading at Karenko. One was *On the Way of a Warrior* about Buddhism. The other was the Bible. "Finished Bible today. Old Testament hard to read and bunch of bull . . . and simple in lots of places. Well read over part of new Testament and Proverbs." Berry especially liked Proverbs: "Spent part of afternoon reading over Proverbs and started over them again—some of them will bear reading frequently."[36]

Underlying the activities that the POWs created to make their circumstances bearable was the constancy of a life of starvation, degradation, and forced labor.

CHAPTER 11

Taiwan POW Camp Karenko
Part 2

COLONEL BERRY'S TIME AT THIS CAMP:
AUGUST 17, 1942, TO JUNE 6, 1943

Diary

On New Year's Day 1943, Colonel Berry began writing a diary every night in the form of a letter to his wife, Alice. Optimism prevailed in his diary—he opened each entry by telling her he had had a good night the night before, even though some nights were not good. He commonly signed the entries with "Nite, Sweet" and often added some kind of endearment indicating he missed her, loved her, and wanted to be with her. He wrote one night, "Night Sweet. Hope to live over it and love you many years longer. We are going to enjoy life from now on believe me, Sweet Patootie, K. L. B."

Longing for news from home, Berry wondered whether Alice was keeping magazines or newspapers for him because "this period of imprisonment is pulling me just that long behind the 8 ball on news and what is going on in the world." He also wanted to know how their children were doing, and whether she was "pushed financially." Looking forward to hearing from his wife, Berry wrote optimistically, "Guess I have many letters from you somewhere but up now our dear Nip friends have not permitted any to get to me. Keep writing however for when they do reach me, they should give me plenty of news to bring me up to date."[1] At Karenko, no one received mail from home.

Berry wrote lengthy entries in his diary about food—especially the lack of it—every single day from January 1943 to August 1945. All the POWs were suffering from inadequate rations and poor quality of food. Many POWs resorted to getting an extra taste of food any way they could, even if it meant stealing from the hogs. Below are just a few abbreviated examples from Berry's daily chronicle that show how intensely he focused on food, including what he ate, how meager the rations provided by the Japanese were, and his resulting loss of weight:

- "We 396 PWs [POWs] got 13 lbs beans tonight in lieu of other vegetables."
- "We in Room 7 always drink our tea in our mess bowls no matter what we have in them and in that way we get all our food in our bellies."
- "Saw Ray O'Day licking the serving ladles today so as not to waste anything."
- "Saw Gen. Seals the other day eating carrot butts at the hog pen—stole them from the hogs like Room 7 [Berry's room] used to do."
- "Poor meals—no potatoes and none came in."
- "I was lucky getting 10 pieces of shins and tendons at noon and 5 at night."
- "Weighed today. PWs average about 5 lbs low. I lost only about 3 from last weighing and now weigh 161."

That his obsession with food was widely shared is evident in the diaries of other POWs. Giving precious food away from his garden became Berry's way of showing friendship and kindness to those around him. He most likely began this practice in the fall, and once he started his diary, he wrote about it. The significance Berry attached to food can be seen in how he precisely recorded each piece of food he gave away:

Mar 8: "Gave Ed Lilly a tomato."
Mar 14: "Gave Wainwright, King and Corkill a tomato."
Mar 15: "Gave Quesenberry, Bunker and Alec Campbell a tomato."
Mar 22: "Steve [General Stevens], Hugh Dumas and Jack Keltner got a tomato."
Mar 23: "Leonard Evans and Howard Frissell got tomatoes."
Mar 25: "Loren Wetherby's birthday—gave him a tomato."

138 CHAPTER 11

Mar 26: "Gave Gen. Funk a tomato."

Mar 27: "Gave Pat Callahan a tomato."

Mar 28: "Gave Ed Cork ripe tom [tomato] and made him eat it."

Mar 29: "Gave Hugh Dumas ripe tom today. Gave Rodman 2 tomatoe [sic] cuttings."

April 27: "Did more garden watering and gave Hugh Dumas 2 Kohlrabis and 2 onions. He is a nice fellow and one of my best friends here."

April 29: "Gave garlic cloves to all my roommates, Brownie, Cork, Hugh and Joe Cottrell also."

Sometimes other POWs gave Berry food or other items in return. From one officer he got a pair of shoes; from another he received carrot tops for his salads. Because the POWs had practically nothing, giving gifts to one another was truly a sacrifice, a deed that made life in the camp a little more humane among them.[2]

COERCED WORK AND FORCED WORK

After seeing the success of the individual squad gardens, Captain Imamura decided that the POWs should establish and maintain a camp farm and garden, in which the POWs would voluntarily work. The camp garden would require garden work from all POWs who were not elsewhere engaged, such as the kitchen, library, or post exchange. Perhaps to motivate the POWs, Imamura assured them that the produce grown in the camp farm would supplement their daily rations. For that purpose, Imamura deducted 10 percent of the POWs' monthly salaries to buy stock, seed, fertilizer, carpenter tools, and building materials.[3]

Imamura's ploy worked—the starving POWs consented to work voluntarily in the garden. Imamura could then boast to his superiors that his POWs were volunteering to work in his camp. In time, however, it became obvious that the only diets being supplemented by the garden food were the guards' diets. The Japanese took almost all of the food, giving the dregs to the POW chefs. Unsurprisingly, prisoners began to complain. In response—and to ensure that working in the camp garden remained voluntary—the commander offered farm workers an extra cup of rice at mealtime on the days they worked. This became known as "workers' rice." Berry did not mind the work because he enjoyed gardening. He also looked forward to the extra bowl of rice. For the high-ranking officers over sixty years of age, the camp garden work proved to be too strenuous, so Imamura relieved them of garden duty. Instead, he gave these officers

We had to sign in and out at night when we went to the Latrine. The liquid diet kept many running.

Benjo Duty. "Our Vigilantes—Night Guardians of the Latrine. We had to sign in and out at night when we went to the Latrine. The liquid diet kept many running." Source: *The Life of a P.O.W. Under the Japanese in Caricature as Sketched by Col. Malcolm Vaughn Fortier*, by Malcolm Fortier.

the undignified duty of tending goats. The American POWs were dismayed to see their Lieutenant General Wainwright tending goats.

Each squad was scheduled for various types of forced work. One of the first duties officers and enlisted men had to perform was to serve as what the Japanese called "vigilant guards." To the POWs, it was just plain "benjo (latrine) duty." Vigilant guards were "members of a nightly detail consisting of one officer and one enlisted man for each hour from 'lights out' until reveille. They were stationed at a table in one of the lower halls leading to the benjo and were equipped with a pencil, a writing pad, and a timepiece. Their duties were: to record the names of the POWs going to the latrine, together with time of going and returning; and to tour the barracks each half hour on the lookout for fire or anything unusual."[4]

According to Colonel Bunker, American generals were excused from this vigilant guard duty early on "because it was customary, in the U.S. service" that they be excused, because they would "lose prestige" by doing such a duty. The always acerbic Bunker remarked at the time, "If they only knew how much more prestige they are losing by their lazy deadbeating!"[5] As a result of generals being excused from "benjo guard," as Berry called it, the colonels and other officers had to perform this duty more often. Berry seemed good-natured about it, and in his diary, he often recorded when he was on benjo guard; on one occasion, for instance, he wrote, "Good night—2 trips & Benjo guard." The two trips meant that Berry had to go to the benjo two times that night himself. In a later entry, he recorded, "Did 9–10:30 Benjo guard and read Proverbs to pass time."[6]

Because their Japanese captors were terrified about a possible fire in the camp, they required the POWs to perform fire drills regularly. The fire equipment on hand included outdated hand-pump mechanisms and leaky hoses. Barely any water came out of the fire hoses, a fact that caused some quiet laughs from POWs who stood at attention watching their fellow POWs conduct the fire drills.[7]

Hoping to impress their superiors, the Japanese required the POWs to "police up" the camp whenever anyone from outside of the camp was expected to visit. "Policing up" entailed, among other tasks, cleaning the camp and—absurdly—cutting the grass with scissors. These preparations took place for days ahead of the visit—even if the visitor was a low-ranking Japanese corporal. When high-ranking Japanese officers came into camp for an inspection, not only did the POWs have to "police up," but they also had to put on their best clothes. Thus dressed, POWs lined the walkway into the camp through which a visitor would walk. Rarely did visitors inspect rooms or the benjo; instead, these Japanese officers met with the highest-ranking POWs. When the president of war inspection came to Karenko, only General Wainwright and other top officials were allowed to speak with him. Berry learned that the leaders of the POWs reported numerous problems to the inspector. But nothing ever changed after these inspections.[8]

At the end of March, another important Japanese official visited Karenko: Colonel Nakano, "Old Sourpuss," the main commandant of all the prison camps in Taiwan. Nakano must not have been respected by the guards because one of the Japanese soldiers said of him to Braly, "Big Noise, Little Do."[9] When the "inspection" was over, Berry wearily described the humiliating charade.

> Got up at 5:30 AM and had fair breakfast and then inspection by Prison Camp Commander. Did not talk with any of us nor come into our rooms. Raised hell with the Camp Personnel I hear. Spent long time going over records of camps and last legal (?) inspection of last permanent detachment. In afternoon had air drill followed by fire drill. All went off well. Then we

were released until 4:40 P when we had to fall in and line walk to bid him on his way. Supposed to wear our shoes but store room locked so went out in clogs. Guard said he saw that shoe room open which was a d—- lie as it was opened just before he came. They not admit mistake. Sent us back for shoes. Reformed and at about 6:10 P the Colonel made his appearance and by 6:20 it was over. Kept us in rain for 1 ½ hours. It was very tiresome and despicable.[10]

HEALTH OF THE PRISONERS

For their health needs, the POWs had a makeshift hospital with only the supplies they brought with them from Tarlac. Lt. Col. Harold Glattly ("Doc"), who had been the surgeon for the Luzon Forces during the battles before being captured on Bataan, served as the camp doctor. Dr. Glattly was the only POW doctor for O'Donnell, Tarlac, and Karenko until the British prisoners arrived at Karenko with a doctor in their midst. Berry described Glattly as a "nice fellow and I believe he is a good doctor—is doing a good job for us and under most trying conditions."[11] POWs visited the improvised hospital with ailments like malaria, dysentery, heart troubles, pneumonia, sprained ankles, bowel problems, and malnutritional edema (beriberi). To ensure the hospital did not become a place of respite, the Japanese did not allow anyone to stay there very long unless their death appeared imminent.

The starvation diet and the withholding of medical supplies by the Japanese resulted in the deaths of three POWs—one British and two Americans—at Karenko. British Maj. Gen. Merton Beckwith-Smith was the first POW to die on November 11, 1942. A few days before he died, he had been severely beaten by a couple of guards. After this beating, Beckwith-Smith came down with a throat infection. Under normal circumstances, his infection might not have been life-threatening. But he was too starved and weak to fight off the infection. Beckwith-Smith's death was followed by a second three months later. The senior American enlisted man, M.Sgt. James Cavanagh, died from the same type of throat infection that had claimed Beckwith-Smith. He, too, was too weak and starved to survive the infection. In March 1943, Colonel Bunker died from malnutrition. Berry wrote about Bunker, "He very sick in hospital with malnutrition which has run into some kind of trouble wherein his kidneys don't function—don't eliminate properly. Fear for his life as he is about 62 years old." In his diary entry March 16, Berry lamented, "Bunker not expected to last the day out—plain case of malnutrition and starvation—terrible thing as so little would have saved him."[12] Bunker died as Berry wrote this diary entry.

Because the predominant meal for the POWs was white rice, their diets lacked sufficient Vitamin B1. As a result, many POWs suffered from badly swollen feet and legs, or beriberi. Berry was among them: "Had a good night—1 trip to Bjo and have badly swollen feet and legs today—pus ran all down into feet. Went to Doc he said lay off the salt for 4 days and do not drink so much tea." Private First Class Matsumura (known as "Grumpy" to the POWs) was a member of the Japanese medical staff who had other ideas for treating swollen feet and legs—early morning dew walking. The POW doctors, like Col. John Bennett of the British Army, were ordered by Grumpy to march beriberi victims outside in the early morning. In response to that treatment, Berry wrote, "Looks funny to see our medico marching our diet deficiency cases around in the dew in the early morning—most ridiculous."[13] With a keen eye for the absurdly amusing, Colonel Fortier portrayed Grumpy's "new cure."

The Japanese were not entirely indifferent to the health of their prisoners, however. They gave out reading glasses in early September and again in March. Many POWs from Bataan had no glasses, so this gesture from the Japanese was helpful. In addition, POWs received typhoid and dysentery shots, which were painful and hard for some POWs like Colonel Bowler to tolerate. But they helped prevent diseases among all in the camp, including the Japanese. Berry wrote: "Got shot again this AM for dysentery and arm very sore this evening. After inoculation required us all to stay in bed rest of day and it was welcome in a way as it first time we able to use beds in daytime since last fall—prevented working in garden however and we almost missed out on salad."[14] A few days later, Bowler, one of Berry's roommates, had what Berry called a heart attack at bango (roll call) and had to be given a shot of adrenalin; he wondered if the shot for dysentery caused Bowler's heart problems.

BAD TREATMENT LINKED TO WAR EVENTS

By January 1943, the Allies had won a few more battles in the Pacific arena, including in New Guinea and Guadalcanal. News of these Allied victories aroused anxiety among the guards at Karenko and in turn led to more brutal behavior toward the POWs. They handed out beatings and "boppings" as they saw fit, with Captain Imamura looking the other way most of the time. "Boppings" included slapping prisoners in the face or hitting them on the head using a bamboo stick. Bad as they were, boppings were milder than beatings. Beatings were all-out savagery—beatings with fists, gun butts, and bayonets. Sometimes, the guards used a bayonet as a slapping weapon on the hands. Berry described some of the brutal behavior linked to war news. He wrote, "The Japanese guard on the 5 AM

The Doctors tried everything for Beri Beri (swollen ankles) and finally resorted to early morning dew walking, to no avail.

Dew Walkers. "Cornell, Ives, Rutherford, Swanton, Stickney Prepare for Maypole Dance. The Doctors tried everything for Beri Beri (swollen ankles) and finally resorted to early morning dew walking, to no avail." Source: *The Life of a P.O.W. Under the Japanese in Caricature as Sketched by Col. Malcolm Vaughn Fortier*, by Malcolm Fortier.

shift at the Benjo knocked Bell, Churchill, Quesenberry, Brawner & Dougherty around with bayonet across back of hands and in face with fists. No one knows why."[15] Two days later, Berry wrote:

> Guard beat hell out of Brig. Lucas last night & Col. Ives today. Said Lucas saluted with 1 hand in pocket & Ives sitting down leaning against tree. Issued orders tonight so that we were not to appear with buttons unbuttoned, hands in pocket or to sit or lie on beds from reveille until evening Benjo. Perhaps new guard is cause of the campaign of terror or perhaps they want to deflate us over some good news which is coming out other than the good news of the Russian steamroller. We are all doing our best to keep heads up as we don't want our heads knocked around. The two officers beaten up were met with fists, gun butts, bayonets & kicked. Disgraceful.[16]

144   CHAPTER 11

Making matters worse, in late January the Japanese heard rumors that their own POWs held in the United States or Australia were not being treated well. They requested that General Wainwright and top British leaders write letters urging the US and British authorities to be kind to the Japanese POWs. At first Wainwright refused to write a letter. Rough treatment of the POWs worsened after he refused, as Berry recorded in his diary:

> Last 24 hours have been hell due to our high-ranking officers not writing letters to our government which Nip. authorities requested. One EM [enlisted man] beaten up for over an hour with a heavy stick. Gen. Wainwright hit, Lt. Gen. Heath who is crippled beaten up badly but Nip. authorities had him identify the guard and they then beat hell out of him. Brig. Duke beaten up pretty badly. Louie [Bowler] slapped twice again last night by member new guard because he was in bed at 8:40 pm which is OK but guard thought otherwise and slapped him twice but did not do it very hard—only face-saving.[17]

It is interesting that the camp authorities meted out rough justice to one of their own guards who had beaten up crippled Lt. Gen. Lewis Heath. Apparently, there was a line the guards could not cross. After several more days of boppings and beatings, Wainwright relented and wrote a letter so the brutal treatment would stop. But it did not. It never stopped.

Berry was the target of mistreatment a few times but not as often as his roommate Bowler, who suffered many boppings over their time together at Karenko. Before he died, Colonel Bunker had written about how the guards looked for behaviors or even non-behaviors that the POWs supposedly did or failed to do, such as saluting. "Starting yesterday p.m. guards ran amok with their petty persecution. Centering on super sensitiveness to saluting. The sergeant came up on front porch and dragged Bowler, Bronner [Brawner] and Berry and a Limey upstairs over to guard house for not saluting him though he was outside on the porch, and they were inside in their rooms with windows closed and did not see him!"[18] The guardhouse was a small, enclosed space with a dirt floor and no windows. Prisoners were often taken there to suffer in the space with no water or food for a day or more as punishment for small infractions.

Over time, it became clear to Berry and others that the cruel behavior of the guards was in response to news unfavorable to the Japanese, such as the rumor that Japanese POWs were being treated badly by Americans. Even more important, the guards' behavior became a somewhat reliable indicator of what news they had learned about the progress of the war. On the very day of the beating

Berry described above, Baggy Pants had told the POWs in Berry's room that the Russians had retaken the cities of Kharkov and Rostov from Germany. In addition, the Nipponese newspapers were full of apologies for the Japanese losses in Burma and at the Guadalcanal. In a bitter irony, the POWs in Karenko paid a heavy price for Allied successes.[19]

In addition to the abuse and brutality handed out to the POWs in response to war news unfavorable to the Japanese, camp leaders began issuing nuisance regulations to assert their control over the prisoners. Berry wrote, "Are now issuing a great many unusual regulations just to pester us and have guards going around frequently to badger & pester us all during the day. Worst regulation is one which makes us halt & put anything out of our hand before saluting even a private in the Nip. army. Act like swell boys when they are frustrated."[20]

BERRY'S PROTEST

In response to the increasing bad treatment and nuisance regulations, Berry began a protest, of sorts. He began to grow a mustache and decided not to cut his hair until the situation improved. As mistreatment of the POWs continued, Berry expressed concern that he would be a target before long. "Looking for them to clout me for having long hair but believe I will wear it for a while longer. Still waiting for better news before I cut it off."[21]

The very night that Berry expressed anxiety about becoming a target of the Japanese guards, the inevitable happened. He wrote on March 2, "Texas Independence Day but no independence for me." He was beaten up that night by a Japanese guard who said he did not salute him when he went to the benjo. Berry had not seen him, or he would have saluted as was the rule. Berry wrote, "Called me out and hit me 4 full arm swings to my left jaw but did not yet floor me. Then had me stand up and hold heavy bucket at both arm's length for 15 minutes or so until new guard came on and let me off. I was beginning to weave badly by that time. Did 2 or 3 other[s] in a similar manner."[22]

Berry's roommate Braly recounted this incident with a little more zest than in Berry's account. "As the reveille bugle sounded Roommate Berry came out cussin' from under his mosquito bar [net]," Braly recalled. "K. L. Berry is a big Texan. He had a fine head of curly hair and a long cowboy mustache. He was a big man, had been a football and track star at the University of Texas, and was probably the only man in camp who could have stood up to what followed."[23]

In Braly's account, Berry's hair and mustache were targets, too, details not mentioned in Berry's diary account. The colorful language Braly used to retell the same event must have come from Berry himself. Braly retold Berry's story:

During the night that bastard "Slant Eye" hid behind the hedge on me so when I came out of the benjo he was waiting. After yelling at me awhile he reached up and pulled my hair a couple of good yanks, then my mustache. Next he started landing haymakers on my jaw. I stood the first three all right but the fourth really rocked me. Not satisfied with that the dirty skunk picked a wooden bucket, filled it with water and made me hold it horizontally in front of me. Fortunately the bucket leaked so about half the water ran out. Even so it got damn heavy before "Slant Eye" went off post at the end of the hour. Soon after the next sentry arrived Stu Wood [a POW interpreter] came out to the benjo and after some talk persuaded him to release me. But if I ever get my hands on that other lousy son-of-a-bitch when this is over, etc., etc.[24]

This incident caught Colonel Mallonee's attention as well. His account added a few more details that were not in Berry's or Braly's accounts: "The reign of terror increased. Berry and Lathrop were forced to hold water buckets at arm's length, with an upright bayonet just under their hands, which was withdrawn with great laughter just before exhaustion. Worse yet was having one's organs 'stirred' with a rifle butt."[25]

As a result of this incident, Braly noted that many POWs "started the nightly practice, in case the benjo sentry was not in sight, of rendering a formal bow to the darkness, regardless."[26] Commenting on this development, Berry contemptuously updated the name the POWs who fought in the Battle of Bataan had given themselves. The "Battling Bastards of Bataan" had been reduced to the "Bowing Bastards of Karenko."

In the days following his beating, Berry and roommates experienced still more harassment. "Guard has just stopped at our window and bawled us out because we did not look around when he knocked and jump to attention fast enuf. Did not hit any of us however." Then the next day, "All in our room got a gun butt dropped on our toes for having shoes on in our room and this PM Brownie and I got it again for same offense. We had just come in the room from outside when guard came in and dropped gun on our toes. He did it gently however and laughed. We will not be caught again. Orders have never been issued against the practice but they make regulations up on the spur of the moment just to catch us I suppose."[27]

The harassment and often brutal behavior by the guards continued unabated, so Berry's long-haired protest continued as well. In time Berry even displayed a little vanity about his looks, as he wrote about his long hair to Alice in his diary. He thought she would like how he looked. He definitely thought he looked good.

"Certainly have a long, glossy, curly head of hair—looks like Buffalo Bill or Gen Custer or someone. Have had several compliments on it. Wish you could see it. Will try to get some pictures of it before cutting it off and will save you a few curls then." His protest had turned into something else that gave him pleasure. In April, still sporting his long hair, he wished again that he could get a picture of himself with his long locks to bring home to her. By May, he jokingly wrote, "We all go on goat guard tomorrow—you ought to see your Taiwanese goat boy—my hair is longer than the goats. Perhaps will celebrate our anniversary and get my curly locks shorn."[28] Even though Berry was quite proud of his luxuriant locks, they came off on May 25, two weeks before their move to another camp. As he had promised, he saved Alice some curls.

WAR NEWS

The war news about North Africa that trickled into the camp captured Berry's attention when he learned that his friend, General Eisenhower, was in command of the North Africa campaign. Berry wrote, "See Ike made Field General so G-2 I bet N. Africa is OK. Also see where he is placed in command of all the Allied troops in N. Africa. Seems he is doing OK also."[29]

This news about Africa and Eisenhower made Berry reflect upon the prediction Eisenhower had made to him at Fort Sam Houston in October 1941 that the "big show" would be in the other direction. On February 28, Berry bemoaned his situation: "Have just read Feb. 17 paper (newest one) and in it FDR's speech is quoted. Surely sounds good to me and I am sure bad to them.... Sure hell to sit here and undergo this harsh & cruel treatment and all the parade of stars pass you by. Believe I would be in Africa with Ike Eisenhower if I had not been so unlucky as to draw this d—- assignment."[30] Perhaps unstated is Berry's recognition that, had he followed Eisenhower's advice and succeeded in changing his orders, he might have been in the parade of stars.

Defeating the Germans and Italians in North Africa would give the Allies an opportunity to invade Europe from Italy, something the POWs wanted to hear badly. Any little bit of good news filled the POWs with hope that the war would be over soon. Berry was almost always optimistic and began to think he would be going home in 1943. "All in all we feel that things are going OK for our side and as long as we are not starved to death or our health ruined we are OK and will be out of here this year."[31]

During the spring of 1943, the United States scored a victory over the Japanese in the Battle of the Bismarck Sea, where the Japanese suffered heavy losses. The British were having on-again, off-again success in Tunisia. And of great strategic and symbolic importance, code breakers pinpointed the location of Japanese Admiral Yamamoto, the mastermind of the Pearl Harbor attack, flying in a Japanese bomber near Bougainville in the Solomon Islands. Eighteen P-38 fighters then located and shot down Yamamoto. His death was a major coup for the United States.[32]

As had become a predictable response, their Japanese captors stepped up their brutal mistreatment of the POWs in the aftermath of these defeats. In his diary, Berry noted numerous acts of mistreatment by the guards in the spring. They beat officers severely for minor infractions, bopped prisoners at the latrine and for lying on beds, and threw rocks at prisoners or made them stand at attention for two hours for not saluting or other reasons the guards invented on the spot.

Given the level of mistreatment visited upon the POWs in the spring of 1943, Berry was confident that the war would end that year. "Sure will put on a splurge with this when this thing is over which we all believe will likely be this year as we [see] signs of a desire for peace in Germany and in Nippon. Might come fast when it comes and suddenly. Believe Germany definitely sees she has probably lost and will want US and Briton [sic] to come in before Russia can get in because Russia will be awful hard and we don't want her in too good a position to dictate too much at the Peace Table."[33] Berry's experience in Siberia had given him clear insight into the danger to world peace that Russia could pose.

Through March, April, and May, the POWs received snippets of news about the war in Africa and Russia, news that helped keep their spirits up despite their circumstances. They grew confident that the war was coming to an end. When the newspapers they read contained no news about Africa or Russia, the POWs nevertheless surmised that the Allies were "kicking hell" out of the Germans. The POWs assumed no news was good news for the Allies, especially because the harsh treatment by the guards continued. That could only mean the Japanese were losing.

One guard nicknamed "Scarface" was the worst. When the enlisted men failed to turn out fast enough for reveille, he made them run around a field and beat them if they did not run fast enough. Bopping of officers continued, and two colonels were forced to hit each other several times. When the guard thought they were not hitting each other hard enough, he punched them both. A Japanese corporal watching the slugfest came over and punched them both, too. The colonels fell to their knees.[34] Beating up defenseless, weak men—at bottom an act of cowardice—must have allowed these guards to feel powerful and in control when in fact their future was looking increasingly dire.

After the Allies defeated the Germans and Italians in North Africa in May 1943, Berry, like all POWs, hoped for an easy, quick establishment of a front in Europe and then a swift march through Italy to defeat Germany. Ever the optimist, Berry wrote, "Nip papers mum on continental landings and we feel sure that it has already taken place and was a success and that by now the Brits-Axis forces are well on their way to dragging out this bloody business within the next 4–6 months in Europe and then for this place."[35]

Like his fellow POWs, Berry was overconfident about how fast the war could end. The Allied invasion of Europe through Italy proved to be neither quick nor smooth. The Allies had some victories in the spring, but so did the Axis powers. When news that victory was not at hand reached the POWs in Karenko, their optimism and hopes fell. Pessimists in the camp expected that they would be POWs for another year. Unfortunately, even the pessimists' gloomy predictions proved to be wrong—the war and their imprisonment would drag on for two more years.

RED CROSS AND MOVING

Along with war news in the spring of 1943, the POWs began to hear two rumors. The first was that Red Cross supplies were coming. The second was that one hundred prisoners would be leaving Karenko. When the Red Cross rumor turned into reality, Berry exulted, "There is a Santa Claus and he came today—brought us 15 tons of Red Cross supplies—books, shoes, 6 sacks of sugar salt, about 1000 individual packages containing sugar, jams, cocoa, chocolate, corned/canned beef, soap, milk, tomatoe [sic] juice and about 6–7 other items."[36] However, disappointment swept the camp when day after day, the Red Cross supplies sat in plain view, undistributed.

On April 1, the second rumor about a potential move became a reality for the generals and high-ranking civilians. They learned that they were to leave Karenko on April 2. In all, 129 POWs left. The high-ranking officers took with them their orderlies, the doctor, the best cooks, the barber, and ten goats. The Japanese told everyone that both groups would be more comfortably housed and better fed with the high-ranking people going to another camp. The move was better for the generals in their new camp, but for those left behind at Karenko life was no better.[37]

The reason behind the move of senior officers and civilians was political. International Red Cross officials had expressed a desire to visit the officers' camp, but Karenko was overcrowded and conditions dreadful. It was not a camp to show off to the Red Cross. By transferring high-ranking officers and officials to another camp, both camps would appear less crowded and would perhaps satisfy the Red Cross that the POWs were being treated humanely.

After the generals left, the POWs learned that the Red Cross supplies had to be divided among three POW camps numbering eight hundred prisoners. Finally, after three weeks of waiting, the Karenko POWs received their Red Cross supplies, the first they had received in over a year. Berry's share consisted of "10 lime drops, chocolate slab, 16 cakes, 2 slab sugar (about 2 ozs), bar soap, 2 oz cheese, 2 oz turkey and tongue paste, small can of tomatoes, marmalade, margarine, 10 crackers, carton pudding, . . . steak and tomatoe [sic] pudding, meat galatine [sic], condensed milk, bacon."[38] Because of their starvation diet, the temptation to eat everything all at once overcame Berry's roommate Colonel Bowler, who ate everything at one sitting and became very ill.

Berry had predicted that the Japanese would not allow the Red Cross supplies to supplement their starvation diet. He was correct. The Japanese used the Red Cross food to reduce the POWs' rations. Berry wrote, "Nip issue of vegetables back to old basis—starvation. Red+ is all that is saving us. General wrote letters about it. Rumored Nips wanted to borrow some Red+ food and are mad because our committee either would not lend or swap or were too slow in giving affirmative answer. Prison groups sore about committees action for we feel it will be reflected in Nips treatment of us."[39]

Fortunately, the post exchange (PX) began receiving additional items to sell. Berry's mood improved with more food and, as always, he wrote about the food in great detail:

> April 24-Forgot to say we had PX issue yesterday. There was enuf so that I got 1 pound Nittoh Black (as good a tea as can be found) 4 bottle Wakamoto [yeast] 10 package of cigarettes (10 in a pack) and 7 Noko cigar. I was lucky at cutting cards and drew the 4th bottle Wakamoto on the ace of spades and the 7th cigar on the King of spades—21 of us drawing. I managed to spend most of my 40 yen allowance. Louie, Bill and I now have enuf tea to do us about 40 days and should have more by that time. I also have about 70 packs of cigarettes and they should last until I can get more. Had ½ of a lb can meat and vegetables ration for breakfast in addition to regular issue. They from weekly Red+ issue. Felt so good this AM that I worked in garden watering and piddling. . . . Glad I am back on my feed as I can enjoy the bountiful (8) repast.[40]

When Berry was weighed in May, he was certain that he had gained ten to fifteen pounds from all the new food. He felt stronger and healthier. But he was disappointed to find that he had gained only four pounds since February.

The day after the generals left for Tamazato, all the squads and room assignments were changed. Berry lost Brownie as a roommate, which he regretted very much, but Brownie was at least in his squad. In Squad 1 there were twenty-one colonels arranged by prison numbers: Aldridge, Amis, and Atkinson were in room one; Ausmus, Balsam, Bell, Berry, Boatwright, Bonham, Bowler, and Braly were in room two; and Brawner, Brezina, Browne, Campbell, Carter, Chase, Cooper, Cordero, Corkill, and Cottrell were in room three. According to Berry, they were very crowded and had no cabinet space. Bill Braly was the squad leader again, but it is unknown if Berry was assistant squad leader this time. As before, each room had its own private mess. Berry tried to get Brownie into his mess, but the others in room two refused to add another mouth to feed. Food was too dear to have another person eating from their bucket. Brownie was "sore" about the whole situation, according to Berry.[41]

The month of May 1943 brought many rumors about moving to another camp because the Allies seemed to be getting closer to Taiwan. There had been several blackouts at night, and the POWs had heard bombing somewhere. Berry wrote, "Rumors flying thick and fast. I am beginning to believe we are going to move somewhere but don't know where. Perhaps off of this island as I have an idea this will become a combat area sometime this year and they will want to get us out of here." Many of the POWs did not want to move because the weather was good, their accommodations were fair, and, as Berry observed, "We will be fed and treated as well here as another place and perhaps better."[42] But what the POWs feared most was being moved to Japan. They did not want to be in the line of fire when the Allies bombed Japan.

The Italian Swiss representative from the International Red Cross, Dr. Pararinci, paid a visit to Karenko at the end of May. Because Dr. Pararinci had a Japanese wife, the POWs at Karenko were very skeptical that he could be trusted. They knew nothing else about him that would be a cause for worry. He promised that letters were coming. However, as noted earlier, no one at Karenko received mail from home. Ultimately, Pararinci's visit did nothing to improve the POWs' lives. Once the Red Cross visit had taken place, most of the generals were returned to Karenko on June 5. However, Generals Wainwright, King, and Moore, along with the British and Dutch high-ranking officers and the governors, were sent instead to another prison camp near Taihoku in northern Taiwan.[43]

Moving day for all the POWs in Karenko finally came, producing mixed feelings of hope and dread. By this time, Berry knew they were going to the other side of Taiwan and not to Japan, but he knew nothing about the new camp. On June 7, they left Karenko at 10:00 a.m. They marched to the port, boarded a steamer named the *Hozan Maru*, and set sail at noon headed north up the eastern coast of Taiwan. They traveled about four hours and entered a bay opposite the town of Suo. Locals came out to the *Hozan Maru* in sampans and slowly gathered all the POWs to take them to shore. Berry wrote, "We debarked on punk."[44]

Later that night, they left the town of Suo by train and rode twelve and a half hours in a very crowded train without any water to drink. The train stopped south of Kagi on the western side of Taiwan, where the POWs boarded a narrow-gage train and traveled for another two hours inland. Tired and dirty, they got off the train at a station that was three and a half kilometers from the village of Shirakawa. Hiking to the new camp, gathering mangos along the way, they wondered what was in store for them.[45]

CHAPTER 12

Taiwan POW Camp Shirakawa
Part 1

COLONEL BERRY'S TIME AT THIS CAMP:
JUNE 8, 1943, TO OCTOBER 9, 1944

While imprisoned at Karenko, the POWs had gotten their hopes up that the Allies would move swiftly in Europe after the North African campaign and win the war in 1943. When the war instead continued, they found themselves in yet another camp, one even more primitive than Karenko had been. It was depressing. Shirakawa lacked fresh, running water, but it did not lack mosquitoes. Numerous pools of stagnant, standing water, perfect for breeding mosquitos, surrounded the camp. The Japanese had abandoned this training camp because of water shortages and the high rate of malaria from mosquitos among their soldiers. Malaria became an ongoing health problem among the POWs at Shirakawa, as it had been for the Japanese soldiers before them.[1]

In addition, the climate was sharply different from Karenko's. Colonel Mallonee noted, "Traveling from bitter cold Karenko to parched, burning Shirakawa in the Formosan southcentral plain was quite a shock."[2] Karenko was on the eastern coast of Taiwan near a port, and Shirakawa was inland and elevated from the western coast. The camp was the fourth POW camp for the prisoners who had experienced the Death March and the fifth camp for those who were captured and held on Corregidor. Spirits were low among the POWs as they entered the compound, wondering how many more months they would be prisoners.

CHAPTER 12

CAMP COMMANDER AND HIS STAFF

Former Camp Commander Imamura from Karenko, no favorite of the POWs, was once again their commander. However, the two guards "Boots" and "Scarface," who were especially brutal to the POWs at Karenko, did not follow Imamura to Shirakawa. As a result, the vicious treatment of the POWs was not as prevalent in the new camp as it had been at Karenko. Surprisingly, Baggy Pants, who had said goodbye to the POWs at Karenko, rejoined them at Shirakawa for a while. Also rejoining them were Corporal Iwai, "Simon Legree," who was in charge of the farm, and Medical Corporal Matsumura, aka "Grumpy." Matsumura was not a trained doctor, but he thought he knew more about medicine than the American and British doctors did. As a result, he would not always authorize medical care even though it was needed. And despite his medical role, he could be brutal to the POWs.[3]

Interestingly, a lieutenant and a private under Imamura's command had both lived in the United States and spoke English. Lieutenant Hioki was nicknamed "Pasadena Kid" by the POWs, because he had lived in Hollywood before the war. Pfc. Bob Yamanaka—nicknamed "Frisco Bob" and the camp's interpreter—had attended Mission High School in San Francisco. He was tougher on the Americans than he wanted to be because he did not want to appear pro-American to the Japanese. The medical sergeant, Nagatomo, nicknamed "Handle Bars" because of his mustache, was one of the better members of Japanese staff according to Berry. He liked Handle Bars and thought he was humane in his treatment of sick POWs.[4]

QUARTERS, ROOMMATES, AND QUARRELS

As usual, the generals fared better at Shirakawa than their lower-ranked comrades. The Japanese had constructed a separate building to house the generals. Their quarters turned out to be fairly nice, according to General Brougher. "Assigned rooms (two generals to each room). Weaver and I are roommates. Looks as if we are going to be much more comfortable here. We even have a chair each in our room, a good table, ample shelf space, and a good reading light."[5]

The colonels were not as lucky. Colonel Braly described their barracks as a "one-story wooden structure with gaping holes in the floors and leaky roofs."[6] Berry went further in his description of the barracks and camp:

> This is regular CCC camp—water has to be boiled as it comes from irrigation ditch and the plumbing is primitive wooded setup. Old wooden, 1 story barracks, partly screened, small blockade surrounded by bamboo

lattice high fence. Fairly crowded. 4 in my room—Braly, Bowler, Galbraith and Berry—room about 12 × 19 feet. Water situation critical—get one cup boiled water each half day but catch rain water and drink it—rainy season now. Rained almost all day. Caught our drinking and dishwashing water from roof drip. Place muddy as a hog pen. Got our straw mattresses today so will not have to sleep on the ladder rung bed tonight.[7]

Col. Nicoll Galbraith became a new roommate for the three Bs—Berry, Bowler, and Braly. The Japanese appointed Braly the squad leader again, and Colonel Galbraith became the assistant squad leader, a choice that suited Berry. Berry was pleased with his new roommate, describing Galbraith as a "nice fellow; now asst squad leader 3rd Squad and one of my roommates and fits in well; seems to have both feet on the ground. well liked by squad." Galbraith generally roomed with colonels whose names began with F or G during his time at Tarlac and Karenko. That he ended up rooming with Berry, Braly, and Bowler is likely owing to the fact that—as Galbraith noted in his diary—he did not get along with one of his roommates at Karenko. This change of scene was temporarily helpful, but eventually his new roommates would also get on his nerves.[8]

In fact, there were more clashes among the POWs at Shirakawa than there had been at Karenko. Depressed and disappointed that the war dragged on, living in crowded conditions on a starvation diet, lacking privacy, and being forced to work made the POWs at Shirakawa increasingly frustrated and angry. Unable to take their anger out on the guards, they took their frustrations out on one another at times. Many of Fortier's caricatures drawn in Shirakawa portrayed instances of his fellow POWs being inconsiderate, or worse, to one another. Annoying behaviors led to many arguments and some altercations that Brougher, Galbraith, and Berry mentioned in their diaries.

Some of Berry's good friends came to blows over minor matters. Virgil Cordero and Ed Corkill had a fist fight right in their room. Their roommates let them fight, but Berry heard about the fight, went to their room, and separated them. Neither man was hurt, but they remained angry for a few days. Berry asked them to shake hands, but Cordero refused. It took him a day to calm down enough to shake hands with Corkill. Perhaps because the cause of the fight was trivial, Berry never indicated what they fought about.[9]

A few days after breaking up the fight between Corkill and Cordero, Berry had to separate another two friends—Gil Bell and Ed Aldridge. Neither one was hurt. Their fight was over food. Bell had dished up the food that night, and Aldridge asserted that the portions were unequal. Bell called him a "damn liar," and fists flew. Berry reflected on the situation that night, "Guess we are all getting 'nerves' with our long harsh imprisonment and about time we getting out of here or some will not be worth a good god damn. Some not worth one now and more of us will get in that unfortunate condition if war not over PDQ." Brougher mentioned this fight, too, "It is a pity how crowded we are together—sitting in each other's laps, stepping on each other's toes, getting on each other's nerves."[10] Quarrels and hard feelings over food portions, water, people making too much noise at night with their clogs, library books, workers' rice, and many other things continued the whole time at Shirakawa.

HEALTH OF PRISONERS

Shirakawa had a makeshift hospital, but just as at Karenko, sick POWs had a difficult time getting into the hospital or staying there. Corporal Grumpy, who was in charge of the hospital as he had been in Karenko, would not listen to the POW doctors in the camp. Many types of health problems were seen in Shirakawa. The prisoners suffered from heat exhaustion, broken arms, sprained wrists and ankles, heart problems, malnutrition, flu, swollen feet, malaria, and bad teeth. Unless these men were dying, they had to rehabilitate in their rooms with help only from their roommates.

During their first week in Shirakawa, nine POWs came down with malaria, and by June 24 there were thirty cases among the 345 POWs and eleven cases among the thirty to forty Japanese guards. Quinine, an effective antimalarial drug, was in short supply, leaving infected POWs (as well as guards) to suffer needlessly. Somehow, Berry never contracted malaria, but POWs who did were in and out of Shirakawa's rudimentary hospital repeatedly. Berry's roommates Braly, Bowler, and Galbraith, as well as friends Aldridge, Browne, and Cordero were among those who went to the hospital frequently for malaria or other problems. In October, Berry wrote that he took Cordero to the hospital for a second time with malaria, "Had hell of a time getting him in as Nips won't let our doctors admit a patient until it approved by and can you imagine—a lance corporal."[11] POW doctors were not allowed to give prisoners quinine for malaria without "the lance corporal's" permission either. Apparently, the Japanese ships were having a hard time getting quinine to Shirakawa because the seas had become more dangerous for their supply ships.

Flu swept through the camp, and Berry's roommate Galbraith came down with it twice. On both occasions that Galbraith was sick, the guards made him sit up all day in bed until 7:00 p.m., at which time he was finally allowed to lie down. Once they refused him entrance to the hospital because his temperature was not high enough. On another occasion, Galbraith was admitted to the hospital with a fever above 102 degrees. Berry wrote, "Hell of a barbarous and uncivilized way to treat a sick man. These fellows have no mercy in their makeup."[12] While many prisoners suffered from one or more of the diseases endemic to the POW camp, Berry was generally healthy during his captivity. He noted a few ailments in his diary. He endured swollen feet, a sign of malnutrition. And diarrhea was a recurring problem for him, owing to the quality of what he was given to eat.

Berry mentioned five deaths at Shirakawa in his diary. The first one was Col. Frank Brezina, who died from heart failure. Berry wrote that the POWs held a Masonic service for him. British Navy Captain Rice, who had been sick for nine or ten months, died in September 1943. Another British officer, Royal Air Force Commodore Selby, died from stomach cancer. British Private Sheppard died from tuberculosis. And on May 9, 1944, Gen. Allan McBride died from a heart attack in his sleep after a very strenuous day working in the hot sun. Berry thought that if some of these men had been seen in a proper hospital or even sent home, they might have lived. He rightly blamed the Japanese for their deaths.[13]

FOOD AND RED CROSS PACKAGES

The food issued by the Japanese at Shirakawa was meager and distasteful, but the POWs did receive bananas and sweet potatoes, two food items they had not received at Karenko. Berry appreciated them because they made him feel that he had filled his stomach with something solid. Meat or any other kind of protein in their rice bowls was almost nonexistent unless there was a special occasion, like Christmas, when they might receive a piece or two of pork in their bowl. Berry lamented, "If we could only get meat and beans with the other stuff we get we would not be faring so bad but we haven't had a dry bean in months and never get over 2 ounces of meat or fish all week—imagine it—2 ounces. That wouldn't make a real sandwich."[14]

From Karenko, the POWs brought a few Red Cross packages with them that they tried to make last for a while. These packages were of special importance to the POWs because they contained things from home, like marmalade. Beginning in June 1943, Berry often wrote in his diary that Red Cross supplies were due the next day in Shirakawa. But they never came. A Red Cross representative who visited in September 1943 said he hoped more supplies would arrive in November. It was not until May 1944 that some Red Cross packages did finally come to Shirakawa—over a year since the last delivery. Nevertheless, when they came, Berry celebrated this big day by enumerating his trove: "Drew Red Cross 1½ packages in PM. The packages are wonderful—cigarettes, soap, sugar, soluble coffee, jam, spam, corned beef, salmon, butter, ham and eggs, pork loaf, prunes, cheese, . . . bouillon powder, Vit. C pills, army chocolate and perhaps other things which I have forgotten."[15]

The items in these packages, along with the post exchange (PX) supplies POWs could buy, became currency among them as they bartered items back and forth to get more of what they wanted. Addicted to smoking, Berry sometimes traded away food for "smokes," as he called cigarettes.

Because there was little else to do in the evenings when they were confined to their rooms before lights out, Berry wrote in great detail about his food diet every day, just as he had at Karenko.

> Usual bad meals but had little fish in soup tonight and few more sorry vegetables. Opened 12 oz. canned corned beef with Bill [Braly] and put 1 oz. in soup each meal. Went fine. Put 1/3 spoon cocoa in soup at breakfast and it wasn't bad. Had PX issue of 10 packs cigarettes today just as I was on my last pack as I had loaned some of mine out. Got Red+ issue of 1 lb. sugar, ½ lbs. salt and a 12 oz. and an 8 oz. can corned beef. Tried to get more but BP [Baggy Pants] would not permit it. Mixed my sugar with cocoa and used ½ spoonful in soup. Tasted swell. What we fix up to eat would kill a hog I believe.[16]

Berry's wry sense of humor was often evident in his diary summaries. His humorous take on his food, along with his grumbling about the starvation diet, helped Berry get through each day. Berry was not alone in his complaints. Everyone groused about the lack of quality and quantity of food. In the generals' squad, two orderlies brought in food in two wooden buckets. Brougher wrote, "I stir steadily to insure uniform consistency while Jim [Weaver] dips with a dipper and ladles into the bowls. And woe unto the soup server who fails to get a fairly uniform level in the bowls! ... We are quite primitive about such things. Hunger is a physical fact that cannot be ignored."[17]

Berry must have complained out loud once too often to his roommates because Galbraith wrote in his diary, "K grumbles at everything—our largest ration cart came in today but he grumbled at the type of produce—also about the food given the pigs and rabbits—also about hot water—and when we received bananas after several days without, grumbled because they were not ripe enough."[18] Unfortunately, grumbling about food was a way of life in the camp, but there was a reason for Berry to grumble about their food being given to the farm animals. The POWs grew the food, and it should have been theirs, especially because the meat from the farm animals was given to the guards.

Berry was a big man who could hardly believe that he weighed thirty pounds less than he should have. He wrote, "Weight 164 lbs last of August and have been eating this swill like a hog ever since and working not too hard about 5 half days a week and today we got our irregular weighing and I weighed 161 lbs—hell of a weight for a 6' 1" man taking not too much exercise if he had decent food and one who for 32 years has weighed right at 195 lbs year after year."[19] Although Berry said he was not taking "too much exercise," in reality he exercised a great deal. Because he was healthy and still relatively strong, he worked every job that came his way, even filling in when his friends and roommates were too sick to work.

Many prisoners who had lost a great deal of weight suffered from malaria and became sickly, a fate that Berry did not want. Two of his good friends, Harrison Browne and Virgil Cordero, were often sick—Browne with migraines and Cordero with malaria. When they were too sick to eat, they gave Berry their food. Curiously, even when they were not sick, they both began to give Berry their workers' rice almost every day. Even though Browne and Cordero were not large men, it is hard to understand anyone giving away food. Perhaps they gave him extra food because he did some of the harder jobs in the camp that they could not perform when they were sick. In any case, even Berry expressed surprise that they gave him food. In his diary, he revealed, "Feel bad eating it in front of my roommates but the fellows who give it to me tell me it is for me and not to give it away."[20] It is no wonder that Galbraith mentioned Berry's grumbling in his diary; he was understandably annoyed that Berry received extra food from friends and still grumbled about food.

To his credit, Berry cultivated a small private garden with Ed Aldridge, Al Balsam, and Nick Carter in his spare time. His roommates, too weak to add more labor to their days to work in a private garden, counted on Berry for some of that fresh food. He and his friends grew potatoes, tomatoes, peanuts, kohlrabi, and radishes. Berry shared food, especially potatoes, with his roommates, the thirty-six other colonels in his squad, and with others outside of his squad. Potatoes were filling and highly desired.

The lack of food began to change in the fall. In November 1943 and into the spring of 1944, food became more abundant—although still lacking in proteins.

ACTIVITIES

Despite their insufferable conditions as prisoners, the POWs maintained some of the same life-affirming activities as they had at Karenko, such as weekly church services and cribbage tournaments. Volleyball was also popular at Shirakawa, but Berry did not participate. Instead, he became a tireless walker, along with General Jones and Colonels Corkill, Browne, and Quesenberry. These were times when Berry could have good one-on-one conversations without others listening in as they would in his room.²¹

Berry also hosted discussion groups, which he called "bull fests," in his room on many Sunday afternoons. Four or five colonels would gather to smoke and discuss the rumors and news snippets about the war. On one occasion, singing replaced the usual "bull fest" as Col. Ed Lilly noted in his diary. "After church this AM, K. L. Berry asked me & a bunch of Texans in his room to sing: 'Home on the Range' and 'The Eyes of Texas are Upon You'—Don't know just what the occasion was—& K.L. did not say."²² Berry was most likely feeling homesick.

As they had done at Karenko, even though it was not condoned, the officers organized another lecture series. Even Berry gave a lecture, recounting, "Bathed today even tho cool & shaved & put on my best bib and tucker & gave 1 hr lecture to young British officers on hunting in China. Hope they did not find it too dry for it was bloodthirsty enuf."²³ Berry was obviously talking about his hunting expeditions when he was stationed in China with the 15th Infantry. The young British officers must have heard Berry's story about shooting the leopard.

The POWs continued celebrating birthdays and Christmas. On July 6, 1943, Berry celebrated his fiftieth birthday in POW style with food, cigarettes, and poetry. He wrote, "My birthday and far from my loved ones but had a good day as these days go. Had many congratulations on my 50th birthday and 3 cigars and 3 packs of cigarettes. . . . Cordero gave me some soup each meal and rice once and Nick Galbraith gave me some rice. I opened can creamed rice and had Roscoe [Bonham] as guest. It was very good."²⁴ His roommates and Colonel Browne even wrote Berry birthday poems, exhibiting both their creativity and their affection for their friend. He loved the poems they wrote (Appendix A).

Because the POWs were held so long at Shirakawa, Berry found himself celebrating his fifty-first birthday there as well. He wrote, "Birthday July 6 1944. My birthday—3rd one as a PW. Had as nice a one as could be expected however—got cigars or cigarettes from Brownie, Ted Lilly, Bill Braly, Nick Galbraith, Ed Keltner, Howard Frissell, Vic Collier. Sugared coffee from Ed Cork and a nice present from PA Brawner and lots of 'best wishes' from many others. Not a bad day considering."²⁵

Christmas 1943 was something special. The Japanese treated the POWs to more food that day. An appreciative Berry wrote, "Had a wonderful duck stew at noon and good bowl rice—grease but usual size. Stew was solid duck and vegetables and wonderful." The prisoners exchanged small gifts and wished each other a happy day. Berry's good mood continued, "Going round wishing my friends 'Merry Xmas' and giving small gifts to Corkill, Keltner, Galbraith,

Bowler, Whitehurst, Pechek and Bitner. Galbraith gave me cigar, Keltner gave me gum and Corkill gave me F Flakes [fish flakes]."[26]

The POWs staged a Christmas show that their Japanese captors enjoyed a great deal. This success led the commander to allow the POWs to set up amateur shows during the first week of each month from February through June of 1944. British and American enlisted men acted in the skits.

Berry's mood remained upbeat as he finished his diary entry that day:

> Christmas '43 is about over and compared with 1942 it has been wonderful—far better treatment and much more kindness and consideration shown us—better food—our countrymen on the march and the Axis on the run. We have gotten no news or rumors but our morale is as high and we know things are on the up and up for us.... We wonder why they are treating us so well and have decided they have received orders from higher up to treat us more kindly and humanely and do we appreciate it.[27]

On New Year's Day 1944, Berry's good friend Brownie surprised him with an act of kindness that truly amazed Berry. He asked Berry to go with him to Yasume Park (a park the POWs built within the compound) and then Brownie pulled out "a 3 oz can cheese and 2 biscuits" and they had a "feast," wrote Berry. Brownie had saved this Red Cross food from Karenko for all those months. Berry could hardly believe that Brownie had saved that food to share with him. He was elated and thankful.

However, later that evening, Berry's upbeat mood from the morning darkened as he grimly reflected: "Jan 1, 1944—New Year's Day 1944 and still in the hands of the Nips. Would have bet all my possessions almost, a year ago that we would not be here now but just at present I wouldn't bet we would not be here Jan 1, 1945. Don't believe we will but it seems as if the anti-axis is moving damn slowly and doing damn little."[28] It was difficult for the POWs to imagine why the war was not moving along faster. He would still be a prisoner in January 1945.

While staging an occasional amateur show relieved the boredom of captivity, reading turned out to be a continuing source of enjoyment and escape for the POWs. Braly noted that "the one alleviating influence for many of us during those distressing months was the Camp Library, and Colonel Freddie Ward and his staff worked untiringly to make and keep it of maximum value. On the shelves we could find text books, biographies, histories, fiction, detective thrillers and poetry."[29] Berry "drew" his first book in July 1943. He became an avid reader at Shirakawa and noted in his diary everything he read (Appendix

B). "Am doing a heck of a lot of reading and not only enjoy it but passing this godawful time more pleasantly thereby."[30]

Generally, Berry did not comment on the things he read in his diary, so when he did, those books were notable. For instance, after reading *The Grapes of Wrath*, he commented, "Vinegar story if I ever read one—one of the nastiest and most vulgar book I ever read." If the book shocked his sensibility, it also gave him something to reflect on besides the rigors of his captivity. After reading *News from Tartary*, he wrote, "I find last book is very interesting and makes me long to visit old China again. You and I must live there a year or so after I retire as there are many things for us to do there and many things for us to buy. Perhaps we will have the money to spend a while there and spend a lot of Jack also." Reading another book about China prompted a similar response, "*Maker of Heavenly Trousers*—a tale of old Peking of about 1913–1920. Makes me long to be there with you seeing the sights and buying pretty things." Clearly, Berry missed Alice and those good times he had enjoyed with her in China.[31]

MAIL FROM HOME

Missing Alice was a theme throughout Berry's diary. His diary entries were always addressed to Alice, telling her about his day, his thoughts, what he ate, his and others' treatment, and rumors about the progress of the war. He wrote about how much he missed her and longed to hear from her. Each diary entry concluded with him wishing her a good night with love. Sadly, Berry's steady affection for Alice was not, it seems, returned in kind. Throughout his captivity—and, indeed, even before—Alice was an indifferent correspondent. The dearth of mail Berry received from Alice at Shirakawa grieved him greatly.

Radiograms and letters began to arrive in August 1943. Thirty-two officers received radiograms, and Berry was one of the lucky ones: "August 16—Truly a Red Letter Day: Got a radio from home through Geneva. Here it is: wife Alice Berry reports all well Celeste with her Tom Senior at college Junior married Phyllis love." Alice had sent her radiogram in January 1943. Remarkably, it was the first piece of news Berry received from her since before he left for the Philippines two years earlier, in October 1941. Despite the long silence, Berry was ecstatic to receive news from home.[32]

His hunger for news is reflected in how much information he gleaned from just thirteen words.

> What a relief lifted from my mind to know that all is well with my loved ones and it told so much in a few words. It meant all well and not starving. Celeste at home with mother and not in Alaska. Bill not killed or seriously

wounded. Tom doing well in school and not drafted as yet. South went back on his word about appointing Tom or else Tom did not want it or failed on his exam and K. L. Jr either graduated or busted out of USM and married Phyllis Boardman [Boarman] and that my baby still loves me. Wonder if you did not send more news but it was condemned at Geneva. Nothing to show whether you received any of my letters except the sending of the radio. Hated to read radio at first as was afraid of bad news. Can't tell when it was sent but believe it was Jan 1943 or later due to Tom being a senior.[33]

Of particular interest in this recounting of Alice's brief message is the significance that Berry attributed to the single word "love" at the end. For him it meant "that my baby still loves me." After receiving this brief message, Berry did not hear from Alice again for another seven months. The fact that other POWs started receiving letters from home in September deepened Berry's disappointment, frustration, and even resentment toward his wife.[34]

On occasion, Berry received important news about his own family from letters sent to other POWs. For instance, Col. Gil Bell received a letter in September 1943 that reported that his son had graduated a semester early in January 1943 from the US Military Academy. Berry knew that Bell's son and K. L. Berry Jr. were classmates at West Point, so the question he had about K. L. Jr.—whether he had "busted out" of the academy to marry Phyllis or had graduated early—was answered. K. L. Jr. had graduated early. Because Bell had received a letter, Berry was hopeful that he, too, would get one soon. Tempering his hope were rumors circulating among the POWs that the Japanese were withholding mail. No one knew whether the rumor was true.[35]

One late October night in 1943 at Shirakawa, a sad Berry reflected on not hearing from Alice since he left for Manila in October 1941. Alice's neglect from that time galled him.

> Baby, 2 years ago 2 days I left you and Celeste.... I remember so well looking out of the car window and seeing you for the last time. Little did I realize that war would come so soon and I would be caught in such a rat trap. ... In all these 2 yrs how I have looked, hoped and prayed for a letter from you but prayers have been unanswered and I can't get over the fact that you did not send me an air mail to be waiting for me when I got to Manila. Why did not you?[36]

In a series of his diary entries about other POWs receiving mail, Berry repeatedly expressed deep disappointment, often with sarcasm or outright anger:

> November 1, 1943—Good bit of mail and radios came in today but none for me.
>
> December 17, 1943—Bill Rutherford got a fine Xmas package today mailed in NYork Aug 27, 43. Contained 11 lbs chow. Wonder if my wife will be as thoughtful and intelligent.
>
> December 24, 1943—Jim Hughes, Dean Sherry, and several others got packages today. Guess my wife has forgotten me. Have you?

Sadly, Christmastime did not bring Berry any letters or packages from home. The New Year was not any better. By February 1944, a frustrated Berry nevertheless got his hopes up when he heard there were 1,500 letters on the way to Shirakawa. Given that such a large cache of letters was coming, the ever-optimistic Berry expected that he would receive several letters from family.

> Work in AM and while at work we heard there were 1500 letters for us American PWs which meant about 7½ letters apiece. We all had high hopes and I thought with a wife, 3 grown kids, 4 sisters and 2 brothers I would at least get my average but when my name was called and I stepped for my large package—Ed Aldridge had already come out with 18—and when I was handed only 2 and only 1 from you [the other from Anne, his sister] guess my chin dropped a foot for my spirits certainly did. It was your letter of Aug 24 [1943] and postmarked Washington. It told me almost nothing and last half of first page was censored. You did not even tell me your address and for lack of news your letter was a masterpiece. Anne gave me more information and I love her for being thoughtful enuf to write. You have to be able to write since February and I feel that you have slighted me very cruelly. Most officers got from 4 to 15 from their wives and many from children and other relatives. Haven't you told my relations where I am?[37]

Overnight he fumed about his lack of mail and became angrier the next day, threatening divorce.

> Heard from Ed Aldridge you live in SA [San Antonio]. Also Texas won Conference in 1942. From Anne that Tom still in college but due to go to Paris Island soon.... Wonder if you ever stop to think I would like some pictures as well as letters. Many got numerous letters running from June 43 to Sept 15/43. Why did you not write more? If you have not written me many times I will never get over it and if you have not sent me packages I will never get over it. One fellow got a package yesterday and many were told in letters that they had sent them. Hope you do not wait until you hear

from me before writing as you do and did in 1941. Will be complaint for a divorce regardless of how much I love you.[38]

Continuing to be upset about the lack of mail and sparse news from Alice, Berry noted the next day, "Several officers got packages today but you know my wife always behind as she was in her letters." He went on to write that he did not like the one letter he did get from her because it was "kinda cool." Still smarting about the lack of mail and feeling bitter, he lashed out again a few days later, "Hope my wife and family open up their hearts and take the little time and write me and send me some clothes, food, news and love. Hell to be in my position and condition and feel your family has forgotten you and perhaps just waiting for your insurance. Am quite bitter tonight so I guess it would be best to close this daily stint."[39] Seeing his friends receive so many letters at a time made him jealous and angry. Even a few weeks later, his frustration and anger had not cooled down.

> Had a nice dream of you last night but a terrible one about you the night before. The latter was about your dereliction regarding writing to me and sending me packages. I feel pretty badly the way I feel you have done me. I fairly boil when the others brag about their 5, 10, 15 letters—yes one got 15 letters from his wife and children and I got one from my wife and not a damn one from a single kid—scabby.[40]

Finally, in April 1944, he received three letters—one from his brother Gene on the 8th, and one each from Alice and his son K.L. Jr. on the 18th.

> Got a great surprise today—letter from K of Jan 43 and one from you of Feb 43. Wanted more but those two were pretty nice. K said you all had heard rumors that I had been made a general. Hope so but I have not been informed. Hope it is true with back pay and rank back to April '42. Proud as hell of Tom and K also but especially of Tom. Hope he makes the grade in the Marines.[41]

In early May more mail arrived at Shirakawa, but, as too painfully usual, Berry did not receive any: "Mail came today and I was one of the very very few who did not get a letter. Many got 4 or more and one fellow got 7 from his family." As before, he learned some family news from others: "Gen Jones son mentioned K and his beautiful wife and stated he had his combat orders and would soon head this way and drop some heavies over in this part of the world. He is a bombardier and believe from what he said that K is also."[42] (In fact, Berry's son K. L. Jr. became a B-17 pilot in 1944, flying bombing raids over Europe.)

Several letters arrived for Berry later in May; he received two letters from Alice and two from his daughter, Celeste; one from his sister Anne; and one from a friend, Mack Hodges. But he was disappointed that his son Tom still had not written a letter. Hoping for more letters in June 1944, he noted, sardonically, that "I have gotten about 10 letters so far from 6 different people and many here have received from 20 to 50. Very illuminating to say the least." In the month of July 1944, more packages and mail arrived in Shirakawa. He waited in anticipation on mail day, thinking he would get a package from Alice. He needed clothes badly: "All my drawers going all to pieces. Will be reduced to a breech about soon." But once again, Berry received nothing. Yet even as he expressed anger over not receiving mail from Alice, there was a part of him that did not want to believe that Alice was neglecting him.

> Big mail day but as usual I was not lucky. One officer got 20 letters and another got 14. Some letters date at March 1944. Why don't I get mail. Ed Aldridge has gotten 21 letters from Bea and I four or is it 3 from Alice. Can't believe it is all her fault however but fear it is partly hers. Can't imagine why my many kids and brothers and sisters don't write more. Also thought I had a few friends who might write.[43]

In early September 1944, the POWs heard that there were 3,300 letters coming for Americans. Berry wrote, "Hope this time that I not only get mail but get my share. I am tired of being treated like an orphan. I want mail and lots of it and mail with good news in it." Sadly, he received only two letters that day, both from his sister Ada. Although he was once again hurt and angry, Berry tried manfully to soften his disappointment by blaming anyone other than his wife and family.

> Why oh why can't I get some mail from my own family. Is it because they don't write. I do not believe that because I feel sure that with a wife, daughter, and 2 sons I must have several letters a month written to me. I would be willing to gamble that at least 100 letters have reached Nip hands for me and why I don't get them is beyond me. This is certainly getting almost unbearable and for God's sake and US and GB and USSR get a hustle on for we want to go home. Afraid if war is not over soon I'll find myself the junior soldier in the Berry family and connections. Ada's letters gave no news except that all of you well. Pray.[44]

For obvious reasons, receiving mail from family was extremely important to the morale of all POWs, so it was not a trivial matter that Berry hardly ever received any mail. It is all the more painful, then, to read in Berry's diary the often-repeated cycle of looking forward to receiving letters from his wife only to be bitterly disappointed time and again. For his part, Berry wrote letters when he was allowed to write, but the Japanese often did not send the letters on. Why Alice so neglected writing many more letters is simply beyond understanding. Letters from home during his fifteen long months at Shirakawa would have helped Berry's attitude while he labored at jobs forced upon him at the POW camp.

CHAPTER 13

Taiwan POW Camp Shirakawa
Part 2

COLONEL BERRY'S TIME AT THIS CAMP:
JUNE 8, 1943, TO OCTOBER 9, 1944
Water Problems and Hard Labor

Lack of fresh water was a major issue at Shirakawa. Carrying water, boiling water, and distributing water were mandatory duties from the POWs' first day at the camp. Water was so scarce that the POWs had to go to extremes to get some. Berry wrote early on that he caught his drinking water from the falling rain. The POWs even tried to bathe in the rain as they had at previous camps. Collecting and carrying water into the camp in the hot Shirakawa sun for the daily use of over three hundred POWs and thirty guards were unbelievably strenuous activities for weak and starving POWs to undertake. Worse still, many, like Berry, had to make several trips every day, because some of their colleagues were too weak and sick from malaria to do the job.

Unexpectedly, in mid-October, the generals in the camp refused to carry in water anymore. As a result, all the other officers had to carry in water for the generals, the Japanese, and for themselves. It is curious that General Brougher and Colonels Braly and Mallonee failed to mention that the generals refused to carry water. It must have been a touchy subject because the generals were the senior officers, and the other officers could not openly speak out against the generals. Berry only mentioned it in passing: "Water detail made 12 trips today. Don't know why they are using so much water than formerly. Now trying to put

The Generals' Squads (Nos. 1 and 2), as a result of a letter protesting carrying water were excused and then we had to carry their water as well as the Nips.

Generals Refuse to Carry Water. "That Water Carrying 'Bugaboo' for Your Health's Sake. The Generals' Squads (No. 1 and 2), as a result of a letter protesting carrying water were excused and then we had to carry their water as well as the Nips." Source: *The Life of a P.O.W. Under the Japanese in Caricature as Sketched by Col. Malcolm Vaughn Fortier*, by Malcolm Fortier.

No 2 squad with PM detail so we have 4 squads to carry water as general and Brigadiers do not carry water." It had to be humiliating to have to carry their generals' water as if they were the generals' servants. Colonel Fortier dared to capture a little about the generals' refusal in one of his drawings.[1]

A few days after the generals' refusal to carry water, an exasperated Berry expressed the humiliation he felt serving on the water detail. "Day of hard work. Water detail of which I was one carried 210 large buckets and Nips took 187 of them. We filled their kitchen vat then officers [Japanese] bathtubs and then to climax it filled the Ems [Japanese enlisted men] bathtub. Was humiliating and galling to the Nth degree. My partner and I carried 14 buckets." To make matters even worse that day, the guards refused to give members of the water detail the

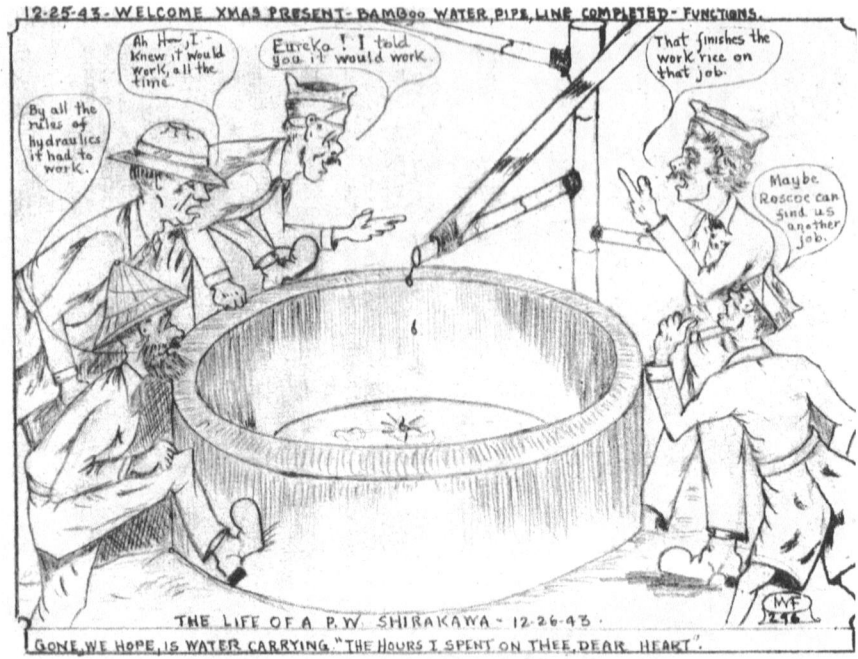

Nips finally allowed us to build a Bamboo pipe line from the spring to our kitchen and water carrying was largely eliminated.

Pipeline. "Welcome Xmas Present—Bamboo Water, Pipe, Line Completed—Functions. Nips finally allowed us to build a Bamboo pipe line from the spring to our kitchen and water carrying was largely eliminated." POWs, left to right, Colonels Sledge ("By all the rules of hydraulics it had to work"), Carter ("Ah, how I knew it work all the time"), Bonham ("Eureka! I told you it would work"), Berry ("That finishes the work rice for this job"), and Aldridge ("Maybe Roscoe can find us another job"). Source: *The Life of a P.O.W. Under the Japanese in Caricature as Sketched by Col. Malcolm Vaughn Fortier*, by Malcolm Fortier.

extra workers' rice promised them. The guards refused the workers' rice again a week later. Berry gave no reason for the guards' behavior. It may have been that the guards were angry with the generals and took the anger out on the colonels.[2]

Finally in December, after six months of carrying water by hand, Col. Roscoe Bonham and Col. Nick Carter obtained permission to build a bamboo pipeline to bring water to the camp. Work on the pipeline began on December 13, and Colonels Berry, [Theodore] Sledge, and Aldridge were invited to join Bonham's "special duty" team to build the pipeline. First, they had to clear a right of way and then dig a ditch for the pipeline. Next, the team had to make elbow joints and an intake tank using very crude tools. Berry wrote that he got blisters from working with "uncivilized" tools.³ The team worked full days on the pipeline from December 14 to December 24. On Christmas day it was finished. It was a wonderful Christmas present for the whole camp.

COERCED AND FORCED WORK

Shirakawa was not unlike Karenko in assigning hard work to the POWs. One of the first arduous jobs the guards mandated was related to the high rate of malaria in the camp. Able-bodied POWs were ordered to hike a mile out of the camp and pull "stink weed" out of the ground by hand. At age fifty, Berry was one of the able-bodied, relatively healthy men. "About 20 of us went out and gathered skunk weed to run off mosquito by burning it. Smoke." This particular weed was the "same plant from which the Chinese made the punk we all used as children to light fire crackers," noted Colonel Braly.⁴

Malarial mosquitoes were so bad that we tried everything, the burning of this stink weed was almost as bad on us as on the mosquitoes.

Stinkweed. "The Stink Weed for Mosquitoes—I mile to Go—Strong Back an Asset. Malarial Mosquitoes were so bad that we tried everything, the burning of this stink weed was almost as bad on us as on the mosquitoes." Source: *The Life of a P.O.W. Under the Japanese in Caricature as Sketched by Col. Malcolm Vaughn Fortier*, by Malcolm Fortier.

The cutting and fetching of stinkweed (or skunk weed) were backbreaking activities in the unbearable heat of June and July. The guards monitored how large a bundle a POW had on his back before he could return to camp with it. Fortier pointed out that POWs who tried to shoulder small bundles of stink weed were stopped by the guards, who then forced those POWs, like Colonels Foster and Bridge, to carry the largest bundles for the mile walk back to camp.

Camp Commander Imamura had enjoyed the profits of the camp garden or farm at Karenko, so he once again mandated that the POWs establish and maintain a farm at Shirakawa. As before, he told the prisoners that the farm food was for them, but the POWs knew otherwise this time. They were not at all happy to work on the farm. But if they did not, their regular rations of food were cut. The work making a farm was hard, especially for POWs who were over sixty years of age. Berry wrote, "Hot day and we worked 5 hours in the broiling sun in the so-called 'PW Farm' carrying grass, breaking clods, etc. We were not pushed but was hard on old men and hot on anyone."[5]

Caught in a no-win situation, Berry worked in the camp garden reluctantly to avoid having rations cut. But at Shirakawa they had to suffer yet another indignity, as Berry noted:

> Now that our garden is coming in they tell us that in order to have more food the Nips will take our 30 yen and buy our vegetables & pay us a lower price than in the market and in that way we will get more vegetables. In other words they will cut down on their issue and we will take our own money and buy our own vegetables before we can eat them. They will probably take the choice ones and make us buy our own poorest vegetables and will likely not get any more food.[6]

Berry was correct in assuming that the Japanese would take the best vegetables and sell back the poorest ones to the POWs. And adding insult to injury, the food they grew was often given to the animals. Berry noted, "Got in 8 baskets of spinach from our garden today but it went to hogs." On another occasion, Berry and Browne walked by the hog pens and found the guards feeding the hogs "fine string beans." Berry was incensed, "Made me boil all over to think these bastards gave the hogs food which we vitally need. We would be much better off if we had no animals if we could have the food the animals get. When we kill a hog we get only about 1½ ounces meat and fat out of it." Nevertheless, Berry worked hard in the camp garden and became "famous" for his sodding skills.[7]

When Lieutenant Hioke, the "Pasadena Kid," took over commanding the camp in the new year, he, like Imamura, endorsed the farm, so the POWs had to

Berry's No. 13's came in handy on the sodding job he was placed in charge of.

Berry's Size 13. "No Special Tools Required—For Once Those Big Feet Are a Blessing. Berry's No. 13's come in handy on the sodding job he was placed in charge of." Source: *The Life of a P.O.W. Under the Japanese in Caricature as Sketched by Col. Malcolm Vaughn Fortier*, by Malcolm Fortier.

keep working on it. Receiving workers' rice drove big, strong men like Berry to do the hard, manual labor of farm work. However, as before, the guards played games with the POWs, denying them the promised extra portion. "Worked hell out of squad 1–6 and EM [enlisted men] and then beat us out of our workers rice—said they had already given it to us." Not to get the extra food after expending so many calories from the hard work was tough on those officers and men, and they lost weight. Even Berry, who often had extra rice from Browne and Cordero, lost weight.[8]

Some officers, who came to be called "Two-Bowlers," received workers' rice without doing hard labor. They worked in areas like the post exchange (PX), the hospital, or in the kitchen, or they had jobs like camp interpreter, personnel officer, commodity officer, maker of clogs, keeper of chickens, etc. Animosity toward these men who received two bowls of rice for less physically demanding work is evident in Berry's diary and in Fortier's cartoons. Berry complained about some of these officers: "Damn shame as their extra rice over ½ small bowl come from our bellies—form of robbery and unfair and dishonest. All laid at foot of personnel director who takes 3 times what he should. Human nature I guess when you are hungry and have the chance. PX crowd gives part of theirs to the generals however which looks very much like that well-known game which we have much of in the army—namely 'Boot Licking.'"[9] Unfortunately, because the personnel director was part of this scheme, nothing would come of complaints about those who took too much rice.

Then there were some POWs who were neither "workers" or "Two-Bowlers." They were, in Berry's word, malingerers—POWs who did not report for work, or POWs who worked less than a full day. About the malingerers, Berry wrote, "[I] feel [malingerers] were taking something of our to which they have no right.... Many of us are quite sore about it. I for one think that they should be 'silenced' by the other prisoners. If they worked only ½ day it would permit someone else to have the other half day and mess then draw 300 grams instead of 150 and each of the two get one work rice issues. Don't know what will become of it." When the guards did find the malingerers in their rooms, they turned them out to work. For Berry, malingerers were in a way breaking an unspoken honor code among POWs: Everyone was in this mess and should do all they could to help everyone survive. Nonetheless, Berry continued working hard labor jobs, doing what he could for others, and always hoping for his second bowl of rice.[10]

As at Karenko, "policing up" the camp for visiting dignitaries was regularly required of the POWs. The work was grueling, especially in the unbearable heat of summer. Berry wrote, "Turned Squads 1, 2 and 3 out this morning and worked hell out of us for 3 hours skinning grass and brush off of a hill just outside the compound. Worked generals and all except about 12 who were considered too weak and sick." Skinning grass was done with scissors, razor blades, and large hoes.[11]

POWs were always subject to the commander's whims. He wanted to have a little park built, supposedly for the POWs. Consequently, he forced the POWs to construct and maintain, or police up, the park. Berry wrote angrily about the park construction: "Took first 4 officers squads today and worked them 4½ hours hard—AM and PM. Going to show us who are the masters I guess. The work we are doing seems to be in making a park on a little hill in the Nip part of the camp—skimming off grass and brush. Absolutely of no use what ever but look like just something to make us work."[12]

As it turned out, the park, named Yasume Park (Yasume means "rest" in Japanese), did prove to be of some use to the POWs. Berry and Browne walked there at times, as did others, and the POWs held church services there in the spring. As the construction of the park neared completion, Berry groused, "Bout got a bamboo fence around our park (?) So I guess we will shortly be ordered up in it so news hawks can get our picture playing in the beautiful park we have built." Berry was correct. The Japanese did have "news hawks" come into the camp and take pictures. A team of photographers arrived in March 1944, "Nip movie picture men," as Berry called them, to film the POWs at work. Their purpose was to make propaganda films to show the American public how well the POWs were being treated by the Japanese. The generals' squad posed willingly for pictures in Yasume Park. Berry's friends Ed Aldridge and Ed Corkill volunteered to have pictures taken of them while doing laundry. Berry defiantly wrote, "Damned if I would!" The next day he wrote again, "Movie men still around today. Made 2 enlisted Basketball teams play and forced PWs to look on. I skipped it but was not caught." Here, Berry showed as much disdain as he dared.[13]

Occasionally the forced labor to which POWs were subjected would leap from humiliating to patently absurd. For example, in the spring of 1944, POWs were ordered to participate in a flycatching campaign. To suppress the fly population, the camp commander ordered all POWs to catch, kill, and turn in as evidence twenty to fifty dead flies each day. Berry commented on this new project, showing his defiance again, "Got orders today that each of us had to kill 20 flies a day. This means 10000 flies a day and they are not here. In fact, we seldom see any. I personally will kill only the ones in my room which I can kill w/o too much effort." One of the POWs, Colonel Sledge, cannily stood by the animals in camp and caught flies off their backs to meet his quota.[14]

There was one forced work assignment that all POWs wanted: planting and gathering peanuts. Braly wrote about his squad's being assigned this duty one morning and how they outfoxed the guard.

> For obvious reasons the choice job of the lot was planting peanuts. And so it was with pleasurable anticipation one April [1944] morning that our Squad 3 received an order to pick up half a sack of shelled peanuts and follow the guard. K. L. Berry shouldered the sack and off we went. We were hardly well started on the planting job when the sentry yelled for me as Hancho and complained (using sign language) that the officers were eating more than they were planting—which was all too true. This was apparently not cricket and we were to desist pronto. It so happened that we had planted, hoed, and harvested a previous crop of peanuts of which we had received exactly nil. Here was a chance for a comeback. Everyone understood the complaint. The only difference thereafter was that we ate more while going down the rows away from the guard.[15]

At the end of their shift that day, Colonels Cordero, Hilton, and Fortier were bopped on the head by the sentry for having peanuts in their pockets. Their punishment seemed minor to them, and they had a laugh about it—they had more peanuts hidden in their clothes. This caper took a dark turn a few days later when Commander Hioke stopped Colonels Worthington and Braddock for their bulging peanut pockets and proceeded to beat Braddock, knocking him down, kicking him, and breaking three of his teeth. The POWs had thought that Hioke was gentler than their other commanders, but this incident proved them wrong.[16]

BERRY'S PROTEST, AGAIN

Despite the starvation diet, forced labor, and lack of mail, Berry remained an optimistic, determined-to-survive prisoner with a sense of humor. Those characteristics, combined with his overall good health, contributed to his survival. An amusing topic that seemed to relieve Berry's frustration and anger about his circumstances was once again his Buffalo Bill hair.

As at Karenko, growing his hair longer began as a statement of defiance toward the Japanese, but in reality, Berry enjoyed his curly locks. After one last haircut at Shirakawa in August 1943, which he said made him look like a "billiard ball," he began his silent protest again. For several months, as it grew, he did not mention his hair in the diary. However, by December, his protesting locks had grown enough to be worthy of mention. "Took a good soap one with a scrub and also shampooed my long curly locks. Beginning to look something like Bill Cody again." On March 10 he recorded an update, "Hair long and curly as it finishes its 7th month—look like Kit Carson, Buffalo Bill, Jim Bridgers all in one. Got a pretty good handle bar mustache also. It is 22 months old. Have trimmed it on ends some so not as long as it should be."[17]

Berry continued to let his hair grow, and in May he wrote, "My hair is nine months old today and quite long and curly. Howard Frissell, overheard Sgt. [Frank] Pechek telling a British officer that I was a typical Kentucky Kurnel and that I was a damn fine officer." Although Pechek got Berry's birth state wrong, his assessment of Berry as an officer was right. Berry enjoyed the compliment. Later that summer, with perhaps a touch of vanity, Berry "had young British officer [Captain Bevin] do my caricature today. Haven't seen it yet but saw his work and it is good." And the next day he announced, "Got my picture this AM and it is pretty good."[18]

Berry the Gardener. Captain Bevan's caricature of Berry with a hoe in his hands made it all the way home in Berry's keepsakes. Source: Author's private collection.

In the end, Berry kept his long locks for twelve months. He wrote on August 10, 1944, "Cut my hair on 10th after 366 days and cut off mustache after 2 years 3 mos and 5 days. Certainly look different." Berry's haircut and shave were such an event that two of his fellow POWs took note of it in their diaries. Ed Lilly wrote, "K. L. Berry who has been allowing his hair to grow for some time and has developed some of the tightest little ringlets (That might well be the envy of any debutante) had it all cut off today. Quite a change in his appearance." Less descriptively, Galbraith commented, "K. L. gets haircut after one year, also shaves off his Texas mustache." The Kentucky Kurnel was gone.[19]

A DIFFERENT KIND OF PROTEST

From time to time, the Japanese commander ordered the officers to write essays on different topics. Several officers mentioned these essays in their diaries. Berry made copies of what he wrote. At first, Berry was compliant, but as more essays were demanded, Berry began to protest in his answers. The first time they were ordered to write, the topic—"what they would do after the war"—caught Berry's imagination, and he wrote a thoughtful piece concerning his health, his retirement, and his dream, owning a ranch.

> When I return to my own country I must undergo a period of hospitalization to regain my health and be operated on to repair two hernias which have developed since the war began and which are now beginning to cause me pain and inconvenience. Then I will be retired from active service due to my defective hearing and age in my grade. Upon my retirement, I will purchase a small ranch near my home town of San Antonio, Texas and raise cattle in a modest way. On my ranch I will enjoy the last years of my life doing things which I have long wanted to do. Among these things I intend to do other than the pleasure of managing my ranch will be a course of reading and study in an endeavor to understand more fully the trials and aspirations of the people of other lands and will work towards a better understanding of the people so that we can all work, live, and enjoy life and peace—free from want and the scourge of war.[20]

Then, the Japanese captors pressed the POWs for information about their thoughts on the war. On November 4, 1943, they were ordered to answer these two questions:

1. Why is the United States fighting Japan?
2. What does the US hope to get out of the war?

Berry answered assertively:

1. The United States of America is fighting in self-defense and to avenge the premeditated, unprovoked and uncalled-for attack of the Japanese Navy on Pearl Harbor while peaceful negotiations were under way between the two countries.
2. The overwhelming defeat of Japan so that the following will take place:
    a. Return of lands conquered by Japan to rightful owners.
    b. Assurance of equal opportunity for all nations in Oriental trade.
    c. Assurance that never again will it be possible for such a war to take place.[21]

Days later, the POWs were ordered to answer two questions about America's treatment of Japanese prisoners. Berry showed his trust in the United States and dared to show his disdain for the Japanese by answering:

> I have no belief whatever in the truthfulness of the two reports which I have been asked to express an opinion. I consider them as just so much propaganda.
> America treats her prisoners strictly as laid down in the Hague and Geneva Convention. She feeds, clothes and houses them well and otherwise treats them with kindness and consideration. They are not put on semistarvation rations nor are they permitted to be knocked and beaten up by the guard personnel. Neither are they harassed with any humiliating and annoying restrictions and requirements.[22]

The Japanese persisted in asking questions. On January 26, 1944, Colonel Sazawa, the senior POW camp commander of Taiwan, asked the POWs to write a short description of "the reality of the most bloody [sic] engagement experienced in the Anti-Japanese battle and the view of the war." With his aversion to the Japanese growing daily, Berry boldly protested in his answer to Sazawa's query:

> I respectfully request that I be not required to write up an account of any battle in which I might have participated in during the present war as I feel that my account might give aid or comfort to the enemy and, if so, would be a violation the Rules of Land Warfare.[23]

CHAPTER 13

MORE PROTESTS, NEW RULES, ALLIED VICTORIES, AND INSOLENT BEHAVIOR

Meanwhile, in the spring, the Allies enjoyed more success in the Pacific, with victories in the Marshall Islands, Truk in the Caroline Islands, New Guinea, and the Mariana Islands. They were also gaining ground in Burma. On the European front, the Allies were still stuck in Italy fighting the Germans, but the British dropped three thousand tons of bombs on Hamburg, Germany, in March. By April, Berry heard a rumor that US forces had landed in the Philippines. The successes in the Marianas, so close to the Philippines, must have given this rumor particular credibility among the POWs—even though it was not true.

As a result of the war news, the camp commander and the guards became more aggressive in their punishing behavior toward the POWs. One morning in May, when a guard kicked British Brig. Gen. Torquil McLeod and a POW from another squad while they were laboring in the camp garden, the whole issue of forced labor came to a head. In response to this outrage, Berry heard that General Brougher had written a letter protesting the treatment and requesting that men over age fifty not work. Berry recorded Hioke's response in his diary: "Camp CO called meeting of squad chiefs this PM and went into matter. Wants list of those who over 50 yrs or other infirmities do not want to work under any circumstances. I am one of those who does not want to work."[24]

Hioke informed the POWs that those who went on record as not wanting to work would be segregated and given reduced rations. Those not working would still have to obey all other regulations. He added, "Remember Karenko." Berry was furious. "In other words the damned SOBs using threats to keep us from going on record as not wanting to work." Despite the threat, almost all of those over fifty were willing to go on record as not wanting to work. They waited to see what Hioke would do.[25]

Berry wrote about the situation in his diary: "About 15 POWs called up and asked if they volunteered to work and if they spoke for squads. One American, Bonham, did not vote and Johnson, another American, said he volunteered. All others said they did not. So tonight we had to sign paper as to whether we volunteered or not. This squad 100 percent against volunteering." Berry meant that all the officers in his Squad 3 over fifty years of age signed the document defying Hioke, declaring they would not volunteer to work. Hioke would have to order them to work. A few colonels in other squads would not sign the document defying Hioke. Berry hoped they would change their minds. The next day, Col. Dorsey Rutherford did change his mind, but Colonels Edwin Johnson, James Hughes, Albert Ives, and Joe Kohn did not.[26]

Berry expressed his disgust with his fellow POWs for not standing together with the other colonels over fifty years of age. "I told Johnson and Hughes that from now on there [sic] were just Colonel 'So and So' to me. Will tell other two if the case arises.... Our men and officers are asking if Parker is going to make an official protest if it. None of us believe he will." Berry, like the other senior officers, lacked faith in General Parker; they wanted him to be a tough leader, protesting and not acquiescing.[27]

Because most colonels over the age of fifty had made it clear that they would not volunteer to work, Hioke came down on all of them "like a ton of bricks" that night. At 1:00 a.m. the bugle sounded for roll call. For two hours and twenty minutes, while the POWs stood for roll call in the hall, the guards ransacked each barracks, taking away their shoes and slippers and their water-carrying poles. And to spite the officers, the next day the Japanese gave the enlisted men extra rations, including sweet potatoes from the camp garden, while the officers were given less food and no potatoes.[28]

By this time, Colonel Sazawa, commander of all Japanese POW camps in Taiwan, had heard about the protest, and he came to Shirakawa. He issued new orders for the entire camp, thinking he had a mutiny on his hands. If anyone disobeyed the new rules, he was put in the dirt-floored, windowless "guard house," with only rice and water for days at a time. His extremely harsh orders were:

- No one to lie down on beds, floor, or the ground between morning and evening roll calls.
- No playing of cards or musical instruments except Saturday after noons and Sundays.
- Yasume Park exercise ground was closed.
- No food from the farm allowed.
- No visiting in other barracks nor through the windows.
- Individual cooking arrangements for boiling coffee water and cooking potatoes had to be destroyed by the POWs.
- Work rice was discontinued.
- No more monthly shows were permitted.
- Chaplains were not permitted to make pastoral calls on other POWs.
- Post exchange stock was curtailed.
- British junior officer magazine, *Raggle Taggle*, was discontinued.
- Several inspections of barracks daily by guards.

- Additional nightly roll calls at the will of the guards started up—sometimes three roll calls during the night.
- POWs on benjo duty were not permitted to read.
- No POW was permitted to salute his senior POW officer.[29]

Alleviating the stress of the Sazawa rules, the POWs heard good war news from incoming British POWs who were being moved north out of POW Camp Heito on Taiwan to Shirakawa in late June 1944. These POWs had been captured in Singapore earlier in the war. They told the Shirakawa POWs that the Allied troops under General Eisenhower stormed the beaches in Normandy, France, opening a second front in Western Europe (D-Day). In fact, the Allies were gaining ground in many places—Rome was captured; the Russians began fighting the Germans in Finland; US Marines invaded Saipan in the Marianas; and US planes flew a bombing raid on Japan, the first since April 1942.[30] Because of the good news from the British, the POWs' attitudes began to shift. As harassment from their captors increased, the POWs matched it by displaying insolent behavior toward the guards. Berry noted these acts of defiance in his diary and recorded one of his own:

> I baited Nip guard at kitchen today. He wanted me to halt and salute but I saluted walking. He slapped me, gave me a lecture, made me about face and go back. He then called to me to return but I made out like I did not know he was talking to me. But just before he exploded I stopped and returned and again saluted walking. We then repeated the same scene but this time I stopped and bowed and everything lovely. If he had known for sure I was baiting him I would have caught hell but what the hell.[31]

Guards tried to spot officers in the act of violating the new regulations by running through the barracks to catch people on their beds during the day. "The bugler ran through barracks and caught several fellows on their bunks. Took them over to guard house and made them slave out in rain naked for 55 minutes. How is that for civilized treatment and right now we are having a flu epidemic," Berry reported in disgust. On another occasion, guards ran through the barracks with rubber shoes on trying to catch officers in bed before 9:30 p.m. Berry showed his defiance, "I fixed up dummy in my bed to fool them but [they] did not come any more. There were numerous other incidents where officers were put in the guard house for several days for minor infractions. They had no windows, slept on dirt floors, had no latrine access, and were given very little food and water."[32]

For some reason, the commander became concerned about the POWs' diaries, which he had previously ignored even though keeping a diary was technically forbidden. The Japanese may have feared that the Americans would invade any day and use the information contained in POW diaries to identify and punish those who had mistreated their captives. In any case, Berry's diary was confiscated by the guards on August 4, 1944. Confiscation of diaries on this date was not mentioned by Braly, General Beebe, or General Brougher, so it may have just been Berry's for some reason. He started a new diary on August 6, thinking his old one was gone, but guards in a future camp returned most of his first diary to him.[33]

Because the officers asserted that they would not volunteer to work, their rations were cut several more times in August and September, leading Berry to say in September, "Today was thinking of what our refusal to volunteer to work has cost us and still the Nips say we are not being punished for not volunteering." As he did often in his diary, Berry couched his complaint in black humor: "Am so hungry could eat south end of a skunk going north."[34]

MOVING

As the war progressed toward Taiwan, rumors about moving the POWs were rife. Although the POWs had not seen a newspaper since April, they remained optimistic that the war events were in their favor based on the punishing behavior of their captors. Berry wrote, "We get a continual laugh out of their efforts to keep us in the dark because the more this happens the more we are satisfied we are knocking hell out of em on all fronts." And General Brougher echoed those thoughts, writing, "Nipponese soldiers forbidden to speak to us—no newspapers—things must be mighty bad for the enemy."[35]

The "knocking hell out of em" moved close enough to Taiwan that the rumors about moving finally became fact. On September 30, orders came in for the guards to get the American, Dutch, and Australian generals ready to move. Twenty-nine of them, along with three orderlies, were rushed out the next day before they could even finish lunch. The Japanese became worried about the diaries and notebooks again, and they started taking away all written materials from the hand baggage before these officers left. General Brougher thought he had lost all his writings, but some of it, not all, was returned at the next POW camp.[36]

A week later, Berry wrote, "Sat. fine day but bad news as we received orders to pack up for a move to a cold climate by boat. Can carry all baggage. Asked if we had a winter shirt and blanket." Berry expressed concern that being moved to Japan, Korea, or Manchuria would put the POWs in mortal danger of attack from their own side. "Japan seems to me only place we could go to tho some think Korea and others Manchuria. We do not like idea as we have to run gauntlet of our subs and planes. Hope we get thru."[37]

As Berry and the rest of the POWs moved out of Shirakawa on October 8, 1944, after almost a year and a half in captivity there, the future was uncertain. Berry's fears about running the gauntlet were justified. Their journey north to Japan on the *Ōryoku Maru* was going to be dangerous. The Allies were flying bombing raids over Taiwan, Okinawa, and Japan, and American submarines were cruising underneath the surface searching for opportunities to coordinate attacks on Japanese ships with the American planes above them. Under these circumstances, traveling in unmarked Japanese vessels anywhere in the East China Sea meant that the POWs could die at the hands of their own countrymen.

CHAPTER 14

En Route to Manchuria

"HELL SHIP" ŌRYOKU MARU

Colonel Berry's Time on This Ship: October 10, 1944, to October 28, 1944

One of Berry's last Shirakawa diary entries recorded the fear that POWs had about American planes and submarines bombing or torpedoing them unknowingly because they would be aboard a Japanese ship not marked as a POW ship. The *Ōryoku Maru* was in Keelung Harbor at the north end of Taiwan when the Shirakawa POWs arrived there on October 9, 1944 (map 11). The story of the harrowing journey from POW Camp Shirakawa to POW Camp Cheng Chia Tun in Manchuria was chronicled by Berry's friends Ed Corkill, Ed Lilly, Malcolm Fortier, and Bill Braly. Berry did not write in his diary during this time.

The POWs were told to get their baggage out for inspection on Sunday, October 8. After the inspection, the baggage was loaded onto trucks. The POWs could hold out from their baggage only the things they thought they could carry on the three-kilometer hike to the narrow-gage railroad platform. At 3:00 a.m. on October 9, the bugler woke them up and they packed up what they were carrying. The guards gave them two rice balls for breakfast and lunch, and the POWs set out around 4:30 a.m. to the railroad platform where they boarded the same rickety train that had brought them to Shirakawa many months before. After the two-hour ride, 250 of them boarded a regular train at 7:30 a.m. that took them to Taihoku at the north end of Taiwan. They arrived at 6:00 p.m. and were immediately loaded onto the *Ōryoku Maru*.[1]

CHAPTER 14

We spend 19 days of Hell aboard this ship, bombed twice by our own bombers, we lay 14 days in the harbor of Quelung before getting started.

Aboard Ship. "We Get Settled (Sardine-Packed) in our Commodious Quarters Aboard Oryoku Maru. We spent 19 days of Hell aboard this ship, bombed twice by our own bombers, we lay 14 days in the harbor of Quelung before getting started. Source: *The Life of a P.O.W. Under the Japanese in Caricature as Sketched by Col. Malcolm Vaughn Fortier*, by Malcolm Fortier.

The POWs were put in the hold of the ship where a double floor had been installed to accommodate more men. They had experienced this setup before on their passage to Taiwan. The port holes were closed, and there were two or three light bulbs for the entire space; as a result, many men sat in semidarkness during the many days that followed. There were no mats to sleep on, and they were given one blanket. It was excruciatingly hot, and there was no ventilation. Even worse, there was no water.[2] On the decks above the POWs were one thousand Japanese civilians who were leaving Taiwan for Japan.

On October 10, Colonel Braly was designated the prisoner OD (officer on duty), which meant all problems big and small were relayed to Braly to fix. One of the first things he had to handle was food distribution. Buckets of rice for twenty prisoners per bucket had to be distributed fairly. Teacups were attached for small amounts of tea, but there was no water to be distributed. Braly tried to organize the officers into groups, but the guards did not wait for him to finish. So new groups naturally formed. Braly remarked, "It now appears that the eating groups of twenty have become our squad organization for the trip."[3] During the hike to the train and then aboard the ship, Berry had settled in with his two best friends, Brownie and Corkill, and was no longer in a tight configuration with his former roommates Braly, Bowler, and Galbraith.

Braly was able to get the ventilation system on, and it worked well for those who were near the blowers. But still there was no water. Dehydration began to set in. No one was allowed out of their own compartment except to go to the benjo in the passageway—four small cubicles designed for small Japanese women. On October 11, the POWs were begging for water, and late in the day, Braly finally managed to get the guards to allow a canteen of water for each POW. Everyone was so thirsty they were "spitting cotton," he said. That evening, the ship moved to the outer harbor, and all hoped they were on their way to wherever they were going. Unknown to the POWs, the Americans had started air raids against Okinawa that day and would turn their attention to Keelung Harbor on Taiwan the next day.[4]

With news of American bombers headed their way, the crew of the *Ōryoku Maru* returned to the wharf at Keelung on October 12. There they evacuated the Japanese civilians from the ship. Meanwhile, the POWs were left in the hold without life preservers or instructions on how to get off the ship. As American planes approached to bomb the harbor, the guards used their machine guns to blast the sky around them. Braly wrote, "It would seem that if hit we are to be drowned like rats in a hole."[5]

The American planes conducted six more raids during the day. Lights were turned off that night, and everyone was left in total darkness to wonder what was coming. The next day, October 13, was much the same; the Japanese machine guns took aim at American planes flying overhead. The POWs crammed into the holds were fortunate that the planes were bombing the other side of the harbor most of the time. That evening, when the raids had ceased for the day, the POWs were allowed to go up on deck for some much-needed fresh air at roll call time.[6]

After three days in the hold of the ship, everyone was filthy, and the hold was rancid. The guards would not allow anyone to use the benjo during the air raids, adding to their misery. Their living conditions were appalling—no baths, barely any food and water, incredible heat, and very close quarters. The next day was quiet, but there was no movement to set sail, so Berry and his comrades continued to suffer the misery of life crammed in the hold. The next day was the same.

The POWs learned that the evacuated Japanese passengers were due to return to the ship in a few days. Once the passengers were back on board, the POWs would no longer be allowed on deck to get fresh air. So Braly arranged with the guards that two POW squads at a time be allowed to come up top to bathe by use of the fire hose before the deck was closed to them. Corkill wrote, "We have not even been given a chance to wash our hands or faces until this a.m. when we were marched onto the public dock and hosed off naked there before the public gaze but there are few people on the streets since the raid." The experience was humiliating, but at the same time the fire hose bath felt good to all. Lilly took his soap with him and wrote that he had a refreshing bath. They were allowed to bathe in this fashion a few more times before the Japanese passengers returned.[7]

A newspaper reached the POWs on October 16, and from it they learned that while they had been sitting in port, Adm. William Halsey's Task Force had conducted air raids and dropped bombs all over Taiwan. That news was good for the POWs to hear, but they also feared for their lives as they continued to be imprisoned day after day on board the ship, sitting targets should the ship be attacked.[8] Elsewhere, American planes attacked Leyte in the Philippines and the Japanese base at Truk, an atoll in the Pacific.

On October 22, the ship moved to the outer harbor once again and seemed to be sailing full steam ahead to the POWs' new destination. The port holes were shut, and the bulkhead doors bolted, so there would be no way out if the ship were torpedoed by American submarines. After a night of sailing, Ed Lilly was in the benjo and had a good look at the shoreline. Enough was visible for him to think the ship was back in Keelung Harbor. He wrote, "A few, including K. L. Berry, do not believe we are, but the large majority think as I do." The majority was correct and Berry sadly wrong. The POWs were right back where they had started the day before. No one knew what had made the ship turn around, but the Japanese must have been given a warning about submarines or air raids.[9]

Everything was clear by October 25, so the ship set sail again. A day later, Corkill wrote, "We think we are south of the southern island [of Japan] now but don't know where our destination is. We still have two destroyers convoying us. A sea plane was with us yesterday a.m."[10] No life preservers were issued to the POWs, but the Japanese guards wore them as they sailed. Several times the engines were turned off as the ship entered mine fields, and a pilot from the sea planes came on board to guide them through the mines.[11] After five days at sea, they docked at Moji, Kyushu, the southernmost island of Japan. They marched to a railroad station and traveled by train to Beppu, a resort town on Kyushu, and the hotel Nitchi Man was their new camp for a short period of time.

Berry and his friends from Shirakawa were amazingly lucky to have survived their trip on the *Ōryoku Maru* from the northern tip of Taiwan across the East China Sea to the Japanese Island of Kyushu. The story of the next trip the *Ōryoku Maru* made with POWs on board is worth repeating to illustrate just how fortunate the Shirakawa POWs were.

After Berry and fellow POWs debarked at Moji, the *Ōryoku Maru* turned around at Kyushu and headed for the Philippines to collect more Japanese passengers who were fleeing the Philippines as well as more POWs who had been in camps on Luzon since 1942. Their story, as told by the Naval History and Heritage Command, is heartbreaking.

> Most of the POWs who boarded *Ōryoku Maru* on 13 December 1944 had been in custody since the Bataan campaign of 1942 and had survived some of the war's most notorious death marches. They were now exceedingly weak and malnourished.
>
> By 8 a.m., [USS *Hornet*] dive-bombers were on the offensive.... According to an eyewitness account on board, it was at this point that *Ōryoku Maru's* escort ships fled the scene and left the liner to its fate. Prisoners in the after hold descended into panic. Several tried to climb ladders to the deck but to no avail. The Japanese guards fired indiscriminately down into the darkness, driving the men away from the hatch.
>
> *Hornet's* bombers returned every 30 minutes or so and then zeroed in for a massive attack in the late afternoon. Japanese gunners continued to defend *Ōryoku Maru*, not least of all to save their own lives, but also because the ship was loaded with as many as 2,000 Japanese civilians in hasty flight from the American advance through the Philippines.... The damage, which included the splintering of the decks, had also introduced air and light into a few areas of the holds. These came as a relief to many POWs once the attacks were over for the night. For others, however, the second night of the voyage was even worse than the first.
>
> At dawn on the 15th, as a handful of POWs housed amidships were in the process of escaping the stricken liner, *Hornet's* planes returned. A series of direct hits caused the bow to flood, and the ship took on a serious list.
>
> Another wave of assault planes arrived overhead. They strafed the deck and dropped 500-pound bombs and multiple rockets that nearly tore the ship apart. Girders buckled; fires engulfed the upper decks. As POWs, bloodied and terrorized, clamored up the hatches from the holds, Japanese sentries shot them dead. Yet the men kept coming, forced out of the holds and into the fires by rising water. Eventually, the chaos, the strafing, and the fires became too much for the Japanese jailors, who abandoned ship.

Ōryoku Maru. *Ōryoku Maru* was a Japanese passenger and cargo ship commissioned by the Imperial Japanese Navy at the beginning of World War II. Early on, it was used to transport Japanese soldiers from Japan to battle sites in the Pacific, and later in the war to transport Japanese citizens, as well as Dutch, British, and American POWs, from Taiwan and the Philippines to Japan. In this photo, American POWs are in the hold of the ship as it is under attack from American planes at Olongapo, Luzon on December 14–15, 1944. Source: US Navy photo, NH 95603 (cropped), US Navy Naval History and Heritage Command.

Left to their own devices, more than a thousand POWs ended up in the water and swam toward land. The Japanese captured them in the water and on the shore—it is unclear whether anyone actually escaped. The guards then herded their pitiful charges to a tennis court on shore for the next stage of their deadly ordeal.[12]

Among the 1,600 POWs who had boarded the *Ōryoku Maru* on December 13 and the one thousand who survived the American attack, swam to shore, and were "herded" to the tennis courts that day was Lt. Col. Ovid O. "Zero" Wilson, one of Colonel Berry's friends from San Antonio who had greeted Berry in Manila in November 1941 before the war started.[13]

Once the Japanese made a new plan for transport, Wilson was put on one of the two hell ships that had arrived to take them to Japan. The two ships were the *Enoura Maru* and *Brazil Maru*. It is unclear which ship Zero Wilson was put on. Another tragedy followed the first when these two ships were docked in Taiwan for a few days. On January 9, another raid by American planes in Taiwan bombed the *Enoura Maru*, resulting in the deaths of approximately four hundred additional POWs. Those who survived the bombing on the *Enoura Maru* were transferred to the *Brazil Maru* with the other POWs. The *Brazil Maru* escaped Taiwan and arrived safely in Moji, Japan, on January 28, 1945. The POWs were imprisoned in Fukota no. 3 POW camp in Japan for three months. It was amazing that Wilson lived to tell the tale. At the end of the war, out of the original 1,600 POWs who boarded the *Ōryoku Maru* in December 1944 in Manila, US authorities could locate only 128 POWs who survived the *Ōryoku Maru*, *Enoura Maru*, *Brazil Maru*, and subsequent Japanese prison camps.[14]

On April 29, 1945, Wilson and a few other soldiers Colonel Berry knew from Bataan were transferred to Mukden POW camp in Manchuria.[15] In Mukden, as detailed in the following chapter, Wilson was reunited with Berry and told his friend about the tragic trip on the *Ōryoku Maru*. The bombing of the *Ōryoku Maru* by American planes is exactly what Berry had feared when he left Taiwan.

POW HOTEL BEPPU

Colonel Berry's Time at This Camp: October 29, 1944, to November 10, 1944

The town of Beppu on Kyushu was a popular vacation area in Japan because of its hot springs and mild climate. The town's five resort hotels were used as the prisoner "camps" for the Shirakawa POWs for their brief time in Beppu. For Berry and others, this was a radical change from the dark hold of the *Ōryoku Maru*. They learned that earlier in the month, the higher ranking American, British, and Dutch officers and officials had stayed in Beppu for a short stop before being taken to Manchuria. General Wainwright had been in that group.[16]

At Beppu, colonels were housed together in the same hotel, the Nitchi Man, and slept on mats on the floor. Corkill wrote about Beppu in his diary, "We were put 4 in a room large enough for 14 at the rate we were loaded in the ship." The rooms were well lit and had good ventilation. The prisoners' mess hall was the hotel's dining room, and they sat on the floor at ten-inch-high tables. Their diet consisted of vegetable soup and rolls. Corkill grumbled a little about the food: "Our meals consist of a bowl of soup or gruel and a bun per meal with 2 tangerines added twice since being here and a salad once. There is not enough salt in the soup to appease our appetites and no sweets except tangerines."[17] Berry likely felt the same way about not having sweets because he craved sweets all the time. After their mealtime that first day, they went to bed around 6:30 p.m. and woke up at 6:30 a.m. Corkill mentioned that no guards were there when they woke up. He wrote, "It is quite a relief to have them out of sight and hearing and feeling."[18]

The latrines were all together in one large room on the ground floor of the hotel. The prisoners had plenty of bathing and washing facilities there, too. Before getting into a tub of hot spring water, they were required to first bathe in cold tap water. Many prisoners who were in poor condition found the hot springs "too enervating" to remain in the hot water for very long.[19]

POWs were not forced to work while at Beppu, but some volunteered to chop wood in exchange for more rations. Berry must have volunteered to cut wood because he was always ready to work for more food. Happily, there were no beatings or harassing treatment during their stay. After five days, there was still no talk of moving on. The place was certainly better than other camps, but Corkill complained, "The biggest trouble with this place is that we get no exercise. We play bridge morning and afternoon."[20]

On November 3, Corkill recorded, "Today the new guard from the camp we are in route to, arrived, so we will be on our way. I am anxious to get there and see what is in store for us for the rest of the war."[21] The rumor was that they were going to Manchukuo (Manchuria), the Japanese puppet state. On November 9, the POWs received more information about their impending move. Braly wrote, "They said we would move out the next day, in two contingents, and we had a three- or four-day trip ahead of us including a ten-hour boat journey."[22] But where exactly in Manchukuo they were going was still unclear.

The next day, the first group left at 5:30 a.m. The second group, including Berry, Corkill, and Brownie, left at 2:00 p.m. Corkill wrote, "We have been divided into sections of 8 officers and 2 enlisted men each. I am in Col Browne's squad with Col Berry, Quisenberry [Quesenberry], Maloney [Mallonee], Hilsman, O'Day, Hoffman and Sgts Parsons and Pacheck [Pechek]." They boarded a train north to a seaport town named Yawata, where ferry boats were ready to take them to Fusan on the Korean Peninsula. Another water journey meant that they faced the possibility once more of being bombed by American planes.[23]

As it turned out, American planes were bombing parts of Kyushu that very day, so the guards told the POWs to put on all their clothing under their life preservers in case they were attacked during their ten-hour-long journey crossing the Korean Strait. It must have been at least some comfort to the POWs that they had life preservers for this leg of their journey.[24]

CHAPTER 15

Manchuria POW Camps

CHENG CHIA TUN POW CAMP

Colonel Berry's Time at This Camp: November 14, 1944, to May 20, 1945

On November 14, 1944, the POWs arrived by train at the small town of Cheng Chia Tun (today Shuangliao) near the Gobi Desert, about 150 miles northwest of Mukden (today Shenyang) in Manchukuo (Manchuria). The POWs were forced to march half a mile to the POW camp of Cheng Chia Tun, which was camp no. 4 in the Hoten POW Camp system headquartered in Mukden. After enduring the extreme heat of Shirakawa, the POWs found themselves relegated to the bitter cold of Manchuria. Limited to the clothing they were able to bring from Shirakawa, the 354 POWs were freezing.[1]

The POWs lined up in an open area in the middle of the compound. At this camp, fortunately, they were not ordered to strip and lay out their belongings for an inspection, as they had been in the previous POW camps. Instead, they wore all the clothing they owned but still suffered from the freezing temperature. They went through the check-in procedures while they waited for Lieutenant Matsumiya, the camp commander, to arrive. According to Colonel Mallonee, Matsumiya finally arrived in a "four-door sedan" with no windshield, windows, or engine. The POWs had to stifle laughter as Matsumiya surveyed the POWs from his car "motored" by "two little Manchurian ponies." The commandant was buried under furs to keep warm, and all the POWs could see of him was his mustache.[2] Naturally, he became known as "Handlebars."

When we arrived in barracks we found these enormous Russian Stoves, with no fire in them. Took us sometime to learn how to manipulate them to satisfaction.

Old Russian Stove. "We meet 'Pachika,' who gives us 'The Cold Shoulder.' When we arrived in barracks we found the enormous Russian stoves, with no fire in them. Took us sometime to learn to manipulate them to satisfaction." Source: *The Life of a P.O.W. Under the Japanese in Caricature as Sketched by Col. Malcolm Vaughn Fortier*, by Malcolm Fortier.

As Matsumiya lectured them, the POWs had time to study their new home. The compound where they stood was surrounded by a wall, and an electric fence stood a few feet inside the wall. This camp included several buildings: the prisoners' barracks, a hospital, the guards' barracks, storerooms, and a cookhouse. The POWs were eager to get to their barracks before they froze to death, but before they could go, they had to sign the standard no-escape pledge.[3]

Once inside the barracks, they discovered that the high-ranking officials and generals were already there. Generals Wainwright, Moore, and King told the newcomers that it was forty degrees below zero outside. Inside their barracks were Russian stoves, which the POWs called "pachikas," standing ready to heat the rooms. For a short time, the POWs had trouble getting the large, old stoves to work, but once they figured the pachikas out, their rooms heated up well, and they were comfortable.[4]

Colonel Berry's squad members changed again at this new camp. Here, he was in a twelve-person room (twenty feet by twenty-two feet) with Colonels Harrison Browne, William Corkill, Marshall Quesenberry, Roger Hilsman, Bob Hoffman, William Braddock, James Gillespie, Stuart Wood, Nunez Pilet, Jesse Traywick, and Clyde Selleck. At last, Berry found himself in the same squad as his good friends Browne and Corkill. Colonel Browne became the squad leader. The guards issued each POW two overcoats, a blanket, a blouse, and a pair of woolen pajama pants. Because it was bitter cold outside, squad members ate, slept, and played cards in their room during the winter months.[5]

At Cheng Chia Tun, the guards returned Berry's old diary to him, but part of it was missing. In anger he wrote, "In checking over my diary I find that the measly catfish bastard at Shirakawa who threw my diary away was told to return it failed to give me the part covering January 1, 1943–February 17, 1943 so I have lost all that." He thought he would borrow Colonel Aldridge's diary to fill in the missing days, but he never did.[6]

Berry's diary entries from the Manchurian POW camps were not as detailed as they had been at Shirakawa and Karenko. Often, he just summarized an entire week. As his time as a prisoner dragged on, Berry may have felt less inclined to write lengthy accounts about his internment. What he did write about were familiar topics—food, lack of mail, forced work, nonwork time, celebrations, and rumors.

In the beginning, the food at Cheng Chia Tun consisted of bread or rolls, beans, and some vegetables. Brougher, who had arrived in October before the colonels, noted, "We are getting beans in large quantity every evening now—too much for some people—some stomachs badly upset—others turning down some of their share and leaving heavy second and third helpings for those of us still taking them (including me)."[7] Once the colonels arrived in November, the portion of beans was diminished because the number of POWs to feed increased. However, the diet was supplemented by Red Cross food that the guards meted out each day. Finding ways to harass the POWs, the guards punched holes in the canned Red Cross food so the POWs had to eat these extras right away and could not save them for later. Berry wrote, "Been here for 25 days and they finally got kind (?) hearted and gave us packs cigarettes each and a 1 lb can powdered milk for 2. Bastards broke seal on can and said we had to turn in can in 7 days. We don't know why they are issuing the Red + an item at a time and with seal broken but it is a damn fool idea but believe it is just to annoy and harass us."[8]

The extra food in December helped the POWs gain weight. Berry noted the change in his own weight in his typically wry fashion: "Beans, Beans and beans with mush and maize meal. Weighed 178 lbs. Thought I had gone up more but scales did not show it. Eating like a horse and feel like am inflated football bladder but can't afford to pass up chow." By March 23, Berry was the heaviest POW in the camp at 189 pounds.[9]

In January, Commander Matsumiya proposed that the POWs plant a garden in the spring. He told them, as had other commanders before him, that the prisoners would get the food from the garden. But the Americans knew that would not be the case, and they told the commander that they would not volunteer to plant a camp garden. By April, Commander Matsumiya approached the American officers again and tried to persuade them to plant gardens. Berry wrote, "CO said all produce for us and he build us a warehouse to keep it in and it to be under a P.W. and Nips not eat any of it. Wants names of all of us who don't want to work tomorrow. Kinda hinted at some retaliation but when called he said there would be not punishment."[10]

Eventually, twenty-eight Dutch and thirteen British officers did volunteer to plant gardens, including Colonel Fortier. But General Parker told Fortier that he could not volunteer because the Americans were standing together, refusing to volunteer. The POWs' defiance would be consequential for them and for Commander Matsumiya, as well. Because he failed to get the Americans to work, Matsumiya was replaced in late April by Lieutenant Ikeda, who understood that his mission was to punish the Americans for not volunteering. Red Cross supplies were withheld and rations cut. The officers lost weight, Berry himself losing fourteen pounds in one month.[11]

Because there was very little forced work, the American officers had more leisure time at Cheng Chia Tun than at previous camps. The reduction in forced work and extra leisure led Mallonee to opine that Cheng Chia Tun was their easiest camp. Berry and Brownie played cribbage and again arranged tournaments for themselves, with Brownie the usual winner. To fill the time, Berry walked more at this camp than previously, sometimes as much as seven miles a day. And as he did at previous camps, Berry led calisthenics. The athlete-coach in him was always present.[12]

Braly and others organized concerts, as before. Others organized follies for entertainment. Bridge games and tournaments took hold, and Berry became a fanatic about bridge. He wrote that he was playing so much he needed to read up on his "Eli." (Berry had made a small book of bridge rules for himself that he called Eli, most likely named after J. B. Elwell, who authored books on the rules of bridge.) Berry and Corkill paired up often. Ed Lilly wrote, "This morning Jack and I played Ed Corkill and K. L. Berry—and beat them."[13] Berry and Corkill played bridge daily. Corkill wrote, "K. L. Berry and I walk 2 or 3 miles around the bull ring each morning and play bridge in the afternoons and read at nights."[14] Reading was a favorite pastime as it had been at Shirakawa. After reading *Crime and Punishment*, Berry offered the following critique: "It was a terrible yarn of Russian life—heavy, morose but interesting."[15] Berry's interest in the novel may have come from his own time in Russia, where he had once lived and had observed the life of Russian peasants for himself.

In April, with slightly warmer weather, Berry and Corkill played golf. Corkill recounted an incident with a guard one day. "This Am [a.m.] I was practicing golf with K. L. Berry. He knocked a ball outside the inner fence. I crawled under and got it. The change of guard came by soon after and the corporal gave me fits about it indicating by pointing his gun at me not to do it again. I won't!"[16]

Along with playing golf, Berry developed an interest in bird-watching because there were hundreds of crows around the camp mating and nesting. Watching crows and other birds flying overhead became a distraction for many other POWs, as well. Some POWs climbed trees trying to kill crows. Berry had always loved bird hunting, but with no gun in hand he settled on counting birds. His interest is reflected in his numerous diary entries mentioning geese, bustards, crows, and sandhill cranes flying overhead. Each day he would detail the types of birds and how many of them flew over. He even wrote "no birds flying" when he did not see any that day. As if this was big news, Ed Lilly even reported in his diary, "K. L. Berry said he saw a flight of 8 or 9 bustards go over."[17] After the war, Berry talked about his bird-watching days at Cheng Chia Tun with Texas friends while he was on a bird hunting trip.

Celebrations of traditions continued at Cheng Chia Tun. On March 2, Berry lamented, "Texas Independence Day and we Texans—Corkill, Keltner, Aldridge, Bonham, [Richard] Rogers, [Louis] Dougherty and Pvt [Leonard)] Evans wanted to celebrate in proper way but could not." Perhaps they sang Texas songs as they had the previous year in Shirakawa. Birthdays were always part of the traditions celebrated. For Corkill's birthday, Berry wrote, "Gave Ed Corkill a surprise birthday serenade on May 4th his birthday. We all avoided mentioning his birthday until then and he thought we had forgotten him." Indeed, Corkill told Clyde Selleck that K. L. must have forgotten his birthday. Berry continued in his diary that Pvt. Philip Hersee sang "Home on the Range" and "Moon over the Gobi," a song that Ed Lilly had written.[18]

Moved by Berry's thoughtfulness, Corkill recorded all the details of the day as if he were writing his wife:

> K. L. Berry pulled a fast one on me yesterday. He failed for the first time to come around the first thing in the morning to wish me many happy returns on my birthday. All day we played bridge together with Cols Browne and Harrison M. H. Quesenberry and still he did not think of it. After supper we walked the bull ring until about 7:15 when Browne called out the window for us to come in and start a bridge game. We said all right and came upstairs. The moment I reached the room I saw a crowd and the camp Orchestra and they struck up the "Happy birthday to you" song. It was a complete surprise. I was flabbergasted. I looked at K. L. and he was grinning and it dawned on me then that he had planned it and it turned out he had started arrangements for it a week or more ago. The orchestra played all evening and included "The Sweetheart of EX." I told them I would certainly let you know that you were remembered in the party. There is no need to tell you how thrilled I was at such a kindly gesture on the parts of the orchestra as well as the audience but most of all to K. L.[19]

Celebrations of any kind at the camps helped lift the POWs' spirits, especially for those prisoners who had not heard from home. In early December, a cache of radiograms reached Cheng Chia Tun. The POWs were excited to get some news because it had been a few months since anyone had received any word from home. As so often before, Berry was not one of the lucky POWs to hear from home. Sad and exasperated, he noted, "About 100 radios [radiograms] came in on 7th but you know old K. L.—he did not get any. Guess my wife has ceased to care for me or she is damned stingy with her funds. Ed Aldridge & J. Keltner got one each." Some mail trickled into the camp during the spring of 1945—but again not for Berry. In the end, he did not receive any mail while he was at Cheng Chia Tun. Feeling neglected again, he wrote, "May 12, 1945—No mail since your August 1943. Practically without clothes, practically barefooted, sockless."[20]

The Japanese allowed Berry and other POWs to write a one hundred-word letter home, but after the Japanese censors redacted much of his letter, Berry did not want to send it. Lilly detailed in his diary, "Yesterday got our postcards for signature.... K. L. Berry found his to be hopelessly garbled and refused to sign it—finally he did though when the Nips threatened to cut him off from the privilege of sending or receiving any messages if he did not. Can't say that I blame him."[21] The Japanese probably did not send Berry's note. It was a common practice for the Japanese not to send the POW letters off. Instead, they read them to see what the POWs were thinking. In any case, no letter from Berry's time at Cheng Chia Tun was found in Alice's possessions after her death.

As there had been at other POW camps, a small medical clinic or hospital was available to the POWs. Several of Berry's friends were in and out of it with the flu, headaches, and just general ill health from lack of food and lack of medical care. One of Berry's friend, Col. Floyd Marshall, who had been sick at the previous camps, did not improve at Cheng Chia Tun. The proper medical care he badly needed was simply not available. Reflecting on his friend's situation, Berry did not think Marshall would make it home. Unfortunately, he was correct.[22]

Rumors, of course, abounded at Cheng Chia Tun. Berry documented the rumors he heard. Rumors were verified or refuted within a shorter time frame than had been the case at previous camps because the POWs had gotten access to some newspapers. Access came at great risk, however. An enlisted man had become friendly with a Japanese sentry who spoke a little English. The enlisted man made a bargain with the sentry. He would give him his watch if the sentry would sneak into the commander's office at night and "borrow" the newspapers so American Colonels Wood and Hoffman could read the papers. Wood and Hoffman then translated the papers and spread the information around the camp. The sentry would then sneak the papers back into the commander's office a few hours later.[23] Hoffman was one of Berry's squad mates, so Berry and others in the squad were among the first to hear what was in the Japanese papers. The POWs were aware that Japanese papers often printed misleading accounts of Japanese successes in the war for their citizens to read. So they reviewed the news with some skepticism.

In January 1945, the POWs heard a rumor that the United States had eleven divisions on Luzon. In fact, the US Sixth Army had invaded Lingayen Gulf on Luzon on January 9. In late January, rumors reached the camp about the war in Europe, as well. There were rumors about big battles in Belgium and Luxemburg (i.e., the Battle of the Bulge) and about the Russian army near Warsaw in Poland. February brought even more encouraging rumors to the POWs. They heard that the Sixth Army in the Philippines had attacked Manila and recaptured Bataan. In addition, they heard that Churchill, Roosevelt, and Stalin had met in Yalta to discuss terms for the end of the war in Europe.

Berry wrote, "Fine rumors and seems to come from good source—Hong Kong bombed about 30 days ago. Taiwan had heavy bombing last few days." In late February, he added, "Papers say that in early Feb many fighting in PI [Philippines] and we in outskirts of Manila and city badly shelled. Russia within 60 miles Berlin and US and E [England] inside Ger [Germany] towards Coblenz. US and E pushed back out of Ger in Dec but now back in. Some US generals lost their stars. 1300 bombers and 500 fighters hit Weimar. Rumored a US armored div [division] and a Chinese Division captured an important point in China. We are digging bomb shelters here as Nips say it is urgently needed."[24] Indeed, the commander and guards sensed that the war was coming to them soon. In the frigid landscape in and around the camp, the commander ordered the POWs to dig trenches and bomb shelters.

In March one of the guards, apparently in a foul mood, walked through the barracks, turning off lights. This behavior led Berry to speculate that the guard had heard the news of the United States victories in the Philippines. As usual, the Japanese guards' treatment of the POWs became rougher when the war news was not good for them. Manila, Corregidor, and Mindanao in the Philippines were all recaptured by the Allies in March. In addition, B-29s bombed fifteen square miles of Tokyo in March, the British liberated Mandalay in Burma, and the Allies in Europe gained ground all around Hitler's troops. Berry wrote on March 29, "Many planes hitting Japan and Taiwan and Germany pushed well in from her borders."[25] The POWs could sense that the war was nearing an end.

On April 12, as the war continued to close in on Japan, Berry wrote about the guards' harassing behaviors and attributed it again to war events. "Nips inspected property today, but it was only to annoy us as they made no check. Took up our bread knives, etc. Bawled out 2 rooms because did not come to attention promptly enuf or remain at attention. Said we had very poor discipline and did not know how to stand at attention and drilled two rooms in it." The guards pointedly reminded the POWs that their treatment could be as bad as they had been treated in Taiwan if they did not obey. Berry wrote, "Said this commandant treated us well and we did not cooperate. That we did not in Taiwan and were treated accordingly and that they treat us the same. They must have had hell knocked out of them some place."[26] Indeed, they had. The Americans were bombing Japan and had sunk the Japanese battleship *Yamamoto*.

As the eventual defeat of Japan became more likely, the POWs responded by increasing their insolent behavior towards the guards. Discipline among the guards began to break down because they knew the war was going badly for them and for Japan. Lilly recounted what a guard told him, "This afternoon new corporal (one of the 3 star privates of the guard has recently been promoted) went into Col. Browne's and K. L. Berry's room and found an iron rod. He picked it up and was taking it away when Col. Browne said 'no, we use it to poke the fire.' The Cpl replied, 'no, you go home this summer.'"[27] As events continued to unfold, the POWs grew more optimistic that the corporal's prediction would come true.

The POWs did not know that President Roosevelt had died on April 12 or that Harry Truman had become the president who would lead the United States to the end of the war on both fronts. Musing about the war news, Berry wrote, "War in Europe seems about over with Berlin capture rather imminent. Bombing in Nippon, Taiwan and fighting still on Okinawa. Many of us believe Germany be out May 1–15th and some believe Japan out shortly afterwards. I think however that she will fight on and be 1946 before they go out unless Russia comes in in which case 6 mos or less from now."[28] Berry did not believe the Japanese would give up easily.

Berry was correct about Germany: Russia reached Berlin on April 21; on April 30 Hitler committed suicide. Germany officially surrendered on May 7, 1945. But this significant news about Germany's surrender had not yet reached the POWs. On the day of Germany's surrender, Berry was thinking not of the war but of his marriage, "May 7, 1945—My wedding anniversary. How I wish I were home with my Sweet Gal. This is the 5th one I have missed with her and the 4th in a row. I believe however that it will be the last and that in 46 we will try to put five into one and have a real day."[29] Filled with optimism that his captivity would soon end, Berry set aside his anger with Alice and instead looked forward to being reunited with his "Sweet Gal." As had been the case from the beginning of Berry's captivity, Berry's frustration with his wife was always mixed with love and longing.

A few days after Germany's surrender, an event known to the Japanese but not yet known to the POWs, Berry noted a change in the behavior of the guards and the camp commandant. "Camp Commandant holding conference after conference with our senior officers. Seems to be easing up and trying to be friendly." The change had tangible benefits for the POWs—each was given seventy-two sheets of toilet paper, along with tooth powder, two packs of cigarettes, a small bar of soap, and a bag of confections. The POWs observed that the Japanese were sending supplies south on trains. Berry interpreted this activity to mean that the POWs would soon be caught up in the Japanese retreat from Manchuria. "Believe Nips are sending everything movable out of this country. Wouldn't be surprised if they moved us."[30]

Berry was correct. Aware of the Russian advance toward Manchuria, the Japanese did plan to move everything and everyone south to be near the ports.

Berry wrote on May 13, "Yes, we were told to be ready to move by 20th. We know not where or when. Some think Mukden, Korea, Japan, Siberia, etc. Also rumored President died April 12th but we do not believe."[31] A few weeks later, the POWs learned that the president's death was a fact but speculated that he had been assassinated by the Japanese. Like many other Americans, Berry underestimated Truman's abilities, writing that he was sorry to hear that Truman had become president.

By May 15, the POWs knew that Germany was out of the war. Berry wondered if Germany's defeat was the cause of more food coming their way again. In a way it was. The Japanese also knew that with the war in Europe over, Russia might now declare war on Japan, and they needed to get out of Manchuria as fast as they could. They did not want to leave any food or supplies for the Russians, so, the POWs enjoyed a little extra food.[32]

On May 20, Berry wrote, "Yes we are moving. Up 4 AM and turned in bedding. Mush, bun and hot water at 7 AM. Fell in at 12:15 PM and marched to station and entrained and given small millet bun. Left at 2:10 PM and reached main NS line at 6:45 PM. Given bun and some got small bit drinking water but I did not. Traveled about 80 miles. Very crowded. Had 11 in our double section where room for only 8. Baggage all over place."[33] The train ride to the final camp, Mukden, was long, covering 180 miles in twenty-four hours. Berry must have been near a window because he said the countryside was beautiful. He also noted the presence of large labor camps.

MUKDEN POW CAMP

Colonel Berry's Time at This Camp: May 21, 1945, to August 20, 1945

Hoten headquarters in Camp Mukden was located in a heavily industrialized area next to an ammunition factory, with a tank factory, airplane factory, and major railyard nearby. There were 1,200 POWs already at Mukden when Berry arrived. He noted that the rest of the group from Shirakawa was there, as well as a few he knew from the Luzon Cabanatuan POW Camp—Lt. Col. Zero Wilson, Lt. Col. Tom Tarpley, Capt. Jack Walker, and Capt. Ernest Brown. He was happy to see them but shocked at their condition, writing in his diary that they had been "thru the mill." Wilson and Walker told Berry about the *Ōryoku Maru* ordeal. Berry gave his Red Cross butter and milk to his four friends, who were in very poor shape. He went on to say, "Most of the PWs from PI [the Cabanatuan POW Camp] killed by disease, our bombs, starvation, suffocation, exposure and murder. Most all of my officers seem to be killed or whereabouts unknown. In fact, of the 35 American officers I had I know of only five of us who are still alive. Others are I hope but cannot say now. Believe some are in Japan."[34]

After he arrived at Mukden, Berry took advantage of services available in Mukden that had not been available in Cheng Chia Tun. He bought grey pajama pants and shirt, navy breeches, and a British uniform coat that cost him thirty yen. Then he had the POW tailor shop fashion a sheet into four pairs of shorts, giving one each to Colonels Aldridge and Phillip Fry and keeping two for himself. He also had his sandals and shoes repaired at the POW shoe shop.[35]

At Mukden, there was little to occupy the POWs' time. Berry and others played a lot of bridge, sunbathed, and walked. He had some long talks with Zero Wilson, Jack Walker, and others he knew who had come to Mukden from Japan after leaving Cabanatuan, Philippines. He was truly touched and shaken by the suffering they had endured at the Cabanatuan prison camp and then on the hell ships. He wrote, "Wish I were a writer and could do justice to the experiences of our officers and men who remained in PI and came up here late last year and this year. It would make lurid reading and would help a long way in forever banning Nips from entering US."[36]

In a touching gesture, Capt. Jack Walker from the 24th Field Artillery Regiment, who had served with Berry on Bataan, wrote Berry a letter expressing his admiration for his senior officer (Appendix C). Walker wrote, "I have had the opportunity of serving under a number of Commanding Officers. But never have had the privilege of serving under one whom I have esteemed more as both a friend and officer."[37] Surprised and humbled by this beautiful tribute to his courageous leadership, Berry wrote, "Jack Walker wrote me very fine personal testimonial letter the other day and totally unasked for or expected."[38]

During the last week of June, the POWs heard they would be getting mail again. Berry grumbled in his diary, "The mail rumor of 3 weeks ago is getting hot. Our men sorted it yesterday and I have 2 letters in long official looking envelopes. Guess from kids or brothers or sisters as my wife seems to have forgotten me. Will call on her to produce some proof that she has written me a reasonable amount—say once a week." Then he wrote, "Got mail on 3rd but you guessed it—not a damn word from my wife and family. Two letters from Ada [his sister]—one July 6 and one July 20, 1944. Stated in one that Tom in Marines but did not give his rank."[39] Those two letters were the last mail he was to receive for the remainder of his captivity.

A rumor reached Berry that although he did not get his star, he would receive brigadier general pay because he had had a command in the war. He really hoped it was true and that it would be retroactive. Then, thinking about his own men, he took the time to write up several citations (Distinguished Service Medal, Distinguished Service Cross, Silver Star, and Purple Heart) for his officers and men of the 1st Division.[40]

On July 6, 1945, he celebrated his 52nd birthday in POW style. Corkill wrote, "Today is K. L. Berry's birthday. I had the mens [sic] quartet sing the happy birthday song and gave him a deck of cards which he needed very much."[41] Berry also described the day.

> Had my 4th birthday as a PW today and not so bad. Beebe, Drake and Pierce sang "Happy Birthday to You" and received congratulations from Jones, Lough, Bluemel, Weaver (Chesterfields), Brougher, Drake, Stevens, Pierce, Funk, Beebe, Aldridge, (pack Nip cigs), Balsam, Bell, Boatwright, Bonham, Braly, Brawner, Browne, Callahan, Campbell, Cordero, Cottrell, Corkill (pack cards), Churchill, Doane, Dougherty (Bun), Dumas (Bun), Foster, Galbraith, Hamilton, Hoffman, Hughes, Johnson, Keltner, MacDonald, Monihan, O'Day, Quesenberry, Rawitser, Selleck, Stansell, Stowell, Uhrig, Curtis, Lowman (Bun), Vachon, Wilson, Walker (bowl pig weed), gave me Chesterfield and Fred Galardi gave me 5 Chesterfields. Pretty nice birthday after all.[42]

The fact that Berry named every person who celebrated with him and everything he received makes clear just how important celebrations were to him and his fellow POWs.

In that same week in July, Berry described a funny incident that happened to Corkill: "Cork given hell by Nip guard because he taking sun bath naked and did not get up and bow when guard came up. Tried to make Cork say he was sorry and ashamed but got nowhere with Cork. Wanted to hit Cork but evidently got orders not to do it."[43] Tired of being harassed by the guards, the POWs demonstrated their *cheekiness* any way they could!

On the serious side, war news about the Battle of Okinawa, April 1–June 22, 1945, reached Berry in July. He learned that two of his friends from Texas National Guard days died in the battles in June. The Japanese had fought ferociously for that small island, and the United States suffered forty-nine thousand casualties, including over twelve thousand deaths. An estimated 150,000 civilians on Okinawa died, and 110,000 Japanese died defending the island. While the sacrifice of American lives was tragic, winning the battle for the island put the Americans within close range of Japan for the planned Operation Downfall, the invasion of Japan, scheduled to begin on November 1, 1945.[44]

The Battle of Okinawa was significant in another respect. The intense fighting and tremendous losses on both sides in Okinawa confirmed fears that the proposed Operation Downfall could become the bloodiest battle of any in World War II. In April 1945, the Joint Chiefs of Staff estimated that Operation Downfall's first campaign, the invasion of Kyushu, could result in 456,000 casualties, including 109,000 dead or missing. The second campaign of the planned invasion was expected to incur thousands and thousands more casualties, and if the Japanese fought as they had in Okinawa, some casualty estimates for the entire operation were in the millions for the Allies.[45]

In April 1945, President Truman learned for the first time about the existence of the atomic bomb. Up to that point, there were no plans in place to use the bomb. Instead, military leaders were focused on the planned invasion of Japan. Dreading the potential loss of Allied lives in an invasion, Truman hoped that the enormously destructive strategic bombing raids that were taking place over Japan during the summer of 1945 would force Japan to surrender, making a bloody invasion unnecessary.[46]

As the Allies closed in on Japan, the commander and guards at Mukden reacted to bad news in their own malicious, weak way by withholding Red Cross supplies and food from the POWs. Food became very poor and scarce as the days went by, and on July 24, Berry wrote that he weighed 164 pounds. "I now weigh 25 lbs less than I did on March 21. Nips thought our men doctored weight so called 30 down for reweighing about 4 days after official weighing and found all had lost more except 2."[47] The end of the war could not come soon enough for the starving POWs.

After the successful test of the atomic bomb in July, Truman, with a new, terrible weapon in hand, issued an ultimatum to Japan to surrender. Japan had no intention to surrender. It was building up its forces on Kyushu expecting an invasion there in the fall. At this point, Truman had no definite plans to use the bomb because Operation Downfall was being organized by General MacArthur and others. Truman sought more information about every military option, including the use of the bomb. A committee was formed to explore the option, and its members came out in favor of using the bomb, with no advance warning to Japan. Truman issued a second ultimatum to Japan in late July, but as before, Japan refused to surrender. Truman gave the approval to use the bomb.[48]

During the first week of August, Berry wrote, "Very poor food all during this period. Same fare—mush and a cornmeal-millet flour biscuit for breakfast, bowl watery soup for lunch and a biscuit and watery soup for supper. Hungry as hell all the time and nothing we can do about it.... Lots of rumors flying but we are still prisoners so don't believe them. Guess we will hit them hard this month however."[49] The US did hit Japan "hard," though Berry could not have known that the United States had dropped the first atomic bomb, nicknamed "Little Boy," on Hiroshima, Japan, on August 6. Still, Japan refused to surrender.

The diaries of Berry, Galbraith, Brougher, and Corkill did not mention anything about the atomic bomb, so the POWs must not have heard about this historic event. Instead, the POWs remained uncertain about the state of the war. On August 8, Russia declared war against Japan and invaded Manchuria. Brougher wrote on August 9, "Air raid alarm this morning—men apparently not going to work in the factories. Wonder what's up?"[50] What was up was that Russia was now in the war, and on August 9, 1945, a second atomic bomb, dubbed "Fat Boy," was dropped over Nagasaki, Japan. With this horror unleashed, Japan surrendered a few days later.

Summarizing the week of August 8–13, during which news of Russia's entry into the war had reached the POWs but news about the atomic bombs had not, Berry wrote:

> Old Russian Bear is in and going great we hear. We heard on the 10th Russia was in and know it for sure now. Hear rumors that she is making powerful lightning drives from the North East & west generally towards Harbin. Nips in a talespin and don't know which way to turn but moving troops thru here. Building trenches around here as if they intended to defend the POW compound. Hear Nips ordered all their families to be sent to Korea by 15th. We are looking forward to future with some dread as we feel our lot will not be very pleasant for a while. We all hope that . . .[51]

With that unfinished sentence, Berry's diary abruptly came to an end. The feeling of dread Berry expressed was based on the fear that the guards would execute all the prisoners in retaliation against the United States if Japan lost the war. That fear was not without some basis. After the war, war crimes investigators found "Order to Kill" documents in Japanese POW camp offices in Taiwan.[52] As it turned out, however, news of the atomic bombs and the subsequent Japanese surrender on August 15 had not yet reached the commander or the guards at Mukden. The POWs were safe for the time being.

CHAPTER 16

Rescue and Freedom

RESCUE: AUGUST 16, 1945

The POWs could not believe their eyes when they saw six paratroopers landing in a field near the Mukden camp on August 16. Brougher wrote, "Wildest kind of rumors all day of end of war. Parachute landing observed near the prison camp about 11:30 AM."[1]

Galbraith wrote, "At about 11 a.m., a B-24 circled around and a number of luggage parachutes dropped. Then six men bailed out."[2] The paratroopers landed about two miles from the camp. There was wild speculation about who these men were. Some thought they were Red Cross representatives. In fact, the men who parachuted in were from the Office of Strategic Services (OSS), which in the future became the Central Intelligence Agency (CIA). They were there to tell the POW camp commander that the war was over, that Japan had surrendered, and that the POWs should be freed. After the Japanese surrendered on the 14th, "there was much concern by top officials that if we did not get to the POW camps immediately, that the Japanese would be likely to kill many or all of the prisoners,"[3] wrote Sgt. Hal Leith, a member of the OSS team.

The OSS knew that some POWs in Japan had already been beheaded after the surrender. Time was of the essence because the OSS believed that the highest ranking POWs, Generals Wainwright and Moore, were in the Mukden camp and they did not want anything to happen to them. When the team landed in the field, they were met by a patrol of Japanese soldiers who had heard nothing about the surrender. Leith, who spoke Chinese, asked if any of the Japanese also spoke Chinese. One did, so Leith was able to explain that the war was

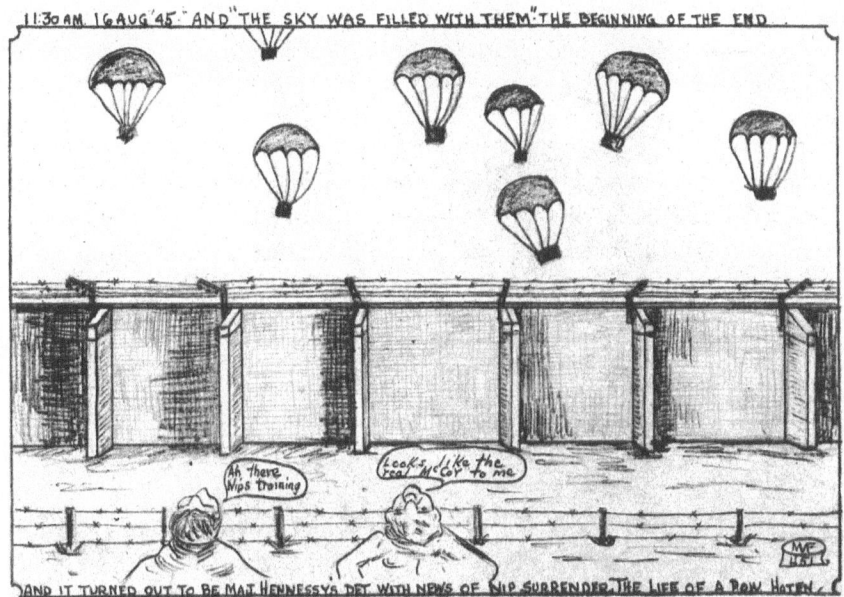

The sight of these parachutes filled many of us with hope, though the more pessimistic loudly voiced their belief that it was only Nip anti-paratroop training.

Rescue. "11:30 AM 16 Aug '45. And 'The Sky was Filled with Them.' The sight of those parachutes filled many of us with hope, though the more pessimistic loudly voiced their belief that it was only Nip anti-paratroop training." Source: *The Life of a P.O.W. Under the Japanese in Caricature as Sketched by Col. Malcolm Vaughn Fortier*, by Malcolm Fortier.

over and that Japan had surrendered. The soldiers did not believe him, and the OSS team members were beaten up before they were taken as prisoners to the Japanese secret police headquarters in the town of Mukden. The commander in Mukden had finally been notified about the surrender, but he had received no orders about what to do next. Seizing the moment, the OSS team asked to go to the POW camp and talk with Commander Masuda about freeing the POWs.[4]

The next day, Japanese soldiers escorted the OSS team members into Commander Masuda's office. Galbraith wrote that some POWs peeked into the commander's office and reported that the parachutists appeared to have the "upper hand." Hearing this, many in the camp thought that the war was over. Others remained skeptical, perhaps not wanting to be disappointed if the war was not over.[5]

Once the OSS team talked with Masuda, he called the highest ranking American, British, and Dutch officers to his office. General Parker was the top

American officer. All were briefed by the OSS team about the Japanese surrender. As soon as he was briefed, Parker made a short statement to the assembled American POWs who were waiting outside:

> There must be no demonstration of any kind. An armistice has been declared between Japan and the United States, Great Britain and China. It is understood fighting still continues between Japan and Russia. For the present we are still under Japanese control and "protection" and will remain within the prescribed limits of this compound. The Japs still have the guns so be careful about starting any disturbance that would bring them into our side of the compound. That is all.[6]

General Parker, like all the other POWs, feared the frightened, armed guards. Mallonee wrote, "One incident offending a scared, trigger-happy moron might set off a blazing massacre."[7] After Parker's short speech, a very excited Corkill wrote on August 17, "The day we have been looking for for 44 months has arrived!"[8]

For the next few days, the POWs followed General Parker's orders, restrained themselves, and kept to their side of the compound. Major Lamar and Sergeant Leith from the rescue team headed 150 miles northeast to the camp where Generals Wainwright and Moore were so they could free them and bring them back.[9]

On August 20, as the POWs waited to be freed, Braly organized a concert and began with the three national anthems—British, Dutch, and American. Galbraith chronicled the wonderful events of the evening in his diary:

> Aug. 20. About supper time a heavy bomber flew low over the camp—pamphlets were dropped stating reps. [representatives] from Am-China Hq would arrive at all P.W. and internee camps and coordinate with Nips for relief, etc. of all—responsibility still remained with Nips for compliance with provisions of surrender terms.... Musical concert at 7 p.m.—started out by mass singing of national anthem. Throats were so constricted that it was difficult to make much of a sound.[10]

Galbraith went on to recount that right in the middle of the singing, several Russian officers appeared at the camp entrance gate. POW Sgt. Walter Hurley translated what one Russian officer announced: "By order of the commander of the Russian Red Army, I declare you from this moment F R E E! I congratulate the United States Forces on its successes. I congratulate the United States, Great Britain and allies on their defeat of Japan."[11]

Understandably, the POWs were jubilant and cheered the Russians. Braly wrote that after the Russian officer set them free, "eleven Jap officers and about

A Russian Staff Officer arrived one evening in the midst of a song fest. He was nearly stampeded but gave a thrilling brief address that warmed our hearts.

I Declare You Free. "6:30 pm 20 Aug 45—Russians Arrive—Break Up our Songfest. A Russian Staff Officer arrived one evening in the midst of a song fest. He was nearly stampeded but gave a thrilling brief address that warmed our hearts." *The Life of a P.O.W. Under the Japanese in Caricature as Sketched by Col. Malcolm Vaughn Fortier*, by Malcolm Fortier.

thirty enlisted men were marched out into the center of the clearing where they laid down their arms including rifles, pistols, machine guns, hand grenades and officers' sabers. The senior Russian officer then selected a pistol from the pile and presented it to Gen. Parker as a trophy."[12]

The Russians then paraded the defeated Japanese before the POWs and marched them out of the compound. In the days ahead, as the former POWs waited anxiously to be transported out of Manchuria, the Japanese were the ones cleaning the camp instead of the POWs—a gratifying sight to the newly freed Americans, Dutch, and British. Because only smaller planes could land at the small, local airport, the sick were taken out first. Then, when Wainwright and the top generals arrived, the generals were transported out. Confusion reigned in the city of Mukden as the Russians took over everything, including

Russian Staff Officer arms our men as guards with Nips arms and they become P.O.W.'s in a thrilling moment in which they were paraded by on their way to guardhouse.

General Parker Receives a Japanese Pistol. "1:00pm 20 Aug 45—Russian Officer Disarms Our Nip guards—Presents Gen. Parker with Nip Pistol. Russian Staff Officer arms our men as guards with Nip arms and they become P.O.W.'s in a thrilling moment in which they were paraded by on their way to guardhouse." *The Life of a P.O.W. Under the Japanese in Caricature as Sketched by Col. Malcolm Vaughn Fortier*, by Malcolm Fortier.

the airport and railroad. The majority of the former prisoners had to wait for the logistics of their transport to be sorted out by the Russians. Meanwhile, American planes dropped food and supplies into the area for the freed men, which was a tremendous comfort to the former POWs, who had all been starved during their last month in captivity. Taking advantage of these provisions, most put on weight before they traveled home.[13]

As the Americans, British, and Dutch waited, they had free time for several days to explore Mukden. Berry's good friend Cordero noted a few events in the days that followed: "The following day, August 21, we were allowed to leave camp. I went for a walk with two other officers [Berry and Corkill] and as we passed through the gates on our way out I drew in a great breath of freedom."[14] He described how the three of them took rickshaw rides into Mukden to sightsee for several days in a row. They brought their own food with them because the American food dropped by the planes was so much better than what they could buy in Mukden.

It was on one of those excursions into Mukden that Berry chanced upon Gerta Bolte, the daughter of an old friend from the 15th Regiment days. Gerta's father and Berry had served together in Tientsin, China, in the 1930s when Berry was a captain. Gerta had been a classmate of the Berry children at the Tientsin Grammar School. When she saw Cordero, Berry, and Corkill in town, she recognized Berry right away and cried out, "Are you Captain Berry; are you the father of the Berry children?" Very surprised to hear his name, he answered that he was. She then explained who she was and told Berry that the Russians who liberated the POW camp had commandeered her house. Then they ordered her to go out and search for whiskey to be brought back to them at her house. Berry was alarmed and told her to be sure not to tell the Russians she was German or else the consequences might be dire. He went back with her to her house and told the Russians—he knew a little Russian from his Vladivostok days—that he would be returning the next morning and if Gerta or her daughter or niece were harmed, he would have the Russians shot. His bluff worked—the next morning, Berry put Gerta and her family on the train to Harbin, Manchuria, where Gerta's husband was. Berry's gallantry, not to mention his chutzpah, was on full display in this small-world encounter.[15]

TRANSPORT HOME

Berry's good friends Corkill and Browne, who had been classified as in need of medical assistance, left on September 5, 1945, on the last airplane out of Mukden. After that last plane departed, plans changed, and the Mukden-Dairen railroad, now under Russian control, had to be used for transporting the bulk of the former POWs out of Mukden.[16]

Meanwhile, the USS *Relief*, a navy hospital ship, left Subic Bay in the Philippines on September 1 and made it to the port city of Dairen, Manchuria, (Dalian, today) on September 8 to welcome the former POWs—now designated as recovered allied military personnel, or RAMPs—on board. Other navy ships that would also transport the RAMPs were making their way into port. But the RAMPs had not yet made it to Dairen. The ships' personnel had to wait a few days in port not knowing when the RAMPs would arrive because the Russians had to get the railroad working and staffed to transport the RAMPs from Mukden. USS *Relief* staff were not allowed to communicate with the RAMPs, only with the Russians.[17]

The day that Berry was to leave Mukden finally came on September 10, 1945. He was among the first group of 759 former POWs to leave by train. Many of Berry's friends were also in this group: Cordero, Bowler, Quesenberry, Braly, Boatwright, Bonham, Lilly, Galbraith, and Fortier, to name a few.[18] The trip to the port city of Russian-held Dairen on the Yellow Sea took twenty-six hours instead of the normal eight hours. But the Americans were okay, not so crowded as on previous train trips, and they had food, water, and places to sleep. They arrived in Dairen on September 11, 1945, at night. The tracks at Dairen were clogged with weeds, so the former POWs had to detrain and walk in the dark down the tracks to the port carrying everything they owned. Braly wrote, "I picked up my bags and violin and stumbled up the tracks into the darkness."[19]

The personnel on the USS *Relief* had waited in anticipation for three days. On the evening of the 11th, they learned from the Russians that the first train had arrived, and the former POWs were walking down the tracks. The USS *Relief* diary chronicled the night:

> When they finally began to arrive, at 2050, the scene was dramatic and poignant. They were heard before they were seen. Then, from out of the darkness beyond the docks, the first of them stepped into the glare of the dock floodlights, first one or two, then ten, then the whole procession. They carried all their worldly possessions on their backs or in duffle bags.[20]

As the POWs came out of the darkness, Braly remembered, "Words can never describe my feelings as I rounded the corner of that warehouse and saw that beautiful, brilliantly lighted U.S. Navy Hospital Ship *RELIEF*."[21] The *Relief* diary noted:

> The Ship's Red Cross lights and green bands had been illuminated, and the Ship started the Stateside music playing over the Ship's public address system—Dixie, The Marine Hymn, Stardust etc. The RELIEF was the first American ship these men had seen in three years; in come [sic] cases, four years.[22]

As the RAMPs came on board, they told the ship's personnel that the first thing they wanted was a shower. "They were provided with soap, towels, and a clean, soft bed. Then, after they were made comfortable a huge steak dinner was served. They had ice cream for dessert."[23] One can only imagine what an extraordinary feeling it was to be treated so well and to be finally going home.

The next day the navy ships left Dairen and headed for Okinawa. Destroyers sailed alongside because the waters were still full of mines. Many mines were

During an inspection tour of the 29th Depot, General Uhl and his assistant, Col William H. Craig, posed with these fifteen full colonels recently liberated from the hands of the Japanese. In the group are, reading from left to right, front row: Col J. C. Hughes, Col R. M. O'Day, Col A. C. Searle, Col J. D. Cook, Col A. W. Penrose, Col A. P. Moore, Col J. W. Thompson and Col H. H. Stickney; rear row: Col R. Bonham, Col W. A. Enos, Colonel Craig, Col J. R. Boatwright, General Uhl, Col J. W. Worthington, Col K. L. Berry, Col V. P. Foster, and Col J. P. Horan.

September 1945 Photo of RAMPS. The photo shows Berry and many of the colonels who made the long trek to freedom with him. General Uhl in the photo was the commander of replacement program. Source: Replacement Command Center for the Recovered Personnel Program, 1945.

destroyed on the way. Frustratingly, in the vicinity of Okinawa, a typhoon hit, so the ships could not go into port, where they were supposed to drop off the RAMPs at Buckner Bay. Instead, the ships had to sail one hundred miles away from Okinawa to wait out the typhoon. Along the way, one of the ships, the USS *Colbert*, hit a mine. Ten RAMPs were killed that day. It is unimaginable to think that these ten men had made it all the way to freedom only to be killed on their way home.[24]

After six days, the USS *Relief* sailed into Buckner Bay, Okinawa. From there, Colonel Berry flew to Manila. At the Replacement Command Center for the Recovered Personnel Program outside of Manila, Berry obtained a new uniform, filled out various forms, and stood with his colleagues for photographers.[25] From Manila, Berry flew to San Francisco, where he underwent health checks at the Presidio Letterman Hospital for several days. After three years and eleven months, Berry was on US soil again.

PART V

A New Lease on Life

"Far and away the best prize that life has to offer is the chance to work hard at work worth doing."

—*Theodore Roosevelt*

CHAPTER 17

Recognition

RECONCILIATION

After Berry was freed by the Russians on August 20, his thoughts turned to addressing the two significant personal and professional disappointments he had suffered during the war and imprisonment—his wife's ostensible indifference to him, and the battlefield promotion that did not materialize.

As a free man in Mukden, Berry could write letters home without Japanese censorship or approval for the first time since April 10, 1942. Smarting from Alice's seeming indifference to him, he apparently wrote her a scathing letter complaining about her lack of correspondence—four letters in three years and eleven months—expressing his fear or belief that she must not love him and had forgotten him.

Alice received his very angry letter and wrote back on September 15, 1945, saying, "I love you, darling, and can hardly wait to hold you in my arms again." Most likely, he received this letter while he was in Manila. She went on to say that she had written him "oodles of letters," and some were returned for being too long, but she said she had rewritten them and sent them again. She asked if he had received her letter sent to Mukden in May, summarizing the last three years.[1] He had not. That the letter Alice says she wrote to Berry in Mukden covered three years invites considerable skepticism about her claim to have been a faithful letter writer throughout her husband's captivity.

Meanwhile, the OSS rescue team "discovered storerooms containing Red Cross parcels and thousands of pieces of mail withheld from the prisoners. This mail was then made available to surviving prisoners within the camp and sent on to those already evacuated."[2] Two very short letters from Alice to Berry dated March 18, 1945, and August 6, 1945, were found in a group of letters Berry had saved in a box for many years after the war. One letter told him she was in Ohio with their grandson, and the other told him she heard his radiogram broadcast from Tokyo. Neither one summarized three years. Perhaps these letters were in the stash found by the OSS team and sent on to Berry after he was back home.[3] It is clear he did not have them before he wrote his angry letter to Alice.

While Berry was in San Francisco at the Presidio Letterman Hospital, he heard from Alice that she wanted him to travel to Ohio instead of San Antonio. She was staying with their son K. L. Jr. at Wright-Patterson Army Air Corps Base. Alice wanted him to meet their one-year-old grandson, Kearie Lee Berry III. He agreed, but when he did not come to Ohio right away, a now anxious Alice sent him a telegram on October 10 asking him where he was. He may have been stalling the reunion, but when he received her telegram, he set off for Ohio the next day.[4]

When Berry and Alice were reunited, it is not known if he brought up the controversy over letters. Conceivably, he had decided to believe Alice. His actions over the next twenty years showed a man busy as ever with work, politics, hunting, and grandchildren who called him "Daddy." However, he did not follow through on his dream of building a beautiful ranch house with his wife for them to live in together someday. Money may have been an issue, or possibly the idea was not as appealing as it had been in captivity. In an interview in 1953, Berry was asked about Alice, and his answer was short and factual: "I still have the same wife after 36 years and she has borne me three children." That statement seemed rather cold in print. However, in spite of Berry's past anger at Alice while he was a POW, they did carve out a new life together where he lived in the limelight, and she took a supportive back seat.[5]

Berry and Alice stayed a few days in Ohio and then flew to Washington, DC, where they visited his brother-in-law Col. Harvey Matthews and his sister Anne. Colonel Matthews worked at the Pentagon and offered to help Berry uncover what had happened to his battlefield promotion. Also in Washington, DC, was Berry's good friend Colonel Browne, who had chosen to go to Walter Reed Hospital to convalesce. Brownie was elated to see his best buddy.[6]

PROMOTION TO BRIGADIER GENERAL

By the end of October, Berry entered Brooke Army Hospital at Fort Sam Houston to seek treatment for several ailments as well as malnutrition. From his

Reunited. K. L. and Alice Berry, October 1945. Source: Author's private collection.

hospital room, he turned his attention to the second matter that had caused him pain as a POW, his failed promotion to brigadier general. He began a letter-writing campaign to secure the promotion that Wainwright had given him on the battlefield in 1942. He enlisted his brothers Gene and John Berry, first cousin Sam H. Riley, and brother-in-law Matthews, as well as many close friends to help in writing letters to congressmen and senators on his behalf. In addition, Berry and General Wainwright exchanged several letters on the subject. Wainwright wrote in November, "My dear Berry: I inclose [sic] herewith copy of a letter I have written to General Marshall concerning your case and I am hopeful that my recommendation may be approved."[7]

What Wainwright meant by Berry's "case" was that Berry's name had been deleted from the recommended promotion list submitted to President Roosevelt in April 1942. Berry's friends were never able to tell him who deleted his name, but they speculated that it was deleted because, in the chaos of surrender in the Philippines, officials in Washington, DC, did not know if Berry was dead or alive. Matthews lamented, "It seems to have been unnecessary to cut your name, the only one from the Philippines, just because Bataan fell. You might have been on Corregidor or might have taken to the hills as a guerilla."[8] Matthews later learned and reported to Berry that Gen. George Marshall decided to promote only those temporary generals whose promotions had been confirmed before the surrender. Thus, Berry's promotion had been deleted. However, Wainwright gave Berry hope in late November when he sent him a copy of the response he got back from General Marshall: "His [Berry's] promotion to the temporary rank of brigadier general is now again under consideration in the War Department. Consideration will also be given to his advance on the permanent list."[9]

Berry's supporters stepped up their letter-writing campaign, writing Texas Senator Tom Connolly and Representatives Lyndon Johnson and Paul Kilday about Berry's case. All three congressmen became interested in Berry's promotion to temporary brigadier general, and they used their influence in Washington, DC, circles to move Berry's promotion along. Colonel Matthews wrote Berry in late January 1946 that he had seen Berry's name in *The Evening Star* Washington, DC, newspaper under the heading "Colonels to be brigadier generals." President Truman had nominated Berry for temporary promotion to brigadier general and sent his name to the Senate.[10]

Everyone was happy to see Berry's name move forward, and on February 7, 1946, Colonel Berry finally received the recognition he had longed for since 1942: his star, the rank of brigadier general, temporary. It was bittersweet, in that he had had to wait so long, but in the end the recognition of his battlefield leadership was now his to savor.

HONORS

During the months Berry was seeking his promotion, he was awarded several notable medals for his leadership in the Battle of Bataan. In the order of action during the war, the first medal was the Distinguished Service Cross, the second highest military award that can be given to a member of the army. General MacArthur had nominated Berry "for extraordinary heroism in action in the Left Subsector of the I Philippine Corps, near Bagac, Bataan, Luzon, Philippine Islands, from 26 January 1942 to 17 February 1942. As Commanding Officer, 3d Infantry, and subsequently of the 1st Division, Philippine Army, Colonel Berry displayed outstanding courage in constantly remaining with the most advanced elements of his command, encouraging them to outstanding efforts during the height of the heroic struggle for the Tuol and Cotar River basin."[11]

The second medal was the Distinguished Service Medal, a military decoration of the US Army that is presented to a soldier who has distinguished himself by exceptionally meritorious service to the government in a duty of great responsibility. Berry was so honored for "Commanding the 1st Regular Division, Philippine Army, in the Philippine Islands from 17 February 1942 to 10 April 1942."[12]

Next, Berry received the Silver Star: "Lieutenant Colonel (Infantry) Kearie Lee Berry for gallantry in action as Regimental Commander, 3d Infantry, First Regular Division (Philippine Islands Army), at Bataan, Luzon, Philippine Islands, on 6 February 1942." Berry was also awarded two other significant medals—the Purple Heart for being captured and beaten by the Japanese on the Bataan Death March and the Combat Infantryman's Badge for being in active ground combat as a member of an infantry after December 6, 1941.[13]

After his promotion, and at the urging of Colonel Matthews, Berry applied for five different army refresher courses established by the Army Ground Forces for ex-POWs who were seeking a return to duty as Berry was. The courses were as follows: Infantry at Fort Benning, Field Artillery at Fort Sill, Antiaircraft Artillery at Fort Bliss, Armored at Fort Knox, and Cavalry at Fort Riley. Thus, from March through June 1946, Berry, with Alice in tow, attended these two-week-long courses, learning about new tactics and equipment used by the army since his capture in 1942.[14]

Distinguished Service Medal. General Wainwright awards the Distinguished Service Medal to Gen. K. L. Berry and Col. Nicoll Galbraith in 1946. Source: Author's private collection.

Although Berry had been recognized several times over for his heroism and leadership and had the prestige of completing five army refresher courses, he was not content. He wanted his promotion of temporary brigadier general to be made permanent, an honor very unlikely to happen in the postwar downsizing of the military. A bill in the US Congress (S. 1533) to promote the other temporary brigadier generals from the Philippine battles to permanent status lacked Berry's name. Because Berry's original promotion to temporary brigadier general did not go through in April 1942, decision-makers in the War Department chose not to put Berry's name forward for permanent status.

Hoping to correct this injustice, Matthews encouraged Berry to write to General Eisenhower, chief of staff of the War Department, about his case. Matthews knew that Berry and Eisenhower were old friends but believed that the chief of staff was too high up in the War Department to be aware of Berry's case. Berry took Matthews's advice and wrote Eisenhower on June 4, 1946. Lyndon Johnson wrote Eisenhower on Berry's behalf on June 6, 1946. Eisenhower's aide wrote Berry on June 7: "General Eisenhower has asked me to write you a response to your letter to him of 4 June. He feels that it would probably be better if you send to him by mail all the facts of your case."[15]

Of course, Berry responded to Eisenhower immediately but heard nothing of note until a month later when Eisenhower wrote:

> 9 July 1946
> Dear Berry:
> I have your recent letter requesting the addition of your name to the officers listed in the "Johnson Bill" and am happy to say that the matter has already been settled in a manner which I trust will prove entirely satisfactory to you.
>
> Acting on the recommendation of the War Department, the President forwarded a request to the Senate on 2 July 1946 that the "Johnson Bill" (S. 1533) be amended to include your name. At the same time, the President nominated you for appointment as a brigadier general of the Regular Army, so that the appointment may be expedited once the enabling legislation has become law. Thus you are assured (subject to favorable action by the Congress) equal consideration with the other general officers who were captured in the Philippines.
>
> With warm regard.
> Sincerely, Dwight D. Eisenhower[16]

Congratulations came Berry's way, prematurely, as it turned out. Everything had seemed all set for S. 1533 to pass before the Seventy-Ninth Congress adjourned in August. Congressman Kilday from Texas was certain of it. But, apparently, the job of getting Berry's name added to the bill was not completed in time before adjournment. As a consequence of this delay, none of the general officers from the Philippine battles achieved permanent status that summer. Kilday promised to try again in the following congressional session.[17]

In the meantime, Berry was eager to be assigned to duty. With General Eisenhower's help, Berry took a brief assignment with the Eighth Army Personnel Headquarters in Dallas before that post was moved into the Fourth Army headquartered at Fort Sam Houston, where General Wainwright was the commander. It was not long before Wainwright then assigned Berry to be the executive officer of the Texas Military District in Austin. As Berry happily fulfilled his duties for the Texas Military District that summer, he watched his chances of becoming a permanent brigadier general come and go once again when S. 1533 died with adjournment in August.

INVITATION FROM THE GOVERNOR

Unexpectedly in December 1946, Beauford Jester, the Texas governor elect, invited Berry to lunch and informed him that he wanted Berry to be his adjutant general of Texas. Jester and Berry had known each other for a long time, both having attended the University of Texas and then the First Officer Training School at Leon Springs in 1917. They were good friends. The governor knew Berry would make a fine adjutant general, eminently worthy of the post.

Berry wrote his good friend Claude Birkhead, former commanding general of the 36th Division, that he was "flabbergasted" and "had never dreamed of such a thing." His mind spinning, Berry continued in his letter that he told Jester that he would be "proud to serve him and the State of Texas," but unless he had the full support of the Texas National Guard, he would not care to take the position. Then, with his mind still racing, Berry mused he would have to retire and be taken off the active list. Still hoping that he would become a permanent brigadier general in the Regular Army, Berry did not want to retire or be removed from the active list. In light of these considerations, Berry told his friend Birkhead that he asked the governor to select someone else, and that Jester had responded that, no, he would not select someone else.[18]

When the word got out at Jester's inauguration that the new governor wanted Berry to fill the adjutant general role, Berry's friends "in the Guard, the Regular Army and others" crowded around him and implored him to take the position.

People thought he would be the perfect Texas role model the state wanted—war hero and famous University of Texas athlete. But it was still a very hard decision for Berry to make. However, once he learned that he "had long been slated to be retired and that only waivers" for his hearing deficit kept him in, he reconsidered Jester's offer. While he could have stayed in the army with waivers, he decided that he would enjoy the position as adjutant general better than what he was currently doing as executive officer of the Texas Military District. So, he accepted Governor Jester's offer and began a three-month-long retirement process that ended his army career on July 30, 1947.[19]

The matter of becoming a permanent brigadier general in the Regular Army was then no longer an issue for Berry. But as promised in the next session of Congress, Congressman Kilday reintroduced the "Johnson Bill" (S. 1533) as H.R. 2993. The bill sailed through the Armed Forces Committee to be put on the consent agenda of Congress before it adjourned in late July 1947. However, when Congress reconvened, the only general officers from the Philippines who were awarded permanent status in December 1947 were Brig. Gen. Lewis C. Beebe and two major generals—George Moore and Berry's good friend Albert Jones.[20]

CHAPTER 18

The Adjutant General Years

A NEW CAREER: ADJUTANT GENERAL OF TEXAS

Brigadier General Berry was released from his duty as executive officer of the Texas Military District on May 2, 1947, and was sworn into his new office as adjutant general of Texas on May 7. Governor Beauford Jester wrote Berry a note of congratulations a few days later saying, "I am glad you have accepted to serve. I want to assure you that it will be a pleasure to cooperate with you and to have your cooperation during this administration."[1] For his part, Berry looked forward to working with his good friend Jester.

In June, Berry was promoted to major general, as dictated by a law passed in the Fiftieth Texas Legislature. The law was in response to the War Department's decision that in a state with an allotment of twenty-five thousand or more National Guard troops, the adjutant general should hold the rank of major general. In 1947 Texas had an allotment of thirty-one thousand troops. General Berry was most happy to become a major general in his new role.[2]

As adjutant general, Berry served as the state's top military official and aide to the governor in supervising the military department of the state. Berry would have one assistant adjutant general, whom he could choose and recommend to the governor for appointment. Some of the responsibilities of the adjutant general's office included "providing military aid to state civil authorities and furnishing trained military personnel from the state's military forces—the Texas State Guard, the Texas Army National Guard, and the Texas Air National Guard—in case of national emergency or war."[3] Over his years as adjutant general of Texas,

Texas Adjutant General. Maj. Gen. K. L. Berry becomes adjutant general of Texas. Source: Author's private collection.

Berry would direct military activities for many thousands of servicemen in year-round programs around the state.

The military side of the new position was of course familiar to Berry, but there was another side to the position that required a variety of talents not necessarily tied to his military proficiencies. In discharging these responsibilities, Berry's honest and affable personality were distinct assets. Soon after becoming adjutant general, Berry was appointed as a trustee of the Terrell Military College, east of Dallas. The college was a nonprofit educational corporation and a class A junior college where students wore uniforms and were called cadets. Berry described himself as a "babe in the woods" in matters pertaining to trustee work but said he would conscientiously perform his duties to the best of his abilities.[4]

Governor Jester appointed Berry to be the Selective Service director for Texas for a few months because he needed him to get the draft machine going.[5] In short order, Berry sent the governor a list of members of Texas' 137 draft boards, which the governor then sent on to President Truman for approval. With that task completed, Berry stepped out of the Selective Service role after six months. As adjutant general, he was expected to be the face of the Texas military all over the state in a variety of settings—county fairs, military and nonmilitary speaking engagements, military promotions, award ceremonies, and military training exercises or maneuvers at all Texas Guard posts. His family saved newspaper clippings from his vast array of appearances over his fourteen years—far too many to be listed or described. Be it said, he was constantly on the go.

In addition to the various duties that took him all over the state, there were two notable charities that Berry agreed to chair to assist in raising funds for their causes. One was the Red Cross, and the other was the March of Dimes. The Red Cross was a charity that Berry held close to his heart. Berry reported that he first experienced what the Red Cross could do when he was stationed in Siberia as a young lieutenant after World War I. There was no question of death by starvation for him in Siberia, but he was thankful for the extra comforts and entertainment he and other soldiers found in the Red Cross packages. Berry's experience with the Red Cross in World War II went well beyond enjoying the comforts and entertainment the organization had provided him years before. He emphasized the lifesaving importance of the Red Cross when he spoke at the kickoff of the Travis County fundraising campaign: "I am confident that it was owing to Red Cross packages of food that many prisoners survived to return home. I can't say enough about how the Red Cross saved us from starvation."[6] His prison diary repeatedly expressed his belief that the Red Cross saved many lives. Advocating for the Red Cross was Berry's way of expressing his gratitude.

Berry's acceptance of the state chairmanship of the March of Dimes caught the eye of Berry's old friend, US Senator Lyndon Johnson, who addressed his colleagues in the Senate saying, "Gen. K. L. Berry Adjutant General of Texas and distinguished military man has agreed to increase his services to our state by accepting the State Chairmanship of the 1955 March of Dimes." Johnson went on to say that many Texas newspapers published the story of General Berry's remarkable career, and he made sure that Berry's story as it appeared in the *Paducah Texas Post* was printed in the Appendix of the June 29, 1954, *Congressional Record*. Johnson also read into the record on the Senate floor Berry's statement accepting the chairmanship: "Having been a prisoner of war for 40 months, I have some idea of the feelings of people who are prisoners of the iron lung. It took an all-out war to get myself and the others released—so let's mount an all-out attack against polio." Having such a prestigious military man using military language to kick off fundraising events contributed greatly to the campaign's success.[7]

Berry was the state chairman of March of Dimes for two years in a row. In publicly accepting his second appointment, Berry again used military turns of phrase as he told those present, "The Salk vaccine represents a major victory in the war against polio. . . . The fight of individuals and children personally afflicted with this disease is long and costly, and I will always be interested in entering combat in the fight against this disease as represented by the program of the National Foundation, which is supported by the March of Dimes."[8] It is clear that his years as a prisoner had made Berry especially sensitive to the plight of many other kinds of prisoners who needed a strong helping hand to survive.

When Governor Jester won a second term as governor, he reappointed Berry as adjutant general. Unexpectedly, one year into his new term, Jester died from a heart attack at the age of fifty-six, Berry's age exactly. Losing his dear friend grieved Berry greatly. Berry flew in the plane that carried Governor Jester's body back to Austin from Houston for a state funeral. Before the pilot landed, Berry asked the pilot to circle over the Capitol building twice for a final salute. It was an unusual and poignant tribute to the late governor by his dear friend. After the funeral service in the Senate chamber, Berry flew with Jester's body to Corsicana, Jester's hometown. Again, Berry had the pilot circle over the city twice, as Berry said, "to give the governor a last chance to look at his home city." Governor Jester's executive secretary, William McGill, wrote Berry afterward and said, "All of us who, like you, were close to him will hold for you even more tender sentiments and affectionate regards as a result of the experiences of these recent days."[9]

After Jester's untimely death, Robert Allan Shivers became governor. He was elected governor in 1950, 1952, and 1954, and he served until January 1957. He and General Berry became good friends, and Shivers reappointed Berry to the adjutant general position each term he was governor.[10]

In the early 1950s, the nation became concerned—some would say obsessed—with the threat that the spread of communism posed to the security of the United States and the world. The outbreak of the Korean War was a tangible manifestation of that threat, prompting Adjutant General Berry to enter the public arena as an advocate for a strong military. He called on Americans to urge President Truman to expedite the buildup of national defenses in the event that the Russians became aggressive in Japan or West Germany. In speeches on the subject, Berry outlined a seven-point plan to make the United States more prepared. He urged Americans to advocate for the following: a universal military training law; increasing the tempo of the Selective Service; building National Guard and Reserve units to full war capacity; building up eight to ten state guard units and sending them to West Germany and Japan; manufacturing modern armaments; increasing the size of the Air Force to one hundred groups;

and ceasing criticism of the president, secretary of defense, and congressmen, instead urging them to take action.[11]

Berry was zealous in pursuing this issue. He wrote his good friend Lyndon Johnson, expressing his concern that the United States was not building up its military fast enough to counter the spread of communism. "Dear Lyndon, How much longer are we going to fool around? This Korean War is the beginning of World War III or World War LAST.... [T]he winner will control things for years and years to come and we must be the winner."[12] Recalling whole regiments of Mongolian people in the Russian Army when he was stationed in Siberia, Berry insisted that the "Reds" the United States were fighting in Korea were in fact veteran Russian soldiers. He concluded his lengthy letter by saying, "Lyndon, you have your feet firmly planted so take a good swing at those slow-acting, fuzzy-minded people up there and jar them out of their inertia. With the greatest sincerity, K. L. Berry."[13]

Two years ahead of all other states, Berry organized a State Guard Reserve Corps. The Reserve Corps would serve in the places left vacant by the National Guard if the National Guard troops were called up for active duty. Truman did send six Army Guard units to the war, but none from Texas. In the end, the Texas Air National Guard was federalized and sent to Langley Air Force Base in Virginia to train for the war. For several years to come, Berry continued to give speeches to Texas audiences advocating for a universal military training law.[14]

Berry found himself embroiled in controversy when his actions to improve the military posts under his leadership in Texas came to the attention of the state auditor, C. H. Cavness. Cavness "questioned the legality of maintaining guest houses as recreational facilities for members of the Texas National Guard and the employees of the adjutant general's department and their families."[15] The place in question was Camp Hulen in Corpus Christie, where Berry was stationed as a young private with the 2nd Infantry. Berry responded to Cavness saying that the guesthouses were for the workers at Camp Hulen, an arrangement that saved money compared to housing them in hotels. He added that he "saw nothing wrong with employees going down there to spend time with their families at their own expense" when the guesthouses were vacant.[16]

Cavness also challenged Berry's decision to build a lake stocked with fish at Camp Mabry. Berry replied that "the lake was a byproduct of excavation work done to get dirt for a steel-fenced parking lot for NG [National Guard] vehicles. We piled some of that dirt up over a [stone] dam so that it could be used as a backstop for our firing range. We have not spent any of the state's money on these projects. And besides I am charged with beautifying the base. I think the lake helps in that respect."[17] General Berry survived the state auditor's challenge with good grace and continued doing what he deemed necessary to improve facilities owned by the Texas Guard.

CHIT CHAT AND THE WAINWRIGHT TRAVELERS

As Berry performed his adjutant general duties, gave speeches on the military, and volunteered for major organizations, he remained in contact with his POW friends. One source of information about former POWs was *The Quan*, a newsletter created after the war so former POWs from Bataan and Corregidor could keep in touch. Upon his return to the United States, Col. Ray O'Day, a former POW, began a column in *The Quan* called "Chit Chat." O'Day "chatted" about the colonels and generals who were imprisoned with him in the Philippines, Taiwan, and Manchuria, noting where they lived and what they were doing. In one column, Berry and his friends Brownie, Cordero, Corkill, and Aldridge were mentioned.

> Virgil Cordero, still in Munich, was to have July traveling in Italy, Switzerland and Belgium in his 1947 Packard Clipper.... Ed Corkill knocked on our door for a short call as he motored home with Mrs. Corkill from a visit in Eastern Arkansas. The old golfer looked good even tho it was hot driving. His home is in San Antonio. He plays golf with Brownie, K. L. and Ed Aldridge. K. L. is putting in a golf course at Camp Mabry, Austin, in order to get more practice and save nickels when he plays with the gang.[18]

O'Day's mention of a golf course was in error. Given the state auditor's scrutiny of his expenditures, Adjutant General Berry would not have considered such a project. In one sense, though, O'Day may have been ahead of his time. Many years later, building a golf course on Mabry land was proposed by the City of Austin. Berry opposed the project, a position that would lead eventually to his forced retirement.

In 1950, Berry and other former POWs gathered for a long weekend in San Antonio at Fort Sam Houston to reminisce and have fun. They called themselves "The Wainwright Travelers" because they had traveled together to so many different POW camps with their leader, General Wainwright. Three hundred "Travelers" came that weekend to Fort Sam Houston, where Wainwright was the honorary president and official host. Speakers at the event included Lt. Gen. Leroy Lutes and Maj. Gen. K. L. Berry. A few photos from that weekend have been preserved. Unfortunately, only some of the "Travelers" have been

Above, Before the Banquet. A large group of "Travelers" posed with General Wainwright. Some are identified. Posing front row, left to right: Gen. Edward King, Col. Eddie O'Connor, Col. Harrison Browne, unidentified, Gen. Lewis Beebe, Gen. Jonathan Wainwright, unidentified, and perhaps, Tom Dooley. Second row: Behind Gen. Wainwright's right shoulder is Col. Nicoll Galbraith. Gen. K.L. Berry can be seen in the back by the window. Source: Author's private collection.

Above, Travelers the Next Day After the Banquet. Standing left to right: unidentified, unidentified, unidentified, Luther Stevens, Ed Aldridge (possibly), Jonathan Wainwright (with cane), K. L. Berry, unidentified, Ed Keltner (possibly), Harrison Browne, Ed Corkill, Eddie O'Connor, and Unidentified. Front row, squatting: unidentified, Kearie Lee Berry III, Miller Ainsworth (not a Traveler but a good friend). Source: Author's private collection.

Facing below, Small Group with Wainwright. A small group of Travelers posed for a photo with General Wainwright the next day. Back row, left to right: unidentified, Luther Stevens, unidentified with pipe, K. L. Berry, Bill Braly, Harrison Browne, and Eddie O'Connor. Front row, sitting: unidentified and Jonathan Wainwright. Source: Author's private collection.

identified. If "The Wainwright Travelers" ever met again before General Wainwright's death in 1953, the Berry family had no record of a meeting.[19]

INVOLVEMENT IN POLITICS FOR AN OLD FRIEND

When Governor Jester was still alive, Berry became active in presidential politics. In 1948, President Truman was up for reelection, but Berry and other Texans had their eyes on Gen. Dwight Eisenhower, Berry's old friend, as the Democratic candidate for the presidential nomination. The idea of replacing Truman with Eisenhower was not far-fetched. Doubting that he could win reelection, President Truman did at one time offer to be the vice presidential candidate with Eisenhower at the top of a Democratic presidential ticket.[20]

In pursuit of this goal, Texas Governor Jester, along with General Berry and the Texas delegation, traveled to the Democratic Convention in 1948 to nominate Eisenhower for the ticket. Eisenhower, at the time serving as president of Columbia University (1948–1950), thanked the governor and Berry but told them he would not accept the nomination if offered because he did not want to seek the presidency. Jokingly, he told Berry that his brothers who were Republicans would disown him if he became a Democrat. Disappointed by Eisenhower's decision, Berry reminded him, "You, being a military man, know that conditions sometimes make you change your mind."[21]

In a historic surprise, President Truman won an upset victory in 1948. However, his second term turned out to be very unpopular because of frustration with the progress of the war in Korea and his firing of Gen. Douglas MacArthur as commander of the United Nation Forces. As a result, Republicans thought they had the advantage and could win the presidency in 1952. Senator Robert A. Taft of Ohio became the frontrunner for the nomination. Some prominent Republicans, like Senator Henry Cabot Lodge of Massachusetts, considered Taft an isolationist, so Lodge and others began to court Eisenhower, who was then commander of NATO forces in Europe (1950–1952). Eisenhower did not agree with Taft's isolationist stance, so he allowed himself to be courted by Lodge.[22]

As Berry had anticipated, "conditions sometimes make you change your mind." Eisenhower did change his mind and ran for president—but not as a Democrat. Meanwhile, never doubting the man and without regard to political party, Berry said to a reporter, "I can't think of one who would make a better president [than Eisenhower]. He wouldn't be the kind of president they could bulldoze. He would pick the best men for the jobs, just like he's always done."[23]

Eisenhower won in 1952, carrying several Southern Democratic states, including Texas. Texas Governor Allan Shivers chose Berry to be the official representative of the State of Texas at Eisenhower's inauguration. Traveling to Washington, DC, Berry brought with him the one-hundred-year-old cow bell that the Texas delegation had previously brought to the 1948 Democratic Convention in Philadelphia when they had tried unsuccessfully to persuade Eisenhower to run on their ticket. The bell, which was owned by the family of Gen. Paul Wakefield, had the name "IKE" in bronze soldered to it. At the inauguration in 1953, Berry presented the bell to President Eisenhower, saying that if the president ever got in a fog, the Texans would come and help him out of the fog. Accompanying the bell was a note saying, "Texas Democrats last November [1952] were finally given the opportunity to demonstrate how they felt about the political situation in 1948 and the man who will be President of our great country on Jan. 20. May God bless him." The president, reportedly, kept the bell in a cabinet in the oval office at the White House.[24]

OFF-DUTY HOURS

For some father-son hunting time together, General Berry and his sons, K. L. Jr. and Tom, purchased a ranch north of Austin. To make the purchase, they secured a loan through the Veterans Land Board, established after the war to help returning veterans acquire property. The ranch, which totaled 729.6 acres, was divided into three surveyed portions with Berry and his sons taking title to one portion each. It was not the type of ranch Berry had dreamed about as a prisoner at Shirakawa where he would have "Jersey black fight chickens," and "dutch ovens for smoking meat," and a beautiful "adobe home" straight out of *House Beautiful* where he could live the life of a cowboy with Alice "coasting down the hill ranching and hunting."[25] Instead, the ranch served as a hunting refuge for Berry and his sons. A small, rustic house on the property accommodated them. It had three small bedrooms, a kitchen, and a large room with a dining table, couches, and easy chairs. It worked well as a comfortable place for Berry to leave the hustle and bustle of the city behind him.[26]

Bruce Reilly, his grandson, said Berry named the ranch the "Lazy K Ranch." The branding iron was a capital K, with an upside-down/backwards L on the start of the K, which Reilly still owns. Sometimes Berry invited friends to the ranch, including retired Lt. Gen. Jonathan Wainwright. Wainwright came to the ranch for a cattle branding event, the day after the Wainright Travelers banquet.

Branding Day. Lt. Gen. Jonathan Wainwright brands a steer. Source: Author's private collection.

In the photo, Wainwright, wearing a tie, can be seen branding a steer with the Lazy K branding iron. Berry is standing next to him, ready to help.

Fixing up the ranch was work Berry enjoyed on weekends. He took his sons and grandsons for a day's work on the ranch when they visited. Grandson K. L. III remembered going to the ranch "quite a few times, him driving very fast to the Berry ranch at dawn for a day's work, but finding time to stop for big, really good breakfasts at a local restaurant!" Grandson K. L. continued, "Some of the work entailed using a coffee can sans gloves to remove the sand from abandoned fence post holes on the ranch—and flushing hundreds of black widow spiders from the holes."[27]

Grandsons Bruce and King Reilly recalled some of their ranch adventures with Berry. "He would take us out to the Berry Ranch out North Highway 183, for weekend shooting, and driving around the ranch 770 acres. Here I would learn to shoot shotguns, 20/16/12 gauge. He taught me to hunt quail and dove, along with lots of shooting at beer cans," remembered Bruce. And King recalled going dove hunting with his grandfather and "his longtime army buddy and fellow Bataan defender and Japanese prisoner-of-war Colonel Harrison C. Browne." After hunting, they would "chow down on 2-J's hamburgers, fries and shakes." Other times, King remembered sitting with Berry "for long hours in freezing temperatures in a deer blind" on the ranch. Berry clearly enjoyed teaching the boys how to hunt, but neither one grew up loving hunting.[28]

Berry became a father figure for the Reilly grandsons in the late 1950s and early '60s. The boys' father was gone most of the time, so they came to live with Berry and Alice at Camp Mabry, the home of the adjutant general. They called their grandfather "Daddy." King explained, "He was a father to me. My own dad was away on military assignments much of my childhood during the late 1950's." Bruce added, "We lived in Camp Mabry, Headquarters of the Texas National Guard—in the largest house on base. My brother and I shared a room together, next to Daddy's office."[29]

As a father would, Berry taught them how to "drive" his station wagon on the gravel roads with one or the other on his lap while he pushed the pedals. He always said "yes" to young King's request to ride "horsey" on his leg while he read the newspaper. He taught them fatherly lessons about how matches were dangerous and how following through on commitments, like going to football practice, was important. He pulled their teeth with a string tied to a door. They also learned about his even temper and quick wit. Bruce told the following humorous story:

> One Thanksgiving, with several guests at the dining table, Lulu (the cook and maid) was bringing out the turkey to the dining table, and she dropped it on the floor. My grandfather quickly said, "That's all-right Lulu, take that turkey back into the kitchen, and bring out the second turkey." There wasn't a second turkey, but Lulu knew what to do....[30]

Berry retained the interest in gardening that had played such an important part of his life as a POW. He maintained a garden at his home at Camp Mabry and another at the ranch. His grandsons recalled that he would "walk along the garden, pick out a good-sized carrot, knock the dirt off on his overalls (he wore overalls a lot for gardening/ranching), then eat the carrot (no washing)." Also, when weeding the garden, his motto was, "Pull a weed, plant a seed." What once had been a necessity to survive became a hobby in Berry's postwar life.[31]

When he was not hiking with his dogs and grandsons after work, Berry could be found sitting outside, alone, drinking a beer with a shotgun in hand. On occasion, King and Bruce joined him outside as he sipped his beers. King recalled that his grandfather would shoot at birds in the trees: "I enjoyed sitting with him at Camp Mabry waiting for defenseless blue jays who tried to eat his figs. I later wondered what the neighbors thought of shotgun blasts going off next door." Bruce added, "During these afterwork beer escapades, Daddy would discourage Mocking Birds (Texas State Bird—illegal to kill) and Blue Jays from eating figs at the huge (40–50 feet) fig tree. Daddy would bring a shotgun with him and shoot these pesky birds out of the fig tree. Of course, the shotgun blast would destroy more figs and branches than a single bird, but Daddy Loved to shoot anything!"[32]

CONTEMPLATION

These scenes, described by his grandsons, evoke Berry's days at POW Camp Cheng Chia Tun where he spent a lot of time counting birds. Berry told a hunting partner friend after the war about watching the birds in Manchuria: "We watched huge flights of wild birds fly over—ducks, geese, and bustards—millions of them, going north. God! If we just had a shotgun! There were lots of crows, too. We tried to pick them off with rocks, but we were too weak."[33]

At Mabry, as he watched for birds in his fig tree, shotgun in hand, he thought about those long days in Manchuria. He may have been weak as a prisoner, but he was never broken. On the contrary, his determination to stay fit had been strong in spite of a starvation diet. After the war, a reporter who wrote of Berry's athletic prowess asked the rhetorical question, "How did the little Nips expect to break a guy like that?"[34]

Exactly. How did he survive on the Bataan Death march when so many others did not? What drove him to overcome the odds stacked against him in the prison camps? Part of the answer was Berry's upbringing and heredity. He came from strong stock. He also had a great sense of optimism, hope, a wry sense of humor, and a can-do spirit—even in the bleakest of situations as a POW, he frequently began his diary entries with "beautiful day." He appreciated life, had a sense of humor, and resolved to survive. And, from his years of athletic and military training, he knew he had the strength and perseverance to withstand punishing physical deprivation if he had enough food. And to have enough food, he worked tirelessly in the camp gardens, his gardens, and on forced labor projects to obtain more rice and vegetables. With food, however bad it was, in his belly, he worked at keeping his strength up by walking daily, doing calisthenics, and even swimming at Mukden. And through it all, he kept a sense of humor. He would not be broken.

Berry's attributes had made him a good trainer and leader of men on the football field and in the jungles of Bataan. He even compared army training and football training: "If a coach can whip a team into condition in six weeks, there's no reason the army can't do it. There are many parallels between football—or any other sport—and war. The army uses 'plays' and highly developed teamwork to capture enemy units or to wipe out a machine gun nest. A coach needs to have talented players, the necessary equipment, morale, and physical conditioning. An army needs the same things."[35]

When he arrived in the Philippines prior to war, he had the conditioning a "coach" needed. And his young, strong, raw troops in 1941 would have been well-conditioned with at least six weeks of training if Berry had had the time to train them. But the war came too fast. Some of his young officers, like Lieutenant Rio, did not graduate from the Philippine Military Academy until after the war had begun. In spite of the lack of time for training and military equipment to fight a war, Berry's young Philippine troops under his leadership and coaching grew into soldiers during the onslaught of battle. They came together, learning the "plays" and tactics of warfare on the battlefield. They fought, lost, and won battles together. He had been so proud of them. In turn, the Philippine government was thankful to Berry.

PHILIPPINE LEGION OF HONOR

In 1955, a number of adjutants general from many states had the opportunity to tour either Europe or the Far East to witness the restoration of war-torn areas. General Berry, who had never traveled to Europe, was sorely tempted to choose that tour but ultimately decided he would return to the Far East with an itinerary that included two days in Manila. Berry let his former officers in the Philippines know that he would be coming, and one of those officers, Abe Asis, who then worked at the Department of National Defense, responded with a full itinerary for Berry's visit that he hoped Berry would be able to accept. For Berry, that meant that he would not be touring with the other adjutants general while in Manila.[36]

In November 1946, Asis had written the following to Berry: "I have submitted to the Chief of Staff of the Philippine Army a token of your brilliant, exemplary, meritorious and courageous devotion to duty while you were with us in the dark days of Bataan."[37] Asis hoped the humble appreciation would be approved someday. Berry did not hear anything further about this "token" of appreciation. However, when he saw the itinerary Asis had created for him, he must have been

Legion of Honor. Lt. Gen. Jesus Vargas awards the Philippine Legion of Honor to Major General Berry. Source: Author's private collection.

struck by the detail for May 21, when he would be awarded "the most belated decoration" while he was visiting Camp Murphy in Quezon City with Asis and others. He had no idea what the belated decoration could be, but he knew it was what Asis had written him about in 1946.

Upon arrival at Clark Air Force Base, Berry told those present, "I feel grand coming back to the Philippines. This is my second home, and I feel just happy and proud to be back here."[38] After the welcome party greeted Berry and the other adjutants, Berry's former officers whisked him away to privately meet Philippine President Magsaysay. His adjutant general friends were to meet the president the next day as a group. From the presidential visit, Berry was taken to the Walled City (in old Manila) for the laying of the wreath on the Tomb of the Unknown Soldier. He had lost many brave soldiers in the battles on Bataan, so it must have been a particular honor for him to lay the wreath. After the ceremony, Philippine Lt. Col. Victor Gomez, a close friend of Berry who had been on the Death March with him, welcomed him into his home for a relaxing visit with several of his former officers before he called it a night.[39]

Early the next morning, Berry was taken to Camp Murphy, where he discovered he was the man of the hour. On the parade grounds, Lt. Gen. Jesus Vargas, chief of staff of the Armed Forces of the Philippines, awarded Major General Berry the Philippine Legion of Honor in the degree of commander. The Legion of Honor was the highest honor the Philippine government could bestow on a person for meritorious service to its country. An overwhelmed Berry wept silently as Lieutenant General Vargas put the medal around his neck.[40]

The Legion of Honor credited Berry for molding his young division into a hard-hitting combat unit. As "hero of the pockets" he contributed to the protracted defense of the Philippines, which delayed the Japanese and gave the Allies time to organize their forces in Australia. After the award ceremony, Vargas and Berry sat together with other officers and government officials, including Philippine Senator Macario Peralta and Gen. Alfredo Santos, and reviewed a military parade held in Berry's honor. In a brief speech after the parade, Berry told his adoring audience that his "brave Filipino soldiers, who lacked training, equipment and were then having their first experience of real war, fought up like seasoned combat men."[41] Berry then paid tribute to Brig. Gen. Alfredo Santos, who had been one of Berry's regimental commanders in the Battle of Bataan.[42] Berry then met many of his former officers and stayed the afternoon to lunch with them and catch up on what they were all doing at that point in their lives. He left for Texas the next day.

In a letter to Lieutenant General Vargas upon his return home, Berry wrote: "Words cannot express my appreciation for the great honor your country conferred upon me by awarding me the Legion of Honor in the degree of Commander, and I will be ever grateful to you for personally making the presentation. My heart was too full at the time to adequately thank you and express my deep appreciation, but I can tell you now that it was the greatest moment of my life and one that I will always cherish."[43]

RETURN TO DUTIES AND MORE HONORS

The news of this prestigious award had reached Berry's friends in Texas, so when he arrived back home, a group of Austinites, organized by former Governor Dan Moody, threw a surprise party in his honor at the Driskill Hotel. The party was a "spontaneous move" on the part of the group to congratulate Berry for the recognition he received from the Philippines the week before in Manila. All fifty people present signed a large scroll commending Berry for "his service to this community and to this country."[44]

Earlier that same day, General Berry began another term as adjutant general, so after the elation of receiving the Legion of Honor and being feted by good friends, it was back to his duties as adjutant general. Among many things Berry was working on was the establishment of the Texas National Guard Officer Candidate Academy School. Berry wanted the school to focus on leadership skills. He felt the Guard would be able to recruit high quality leaders in the grade of second lieutenants because of the academy's emphasis. The Officer Candidate School finally came into being in 1959.[45]

Berry's work as adjutant general was recognized by the Texas Heritage Foundation in 1958. At a ceremony at Fort Hood, the foundation's president, Maj. Gen. Paul Wakefield, and Governor Price Daniel awarded Berry the Texas Distinguished Service Medal. Maj. Gen. Carl Phinney also received the prestigious medal the same evening. "The medal and certificate of award was created to commemorate the 120th anniversary of the signing of the Texas Independence at Washington-on-the Brazos and bears the names of the Alamo, San Jacinto and Goliad."[46] In presenting this medal to Berry, Wakefield declared to all present:

> It would be difficult indeed to select from the star-studded military history of Texas' immortals a profile to draw a comparison with the military achievements of Gen. K. L. Berry. At the fall of Bataan and the death march and the prison camp beyond and in WWI, we have here a Texas professional soldier who would feel comfortable in the presences of William Barrett Travis, David Crockett and the immortal Sam Houston.[47]

In February 1959, Governor Price Daniel appointed Berry to his seventh term as adjutant general. His appointment was celebrated by a small group of UT professors, deans, and directors, as well as Texas military officers, all of whom came to his office to congratulate him. Among those present was Clyde Littlefield, UT coach and his longtime friend. Littlefield's presence was special—he and Berry had played sports together at the university from 1912 to 1916 and were teammates on the famous 1914 football team. Littlefield had also been Berry's track coach when Berry returned to college at age thirty-one.

Berry received additional honors, some official and others more personal, in 1959. One personal honor deserves special notice. President Eisenhower was enroute to Mexico for an official visit, and his itinerary called for an overnight rest stop at Bergstrom Air Force Base in Austin. Although the president did not want any fanfare as part of his layover in Texas, government officials nevertheless decided to have a small welcoming party for the president that would include only Bergstrom officers, Governor Price Daniel, and Austin Mayor Tom Miller. Not even the governor's wife and son were permitted to attend. After the handshakes and welcomes from those present, President Eisenhower looked around the tarmac for his old friend, Gen. K. L. Berry, asking for him by his first name, K. L. The president's request to see Berry caught everyone off guard, because he had not been invited to the welcoming committee.[48]

Covering their embarrassment at this oversight, officials quickly found General Berry, who was on the grounds of the airfield, and brought him to the airstrip apron to greet the president. Later that day, when Eisenhower was settled in his quarters at Bergstrom, he met with only three visitors: Governor Daniel, Mayor Miller, and Maj. Gen. K. L. Berry. The politicians spent thirty minutes, listening as the president and Berry reminisced about their days at Fort Benning when Berry coached the football team and Eisenhower was his assistant.[49]

In October, Berry received two additional and very prestigious awards. The first was the Distinguished Service Medal (DSM) of the National Guard Association of the United States, the National Guard's highest award. The medal was "only the 43rd ever awarded anywhere," and General Berry was the first Texan to receive it.[50] General Berry's service to the state and the nation over a long period of time fit the criteria for the DSM perfectly.

> In order to be awarded the DSM an individual must have distinguished him/herself in particularly outstanding service to the United States Government, any of the Armed Forces of the United States, including the National Guard or NGAUS. Outstanding performance of normal duty alone will not justify the award of this medal. An individual must have

Longhorn Hall of Honor. Football teammates from 1914 and longtime friends Gus Dittmar (on Berry's right) and Clyde Littlefield (on Berry's left) congratulate Berry on his induction into the Longhorn Hall of Honor on November 13, 1959. Source: Author's private collection.

offered exceptionally outstanding contributions over an extended period of time to clearly merit the award of this medal. Contributions to state associations or to the state National Guard may be considered, but this award is intended to recognize national level contributions.[51]

The second honor for Berry in October came from the athletic side of his life. On October 24, 1959, the University of Texas named Berry as one of four of its all-time athletic stars who had been selected to the Longhorn Hall of Honor. The announcement about Berry in the *Austin American* newspaper read: "General Berry, who lettered in both football and track, has one of the most unique records in intercollegiate athletics. Captain of the 1915 team, he entered the service shortly thereafter, and did not return to the University until 1924, gaining all-Southwest Conference honors after a nine-year absence from the college gridiron."[52] Berry formally received this award at a dinner on November 13, 1959, with his good friends Gus Dittmar and Clyde Littlefield by his side.

THE END OF AN ERA

The years after the war had brought K. L. Berry gratifying recognition—he was a war hero and football legend. He was a friend of President Eisenhower. And he had given distinguished service in his long tenure as adjutant general of Texas. Nevertheless, in early January 1961, a rumor was circulating that Governor Price Daniel would not nominate Berry for an eighth term when his term expired on February 11, 1961. Because the adjutant general serves at the pleasure of the governor, Daniel had every right not to appoint Berry to serve again if he so chose. Yet Governor Daniel did not tell Berry directly that he wanted to appoint someone else. Instead, Daniel told Berry that he would not have the votes in the Texas senate to confirm him if Daniel put his name forward. Daniel's stated reason may well have been a way to avoid taking responsibility for his decision not to reappoint Berry. Privately, many of Berry's influential friends had done an unofficial count and found that Berry would be approved if the governor sent his name forward.[53]

If Daniel's rationale was not a dodge, opposition to Berry's reappointment may have come from the state senator from Travis County, Charles Herring, where Berry lived on Camp Mabry. To secure a positive vote on Berry's reappointment, Herring would have had to endorse Berry on the floor of the senate in the Texas legislature. Herring, however, opposed Berry's reappointment because Berry did not support Herring's bill in the state legislature to authorize building a public golf course at Camp Mabry. The issue of building a golf course at Camp Mabry had been simmering even before Governor Daniel appointed Berry for his seventh term in 1959.

A few years earlier, when Governor Daniel was a US senator from Texas (1952–1956), Berry had been able to get a bill through the US Congress to deed 189 acres of federal land adjacent to Camp Mabry to the State of Texas for use by the National Guard. Berry intended to use the acreage to improve training of his troops. By 1959, the City of Austin had expressed its desire to lease that same acreage for a golf course, and Daniels, now serving as governor of Texas, supported the golf course idea. Berry was strongly opposed to the idea. During Berry's seventh term in 1959, Officials in Washington, DC, sent him two questionnaires about the use of the 189 acres. Berry made it clear that the land was needed for training and a future helicopter site for the Texas National Guard. Berry later wrote to a friend that he had responded to the questionnaire "truthfully and officially" regarding the golf course matter. His opposition to the proposal may well have contributed to the failure of a bill in the US Congress concerning this National Guard acreage, "which had been proposed to permit the city of Austin to have the Camp Mabry land for 75 years for a golf course."[54]

The failure of the bill was an embarrassing blow to Governor Daniel, who had promised the golf course to the City of Austin. Understandably, Daniel was not happy that Berry had responded to the two questionnaires as he had done.

In a private letter to attorney Eldon Young on February 9, 1961, Governor Daniel explained why he was reluctant to reappoint Berry to another term as adjutant general in 1961. Governor Daniel stated that "serious opposition arose in the [Texas] Senate two years ago, and in order to obtain his confirmation, some of us working in his behalf represented that General Berry intended to retire after that term. In view of these commitments and present circumstances, I think it would be a mistake to insist on another reappointment. I know this could only result in embarrassment to General Berry and to all who have his interests at heart."[55] The "serious opposition" to which Daniel referred was about Berry's opposition to the golf course proposal. Apart from Daniel himself, it is not known who else "represented that General Berry intended to retire" after his seventh term.

As the governor delayed nominating Berry to an eighth term, he received literally hundreds of letters from all over Texas asking the governor to nominate Berry again. Even the American Legion and the Texas Guard Association lobbied the governor to reappoint him. Retired Col. John Roehm, who was president of the retired officers' association of over eight hundred members, wrote Governor Daniel expressing his concern that Berry had not been reappointed because of his opposition to the golf course: "We do not know the merits of the case but we are certain Gen. Berry's position was, as it always has been, absolutely honest and ethical. These two qualities we cannot afford to jeopardize in public office and I am sure you are making a thorough investigation into the matter."[56] General Berry was "eternally grateful" to all those supporters who wrote letters; hundreds of them came from his longtime friends from the 1917 First Officers Training Camp. He wrote thank you letters to each and every one of them.

The days and weeks dragged on with no decision on Berry's reappointment. Meanwhile, the golf course discussion heated up. Governor Daniel felt political pressure from members of the state senate to persuade his top general to sign the lease for the golf course. Berry refused to sign the lease agreement for the golf course as it was written because it did not reflect recommendations he had made for an acceptable resolution to the issue. Berry's refusal to sign handed Governor Daniel the excuse he needed not to reappoint him. After months of inaction on Berry's reappointment, Governor Daniel decided in May not to reappoint him. Disappointed, but a true military man to the core, Berry said in response to the governor's decision, "I had hoped to get the appointment, but the boss man has spoken."[57]

On June 30, 1961, his last day in office, Berry sent a farewell message to all his constituents in the Texas Guard:

> As I arrive at the end of my tenure of office as The Adjutant General of Texas, I wish to express my gratitude to all members of the Texas National Guard for the splendid support you have given me. I can assure you that it would not have been possible for me to carry out my duties if I had not had the help of so many capable men and women who have served the Texas National Guard with such dedication. The wholehearted devotion to the best interest of the services which you have shown has produced results of the greatest value to all of us.[58]

Berry went on to say he regretted closing the "most exciting chapter" of his life but that he had great satisfaction to have contributed to the defense of the country. He concluded his message by saying:

> I know you will continue to keep bright the torch of your dedication to the Guard and the Nation and hold it high. I am confident you will give my successor, General James E. Taylor, the same full measure of loyal support which you have always given me.[59]

Politics aside, Gen. K. L. Berry was and continues to be the longest-serving adjutant general of Texas. Governor Daniel noted this fact about Berry as he appointed his successor. "Gen. K. L. Berry had one of the most distinguished military careers of any Texan and his service as adjutant general has been the longest and most outstanding in the history of our National Guard. Upon retirement, he will be breveted a lieutenant general and appointed chairman of the governor's state guard advisory board."[60]

CHAPTER 19

An Honorable Life

As General Berry's final term as adjutant general drew to a close, in Washington, DC, President Eisenhower finished his second term, and John F. Kennedy was sworn into office as the 35th president of the United States. Berry's good Texas friend Lyndon Johnson became vice president. New challenges for a new generation lay ahead: Troubles in Cuba and Southeast Asia were brewing, as communism gained ground in both areas of the world; the Russians began the construction of the Berlin Wall, closing off the communist sector of Berlin from the democratic sectors of the west; Freedom Riders in the United States, advocating for equal rights for Black Americans, rode through the South and were attacked by people opposed to equal rights for all; and the new president called on the United States space program to send a man to the moon.

New challenges lay ahead for General Berry, as well. The University of Texas, his alma mater, sought out this well-known and highly regarded man to be president of its newly established Forty Acres Club. Berry wrote to his cousin Sam Riley in June, "I'm going to become connected with the new Forty Acres Club when it opens this fall so I will have my day pretty well filled. The Club will be five stories and a real swank affair. It caters to University of Texas faculty members, ex-students and Friends of the University. We already have some 1400 members and hope to open in December."[1] To his friend General Holt Atherton, he wrote, "Perhaps you would want to join as we will not only have a 12,000 square foot club floor with a fine kitchen and dining room but we will have two floors of hotel rooms available for members."[2] He was already recruiting members.

Berry was happy to accept the honor to be its first president, saying:

> I was proud of the University when in my freshman year there were 1800 students. Today, there are 18,000 students. The choice of the name "Forty Acres Club" is very meaningful to early graduates of the University of Texas—it was almost exactly one mile around the original 40 acre campus. Early University of Texas track teams would work out by running the "40 acres." The name has remained a part of the University and today describes the very heart of a University campus that has expanded many times the original acres.[3]

Berry understood how significant it would be for the future of the Forty Acres Club to claim one of the most famous people in the world as an honorary member. To that end, Berry wrote to his old friend and newly retired US president, Dwight Eisenhower, asking him to be the Forty Acres Club "No.1 Honorary Member." Eisenhower accepted with pleasure. How could the club not be a success with Berry and Eisenhower in the mix? The original Forty Acres Club expanded over the years to become the Forty Acre Society, the University of Texas Foundation and fundraising arm of the university. Berry would have been amazed and pleased to know that with the support of the Forty Acres Club, the Austin campus has grown to over fifty thousand students today.[4]

Once he retired, Berry and Alice were required to move off the Camp Mabry base, their home for fourteen years. Even though he had strong ties with San Antonio, Berry decided to stay in Austin so he could fulfill his new duties as chairman of the governor's State Guard Advisory Board and president of the Forty Acres Club. In addition, he wanted to be close to his precious hunting ranch in Burnet County. He and Alice had too many friends and irons in the fire in Austin to return to San Antonio. The couple chose a house two blocks away from Camp Mabry.

Although Berry had made a graceful segue into retirement with his new responsibilities, the stress of the previous several months overcame him on September 22, 1961. He was taken to Fort Sam Houston Brooke General Hospital and admitted for hypertension. His condition worsened as the hypertension developed into heart problems. By late October, he had shown improvement, but no one knew when he would be able to go home. He stayed in the hospital for several weeks. As he worked on getting healthy, he stepped down from being president of the Forty Acres Club.[5]

Berry finally came home in December just in time for Christmas. His son Col. K. L. Berry Jr., stationed at Stewart Air Force Base in New York, who was understandably concerned about his father, brought his family for a holiday visit a few days after Christmas. In January after Berry's family returned to New York, he had a heart attack and was hospitalized at Brackenridge Hospital's intensive care unit in Austin. He remained in critical condition for several weeks and was not expected to live. But Berry defied the odds and returned home in the spring.

As Berry convalesced at home, he heard good news from his son K. L. Berry Jr. Kaye Jr., as his family called him, became base commander of Paine Field in Everett, Washington. Being a base commander was seen as an essential step toward becoming a general, a goal that forty-three-year-old Colonel Berry had set for himself. Senior Berry was very proud of his son and hoped he would get his star someday. Tragedy derailed both father's and son's dreams. Two months after becoming base commander, Kaye Jr. perished in an airplane crash on July 29, 1962. The heartbreak of losing their first-born son affected General Berry and Alice in immeasurable ways. Although their son had been in the Air Force, the Berrys decided to bring K. L. Jr. home to Texas and bury him at Fort Sam Houston, an army base instead of at the Air Force Academy in Colorado Springs. Hundreds of General Berry's friends attended the sad event. A despondent and declining Berry went home to Austin where he and Alice settled into quiet life, all the while monitoring his health.

A year later, the Berrys' peaceful life was unsettled by events in the wider world. In November 1963, an unthinkable calamity struck the nation when President Kennedy was assassinated in Texas. For days, Americans were glued to their television sets watching news clips of Kennedy's motorcade and assassination. Adding to the horror was the murder of Lee Harvey Oswald, Kennedy's assassin, by Jack Ruby on live television. It must have been with very mixed feelings that Berry watched with the rest of the world as his old friend Lyndon B. Johnson was sworn into office as the 36th president of the United States in the immediate aftermath of the assassination. The nation watched the peaceful transfer of power in the wake of an enormous tragedy.

In 1964, an ailing Berry watched from his living room as President Johnson's presidency became entangled in Vietnam fighting communism. Berry had always worried about the spread of communism, so he most likely approved of Johnson's attempt to stop the communist takeover of Vietnam. Also, at the same time, Johnson worked on getting Congress to pass the Civil Rights Act and other Great Society ideas. Segregation was ingrained in the Texas way of life, so it must have been with admiration that Berry watched Johnson pursue desegregation

1914 Revisited. The reunion of the 1914 Longhorns at the Forty Acres Club, October 30, 1964. Back row, left to right: K. L. Berry, H. H. Neilson, Coke Wimmer, Bert Walker, Len Barrell. Front row, left to right: Clyde Littlefield, James H. Goodman, Gus Dittmar, W. S. Birge, Alva Carlton. Source: Author's private collection.

on a national scale. Johnson had already integrated the solidly white Forty Acres Club when he brought his staff to a New Year's Eve party there shortly after becoming president. One of his staff members was an African American woman. From then on it was an integrated club.

In October of 1964, disregarding his weak condition, Berry attended a special occasion—the fifty-year anniversary and reunion of the famous 1914 University of Texas football team. During halftime of the game between UT and Southern Methodist University (SMU), the 1914 football team members, all in their late sixties or early seventies, walked onto the field to be recognized as one of the university's best teams. UT dignitaries briefly recounted the history of their victorious season. Berry's grandson, King Reilly, remembered, "I got to meet Len Barrell, whose 121 total points scored in the 1914 season was the Horn single-season gridiron record until 1997! The Billingsley Report Ratings, recognized by the NCAA, declared the undefeated Horns won the 1914 national title."[6] Many years later, the team's achievements came to be more fully appreciated when the university's H. J. Lutcher Stark Center for Physical Culture and Sport showcased the team in a museum exhibit. After the UT-SMU game, the teammates met at the Forty Acres Club for dinner and photos. They posed as a team, happily assuming their football stances one last time.

In April, Berry suffered another heart attack and was rushed the eighty miles to Fort Sam Houston's Brooke General Hospital. He died five hours later with Alice by his side. Gen. K. L. Berry was buried on April 30, 1965, next to his son Col. Kearie Lee Berry Jr. in the Fort Sam Houston National Cemetery. Texas Governor John Connally honored Berry by issuing this statement:

> Every Texan who knows of K. L. Berry's service to his state and nation will mourn his passing. He will be remembered as one of our most famous athletes, courageous soldiers and distinguished public officials.[7]

Epilogue

The accolades for my grandfather continued after his death. On May 4, 1965, just one week after he passed away, the Texas House of Representatives passed Concurrent Resolution 110, introduced by Richard Slack. It read as follows:

> WHEREAS, Taps sounded across the State on Tuesday, April 27, 1965, on the death of one of Texas's most beloved and distinguished military heroes, Lieutenant General Kearie Lee Berry, 72-year-old retired general who survived the infamous Bataan Death March and led Texas National Guardsmen as the Adjutant General of Texas for 14 years; and
>
> WHEREAS, General Berry was a native of Denton and first served his country in the Mexican border action in 1916, a tour which interrupted his studies at The University of Texas; and
>
> WHEREAS, During his college years he was a versatile athlete, lettering at the University in football, track and wrestling, and holding the Southwest Conference heavyweight wrestling championship in 1915 and 1916; he was named to the Longhorn Hall of Honor in 1959 by former and present lettermen at the University; he also served as vice-president of the University's student body in 1915–1916; and
>
> WHEREAS, After he was commissioned in the Regular Army, he served a series of overseas tours in Vladivostok, Siberia, the Philippines and Hawaii, before returning to the United States in 1921; at the outbreak of World War II, General Berry won fame as Commander of the 1st Regular

Division (the Philippine Army) in the Japanese siege of the Islands, and his division fell with the surrender of Bataan on April 9, 1942; and

WHEREAS, For 40 months he was a prisoner of the Japanese in Luzon, Formosa, Kyushu and Manchuria; his outstanding miliary service brought many decorations—the Silver Star, Purple Heart, Distinguished Service Cross and Distinguished Service Medal—but his greatest pride was in the love and respect held for him by those under his command; and

WHEREAS, Three governors appointed General Berry to seven successive terms as Adjutant General—from 1947 to 1959—and this last tour ending with his retirement in Austin, left an indelible mark of accomplishment in state service matched only by the unstinting devotion and service to his country which he gave during 30 years of active Army duty; and

WHEREAS, He is mourned by the thousands of soldiers and guardsmen who served under him; and by a grateful State and Nation, as well as by his wife, Alice; his daughter, Mrs. William Reilly of Germany; his son, Tom Berry of Houston; his sisters, brothers and grandchildren; and

WHEREAS, Members of this Fifty-ninth Legislature, for themselves and on behalf of the people of Texas, wish to express appreciation for the outstanding service of this great military leader, devoted public servant and friend; now, therefore, be it

RESOLVED by the House of Representatives of the State of Texas, the Senate concurring, That the Texas Legislature by this Resolution pays tribute to a courageous soldier and distinguished public official, Lieutenant General Kearie L. Berry; and, be it further

RESOLVED, That copies of this Resolution be prepared for members of his family as an expression of sympathy from the Legislature and the people of Texas, and that when the House of Representatives and the Senate adjourn this day that they do so in memory of Lieutenant General Kearie L. Berry.[1]

Today, one of those copies of the resolution mentioned above hangs proudly in my office where I wrote this biography. Next to the resolution is a photo of my grandfather in his prime.

In addition to the concurrent resolution, the National Guard Association of Texas established an award for heroism in my grandfather's name—The General K. L. Berry Award. The criteria for the award were as follows:

> The General K. L. Berry Award is a reflection of heroism involving risk of life or personal safety. Major General Berry was a highly decorated veteran of WWII and the first post WWII Adjutant General of Texas. He was a prisoner of war of the Japanese until the end of WWII.

ELIGIBILITY:
- Member or former member of the Texas Military Forces.
- Member of NGAT.
- For a heroic act at the risk of life not in the line of duty.[2]

Fifty-two years after Berry's death, while I was conducting research for this book, he received another recognition. On November 10, 2017, Berry's athletic legacy was honored by his alma mater in a story that appeared in the *Standing Sentinel*, a publication of the athletics department at the University of Texas.

> One more time Saturday, Darrell K Royal-Texas Memorial Stadium will pause to remember its purpose. When it was dedicated in 1924, the stadium was named "Texas Memorial Stadium" in honor of those Texans who had died in World War I—termed, at the time, "the Great War." It was, they said, "the war to end all wars."
>
> But it did not turn out that way. There would be other wars, and other Longhorn athletes who would serve bravely in them. In 1977, they rededicated the stadium to all those veterans who served the United States of America in any foreign conflict. When Darrell Royal's name was added to the stadium in 1996, he agreed, but insisted it would be only with the stipulation that a stadium veterans committee be appointed to maintain the original standards and purpose of the facility's initial charge.
>
> Each season, on or about Veterans Day, we pause to remember those who served. And while each has their own story, and each story has contributed to the legacy of the stadium, it would be hard to match the odyssey of Kearie Lee Berry.

The story went on to chronicle Berry's remarkable life as an athlete, as a soldier, and as a war hero for the Veterans Day crowd at the football game. The crowd was asked to remember and honor him.

> The message K. L. Berry and the others left us is that the only way to combat fear is with faith; the only remedy for despair is resolve. And the only answer to depression is hope. And it is in that space that we honor, celebrate, and remember.[3]

It has been both an honor and a revelation for me to tell the story of the man who was both my grandfather and someone whose legacy of contributions to his university, his state, and his nation will secure him a place in history. May all his descendants live with as much optimism, sense of duty to country, and zest for life as K. L. Berry had.

Dana Berry Frazee

APPENDIX A

Birthday Poems to K. L. Berry

Birthday Poem from Room 7

   The birthday of our friend K. L.
   Is just the day to wish him well,
   And hope the 50 next in line,
   Has him well fed and feeling fine.
   That when his next one rolls around,
   Back in the States he will be found.
   And in the shape as when he played
   On Texas, as a young gay blade.

Birthday Poem from Harrison Browne

   The smoke of fifty years went dim
   The ardor of a hunter's vim,
   If he can fill his frame with chow
   And get his guns, he'd pack right now.
   So smoke and rest until we get
   The trips we've planned supplied and set.

A Second Birthday Poem from Harrison Browne

   A birthday greeting I would send
   A greeting bright and merry
   To one who's been a damn good friend
   Here's luck to K. L. Berry

APPENDIX B

Books Read by K. L. Berry While Interned as a Prisoner of War

MAGAZINES	
Title	**Date, if noted**
Readers Digest	1939–1941 several issues
National Geographic	several issues
*World's Digest	
BOOKS	
Title	**Author, Year**
Padlocked	Rex Beach, 1925
The Eagles Gather	Taylor Caldwell, 1940
Ripley's Believe it or Not	Robert Ripley, 1929
Give Me Death	Isabel Briggs Myers, 1934
Girl on His Hands	Maysie Greig, 1939
The Smuggler's Cave and Other Stories	Grace Stebbing, 1895 or 1923
*Some Experiences of a New Guinea Magistrate	Charles Arthur Whitmore Monckton, 1920
Boon	H. G. Wells, 1915
Book of Short Stories	Arthur Conan Doyle, unidentified year
Why Britain is at War	Harold Nicholson, 1939
No Other Tiger	A. E. W. Mason, 1927

continued

Title	Author, Year
A Sketch of Chinese Arts and Crafts	Hilda Arthurs Strong, 1923
The Temple of Costly Experience	Daniele Vare, 1939
In Malay Forests	George Maxwell, 1907
*News from Tartary	Peter Fleming, 1936
Last Flight	Amelia Earhart, 1937
The Glorious Adventure	Richard Halliburton, 1927
Return via Dunkirk	Gun Buster (pseudonym), John Austin, 1940
*Mr. Emmanuel	Louis Golding, 1938
Grapes of Wrath	John Steinbeck, 1939
Boon's Geography (perhaps, *A Commercial Geography of Foreign Nations*)	F. C. Boon, 1901
A Passage to India	E. M. Forster, 1924
*The Maker of Heavenly Trousers	Daniele Vare, 1935
Travels in Tartary	Peter Fleming, 1941
Men Against Death	Paul de Kruif, 1932
Huckleberry Finn	Mark Twain, 1884
*Night in the Hotel	Eliot Crawshay-Williams, 1931
Inside Asia	John Gunther, 1939
Dew and Mildew: Semi-Detached Stories from Karabad, India	P. C. Wren, 1912
Evolution (perhaps, *On Evolution: The Development of the Theory of Natural Selection*)	Charles Darwin, 1859
The Red Pony	John Steinbeck, 1937
Of Mice and Men	John Steinbeck, 1937
The Man Who Did the Right Thing	Sir Harry Johnson, 1921
The Yellow Poppy	Dorothy Kathleen Broster, 1920
Walden	Henry David Thoreau, 1854
Black Blood	Unidentified author or date
Lost Horizon	John Hilton, 1933
Unidentified Rudyard Kipling Book	Rudyard Kipling
Corvars	Unidentified author or date

Selected Poems of American Poets	Unidentified book or date
Crocus	Neil Bell, 1939
*A Warning to Wantons	Mary Mitchell, 1934
Will Love and Irving	Unidentified author or date
Book on botany	Unidentified author or date
The Story of San Michele	Axel Munthe, 1929
Cody's little book on "Grammar" (perhaps, *The Art of Writing and Speaking the English Language: Dictionary of Errors*)	Sherwin Cody, 1905
The Gentleman in the Parlour: A Record of a Journey from Rangoon to Haiphong	Somerset Maugham, 1935
The Treasure Train	S. A. Van Patten, 1916
Count Belisarius	Robert Graves, 1938
Wuthering Heights	Emily Brontë, 1847
Maneuver In War	Charles Andrew Willoughby, 1939
The Next World War (perhaps, *Between 2 Wars? The Lessons of the Last World War in Relation to the Preparations for the Next*)	"Vigilantes" (K. Zilliacus), 1939
Treasure Island	Robert Louis Stevenson, 1883
Kidnapped	Robert Louis Stevenson, 1886
Typee	Herman Melville, 1846
Restless is the River	August Derleth, 1939
*Crime and Punishment	Fyodor Dostoevsky, 1867
The Bible	

*Books K. L. Berry especially liked

APPENDIX C

Letter from Capt. Jack Walker to K. L. Berry

June 17, 1945
K. L. Berry
Colonel, USA

Dear Sir:

During my eight years active duty with the regular army I have had the opportunity of serving under a number of Commanding Officers. But never have had the privilege of serving under one whom I have esteemed more as both a friend and officer. It was a joy and inspiration to have been on laison [sic] duty with you in your capacity as regimental and then division commander. It was with seeming disregard of your personal safety that you often visited the front lines during times of conflict. From the officers and men of the division I constantly heard praize [sic] of your leadership. Many times I have seen you bolster the courage, spirit, and morale of the men by your leadership and thereby secure a position that would have otherwise been lost. Indeed it is my opinion that without your leadership the division would have been a mob of armed rabble rather than a determined fighting unit. Also the hardships we have undergone as prisoners of war have been the test that could not be passed by many, but your leadership and friendship are still desired by all. It is my hope and desire to have more commanding officers like you.

Jack. K. Walker
Capt. 24th F.A. (PS)
Ln O 1st Reg. Div. I Corps

NOTES

CHAPTER 1

1. Henry F. Graff, "Grover Cleveland: Domestic Affairs," The Miller Center, University of Virginia Miller, accessed October 11, 2022, https://millercenter.org/president/cleveland/domestic-affairs.

2. Henry F. Graff, "Grover Cleveland: Impact and Legacy," The Miller Center, University of Virginia, accessed October 11, 2022, https://millercenter.org/president/cleveland/impact-and-legacy.

3. *1850 United States Federal Census*, US Census Bureau, National Archives in Washington, DC; Records of the Bureau of the Census, record group 29, series M432, residence date 1850; Home in 1850: Summerfield, Dallas, Alabama, roll 4, pg. 322a. Ancestry.com website, accessed February 20, 2019; *1860 United States Federal Census*, US Census Bureau, National Archives in Washington DC, Records of the Bureau of the Census, record group 29; series M653, residence date: 1860; Home in 1860: Division 1, Montgomery, Alabama, roll M653_19, pg. 249; Family History Library Film: 803019, Ancestry.com website, accessed February 21, 2020; *1870 United States Federal Census*, US Census Bureau, National Archives in Washington, DC; Records of the Bureau of the Census; Census Place: Township 20, Elmore, Alabama; roll M593_15, pg. 139A, image 470; Family History Library Film: 545514, Ancestry.com website, accessed February 21, 2020. "General Berry, Pacific War Hero, Born on a Farm near Bolivar," *Denton Record-Chronicle*, August 3, 1953.

4. John Riley studied medicine at Ohio Medical College, not the University of Pennsylvania. Margaret Mayer, "World War I Buddy of Jester: Death March Survivor New Adjutant," *Austin American-Statesman*, May 7, 1947.

5. Viola R. Berry, *The Alamo and Other Poems* (News Publishing Company, 1906) 105–114, 140.

6. Berry, *The Alamo and Other Poems*, 105–114, 140.

7. Mike Cochran, "A Brief History of Denton County," The History of Denton, Texas, website, accessed April 13, 2019, http://dentonhistory.net/.

8. Cochran, "A Brief History of Denton County."

9. "General Berry, Pacific War Hero, Born on a Farm near Bolivar," *Denton Record-Chronicle*, August 3, 1953.

10. Dan Oko, "Teddy Roosevelt, San Antonio, and the Birth of the Rough Riders: Texas Cowboys in Cuba," *Texas Highways*, November 2019, https://texashighways.com/culture/history/teddy-roosevelt-san-antonio-and-the-birth-of-the-rough-riders/.

11. Sidney Milkis, "Theodore Roosevelt: Foreign Affairs," The Miller Center, University of Virginia, accessed November 24, 2022. https://millercenter.org/president/roosevelt/foreign-affairs.

12. E. Dale Odom, "Denton, TX (Denton County)," in *Handbook of Texas Online* (Texas State Historical Association,) updated September 9, 2020, https://www.tshaonline.org/handbook/entries/denton-tx-denton-county.

13. *The Bronco, Yearbook of Denton High School 1911* (Denton High School, 1911) 33, accessed through The Portal to Texas History website, https://texashistory.unt.edu/ark:/67531/metapth743014/?q=1911%20Denton%20High%20School%20yearbook; "General Berry, Pacific War Hero, Born on a Farm near Bolivar," *Denton Record-Chronicle*, August 3, 1953.

14. *The Bronco, Yearbook of Denton High School 1911*, 83.

15. "General Berry, Pacific War Hero, Born on a Farm near Bolivar," *Denton Record-Chronicle*, August 3, 1953. Berry wrote most of what appeared in this article and submitted it at the newspaper's request. The paper took almost verbatim what Berry wrote about himself; "horny-handed" quote is from K. L.'s personal papers. It must have been and still must be difficult to be Kearie Berry. Kearie Lee Berry Jr. called himself Kaye and Kearie Lee Berry III calls himself K. L. A great nephew, Kearie John Berry, is called K. J. Only one namesake uses the full Kearie Lee Berry in her name, and she is Kearie Lee Berry IV, great-granddaughter of Kearie Lee Sr.

16. "Woodrow Wilson," The White House website, accessed December 10, 2022, https://www.whitehouse.gov/about-the-white-house/presidents/woodrow-wilson/.

17. "General Berry, Pacific War Hero, Born on a Farm near Bolivar," *Denton Record-Chronicle*, August 3, 1953.

18. *Cactus Yearbook, 1913* (University of Texas at Austin, Texas Scholar Works, University of Texas Libraries) 225, 227, 229, http://hdl.handle.net/2152/61703.

19. "1914: A Perfect Season," University of Texas Stark Center Online Exhibit, accessed May 9, 2019, https://www.starkcenter.org/1914-a-perfect-season/.

20. *Cactus Yearbook, 1915* (University of Texas, at Austin, Texas Scholar Works, University of Texas Libraries) 78, http://hdl.handle.net/2152/24283.

21. "1914: A Perfect Season."
22. "1914: A Perfect Season."
23. *Cactus Yearbook, 1916,* (University of Texas, Texas Scholar Works, University of Texas Libraries,1916) 323, 281. http://hdl.handle.net/2152/23763.
24. *Cactus Yearbook, 1916,* 323.
25. *Cactus Yearbook, 1915,* 100.
26. *Cactus Yearbook, 1916,* 305.
27. *Cactus Yearbook, 1916,* 262.

CHAPTER 2

1. Robert C. Overfelt, "Mexican Revolution," in *Handbook of Texas Online* (Texas State Historical Association), updated December 2, 2020, https://www.tshaonline.org/handbook/entries/mexican-revolution; Arnoldo de León and Robert A. Calvert, "Segregation," in *Handbook of Texas Online* (Texas State Historical Association), updated January 27, 2021, https://www.tshaonline.org/handbook/entries/segregation.
2. Mitchell Yockelson, "The United States Armed Forces and the Mexican Punitive Expedition: Part 1," *Prologue Magazine,* Winter 1997, National Archives and Records Administration (NARA), accessed September 2, 2021, https://www.archives.gov/publications/prologue/1997/winter/mexican-punitive-expedition-2.html; Fritz L. Hoffmann, "Villa, Francisco [Pancho]," *Handbook of Texas Online* (Texas State Historical Association), updated: July 2, 2016, https://www.tshaonline.org/handbook/entries/villa-francisco-pancho.
3. Yockelson, "The United States Armed Forces and the Mexican Punitive Expedition"; Alexander F. Barnes, "On the Border: The National Guard Mobilizes for War in 1916," *Army Sustainment Magazine,* March–April 2016, accessed December 12, 2022, https://www.army.mil/article/162413/on_the_border_the_national_guard_mobilizes_for_war_in_1916; Leon C. Metz, "Fort Bliss," *Handbook of Texas Online* (Texas State Historical Association), updated October 3, 2019, https://www.tshaonline.org/handbook/entries/fort-bliss.
4. Vaughn R. Larson, "Wisconsin Guard Protected Mexican Border a Century Ago," Wisconsin National Guard, June 24, 2016, https://ng.wi.gov/news/16063; Lonnie J. White, "The 36th Division in World War I," 1999 Military History Associates, Inc. 1984, http://www.texasmilitaryforcesmuseum.org/36division/archives/wwi/white/chap1.htm.
5. K. L. Berry to Howard V. Ratliff, August 15, 1955, in the author's private collection.
6. Larson, "Wisconsin Guard Protected Mexican Border a Century Ago."
7. Berry to Ratliff, August 15, 1955.

8. "Texas National Guard," *Historical and Pictorial Review: National Guard of the State of Texas, 1940*, 44, accessed through the Portal to Texas History website, crediting Boyce Ditto Public Library, https://texashistory.unt.edu/ark:/67531/metapth833790.

9. Berry to Ratliff, August 15, 1955.

10. White, "The 36th Division in World War I."

11. S. W. Pope, "An Army of Athletes: Playing Fields, Battlefields, and the American Military Sporting Experience, 1890-1920," *The Journal of Military History*, 59, 3 (1995): 436, 442, 446, https://doi.org/10.2307/2944617.

12. Pope, "An Army of Athletes."

13. Pope, "An Army of Athletes."

14. Berry to Ratliff, August 15, 1955.

15. "Winter Texan Wednesday: The Story of the 2nd Infantry Football Team." Museum of South Texas History, February 6, 2019, https://mosthistory.org/events/winter-texan-wednesday-the-story-of-the-2nd-infantry-football-team/.

16. White, "The 36th Division in World War I."

17. *The Reveille*, vol. 1, no. 35, edited by G. B. Wilson, Company B, Third Texas. (Office of Nueces County News, November 27, 1916).

18. David Trask, "The Entry of the USA into the War and Its Effects," in *The Oxford Illustrated History of the First World War*, edited by Hew Strachan (Oxford University Press, 2014), 238–239.

19. White, "The 36th Division in World War I."

20. White, "The 36th Division in World War I"; Barbara W. Tuchman, *The Zimmermann Telegram: American Enters the War, 1917-1918* (Random House, 2014), 164–167.

21. Art Leatherwood, "Leon Springs Military Reservation," *Handbook of Texas Online* (Texas State Historical Association), updated March 1, 1995, https://tshaonline.org/handbook/online/articles/qbl06--leatherwood.

22. "World War I: Short History of the 90th Division at Camp Travis," The 90th Division Association website, accessed May 21, 2019, https://90thdivisionassoc.org; Gus C. Dittmar, *They Were First, Recollections of The First Officers Training Camp of Leon Springs, Texas, May 8 to August 15, 1917* (The Steck-Warlick Company, 1969), 4.

23. Dittmar, *They Were First*, 6.

24. "The Defining Role of the National Guard in WWI," National Guard Bureau Historical Services, August 7, 2017, https://www.army.mil/article/191849, accessed May 21, 2019; "First Officers Training Camp," Marker 11744, Texas Historic Markers website, accessed June 13, 2019, https://texashistoricalmarkers.weebly.com/first-officers-training-camp.html; *US Select Military Registers, 1862–1985*, K. L. Berry, Ancestry.com. website, Lehi, Utah, Accessed February 20. This collection was indexed by Ancestry World Archives Project contributors from the original

data at US *Military Registers, 1902–1985*. Salem, Oregon: Oregon State Library, 2019. "World War I Buddy of Jester: Death March Survivor New Adjutant," *Austin American-Statesman*, May 7, 1947.

CHAPTER 3

1. "World War I Draft: Topics in Chronicling America," Library of Congress Research Guides website, accessed May 23, 2025, https://guides.loc.gov/chronicling-america-wwi-draft.

2. "The Formative Period: Formation of the Division and Training at Camp Travis," 90th Infantry Division Association website, accessed May 21, 2019, http://www.90thdivisionassoc.org/90thDivisionFolders/mervinbooks/WWI90/WWI9002/WWI9002.htm. The record of Lieutenant Berry's posting to California is in the Texas Military Forces Museum, Camp Mabry, Texas. It was sent to the author by Lisa Sharik, deputy director of the museum, in 2020.

3. "21st Infantry Regiment," 25th Infantry Division Association website, accessed November 14, 2021, https://www.25thida.org/units/infantry/21st-infantry-regiment/.

4. *Passenger Lists 1910–1939*, US Army Transport Service, K. L. Berry, June 5, 1919, National Archives at College Park; College Park, Maryland; Records of the Office of the Quartermaster General, 1774–1985; record group 92; roll or box 571, Departure Place: San Francisco, Ancestry.com, accessed February 20, 2019.

5. *Passenger Lists 1910-1939*, US Army Transport Service, K. L. Berry, 1920, National Archives at College Park; College Park, Maryland; Records of the Office of the Quartermaster General, 1774–1985; record group 92; roll or box 335, Ancestry.com website, accessed on June 5, 2025.

6. Alexander F. Barnes and Cassandra J. Rhodes, "Logistics in Reverse: The US Intervention in Siberia, 1918–1920," in *US Intervention in Siberia and Northern Russia 1918-1920*, Progressive Management Publications, 2019, 37–46; Gibson Bell Smith, "Guarding the Railroad, Taming the Cossacks: The US Army in Russia 1918–1920," *Prologue Magazine*, found on the website of the National Archives and Records Administration, accessed September 9, 2019, https://www.archives.gov/publications/prologue/2002/winter/us-army-in-russia-1.html.

7. O. Edmond Clubb, *Twentieth Century China* (Columbia University Press, 1964), 88; Erick Trickey, "Forgotten Doughboys Who Died Fighting in the Russian Civil War," *Smithsonian Magazine*, February 12, 2019, https://www.smithsonianmag.com/history/forgotten-doughboys-who-died-fighting-russian-civil-war-180971470/; Robert L. Smalser, "The Siberia Expedition 1918–1920: An Early 'Operation Other than War,'" in *US Intervention in Siberia and Northern Russia 1918–1920*, (Progressive Management Publications, 2019), 64.

8. Smith, "Guarding the Railroad, Taming the Cossacks"; "Wolfhounds," Wolfhound Alumni website, accessed November 2, 2021, http://www.vn.alumni.wolfhoundsonline.org/images/siberia.htm; "An Army With No Country: the Czechoslovak Legion in Europe," on the website of Czech Center Museum Houston, November 11, 2021, https://www.czechcenter.org/blog/2021/11/11/an-army-with-no-country-the-czechoslovak-legion-in-europe.

9. Trickey, "Forgotten Doughboys"; Smith, "Guarding the Railroad, Taming the Cossacks."

10. Barnes and Rhodes, "Logistics in Reverse," 39; Smith, "Guarding the Railroad, Taming the Cossacks"; Jaime Bisher, "American Expeditionary Force-Siberia (AEFS)," Cossack Warlords of the Trans-Siberian on Weebly website, 2016, https://cossackwarlords.weebly.com/american-expeditionary-force-siberia.html, accessed November 5, 2021.

11. Smalser, "The Siberia Expedition 1918–1920," in *US Intervention in Siberia and Northern Russia 1918-1920*, Progressive Management Publications, 2019, 74; Richard K. Kolb, "Walking on Eggs Loaded with Dynamite," *Veterans of Foreign Wars*, February 1991, 14–17; Barnes and Rhodes, "Logistics in Reverse," 40.

12. Christine L. Putnam, "AEF Siberia," Doughboy Center: The Story of the American Expeditionary Forces, accessed October 20, 2021, http://www.worldwar1.com/dbc/siberia.htm.

13. Putnam, "AEF Siberia"; Smalser, "The Siberia Expedition 1918–1920," 76; Kolb, "Walking on Eggs," 16.

14. Kolb, "Walking on Eggs,"15; "A Brief History of the 27th Regiment," The Wolfhound Pack, US 27th Infantry regimental Historical Society website, accessed September 27, 2019, https://wolfhoundpack.org/history-of-the-regiment/#brief_history.

15. K. L. Berry, Photo Captions, 1919. Several of Berry's photos had text on the back of them for Alice to read and learn about Siberia, in author's collection.

16. Berry, Photo Captions.

17. Clubb, *Twentieth Century China*, 91–92.

18. Berry, Photo Captions.

19. Berry, Photo Captions.

20. Barnes and Rhodes, "Logistics in Reverse," 39, 44; Kolb, "Walking on Eggs" 16; Michael C. Grieco, "Making Sense of the Unknown: The AEF in Siberia," *Monograph*. (US Army School of Advanced Military Studies, US Army Command and General Staff College, Ft. Leavenworth, Kansas), 2018.

21. Trickey, "Forgotten Doughboys,"

22. Barnes and Rhodes, "Logistics in Reverse," 45; Kolb, "Walking on Eggs," 17.

23. Clubb, *Twentieth Century China*, 93. Other sources note that Kolchak was handed directly over to the Bolsheviks by the Czechs.

24. *Passenger Lists 1910-1939*, US Army Transport Service, K. L. Berry, 1920, National Archives at College Park; College Park, Maryland; Records of the Office of the Quartermaster General, 1774–1985; record group 92; roll or box 335, Ancestry.com website, accessed on June 5, 2025; Bisher, "American Expeditionary Force-Siberia (AEFS)."

CHAPTER 4

1. Richard W. Stewart, ed., "The United States Army in a Global Era, 1917–2008" American Military History vol. II, second edition, 59, Center of Military History United States Army website, accessed May 24, 2025 https://www.armyupress.army.mil/Portals/7/educational-services/military-history/american-military-history-volume-2.pdf.

2. S. W. Pope, "An Army of Athletes: Playing Fields, Battlefields, and the American Military Sporting Experience, 1890–1920," *The Journal of Military History* 59, no. 3 (1995): 441, 436, accessed October 19, 2019, https://doi.org/10.2307/2944617.

3. Pope, "An Army of Athletes."

4. Invitation to the ball. Author's private collection.

5. "Schofield Barracks Army Base Guide," Military.com Network, accessed October 19, 2019, https://www.military.com/base-guide/schofield-barracksfort-shafter.

6. David Greenberg, "Calvin Coolidge: Domestic Affairs," The Miller Center, University of Virginia, accessed October 9, 2022, https://millercenter.org/president/coolidge/domestic-affairs.

7. *Longhorn T.* October 1, 1924. The *Longhorn T* was a newsletter about athletics in 1924. This information was found in a box with Alice Berry's scrapbooks at the Lutcher Stark Center at the University of Texas in Austin.

8. "Five of 'Doc' Stewarts's Longhorn Football Aces," *Fort Worth Record Sports*, October 14, 1924; Lloyd Gregory, "'Red' Sprague, Jim Marley and K.L. Berry Picked for A—Southwestern Eleven", *Austin Statesman*, November 30, 1924.

9. UT Certificate, Box contents at H. L. Lutcher Stark Center University of Texas.

10. "An Academic Certification for Berry, Kearie Lee 07/06/1893," University of Texas at Austin, Office of the Registrar, in author's collection; "Grandmother Takes Work at University," *Austin American-Statesman*, March 26, 1925, 3. In 1924, Berry's mother, Viola, earned her BA at Sam Houston State Teachers College and was attending UT in 1925 to earn a master's degree.

11. Sam Kinch, "Texas Adjutant General One of Eisenhower's Greatest Fans," *Fort Worth Star-Telegram*, June 15, 1952.

12. "The Early Years at Fort Benning," *Fort Benning: Home of the Infantry* booklet, chapter 3, p. 12. Chattahoochee Valley Libraries website, https://dlg.galileo.usg.edu/chattahoochee/do:cvl159.

13. "President Herbert Hoover," Herbert Hoover Presidential Library and Museum, National Archives, accessed December 10, 2022, https://hoover.archives.gov/hoovers/president-herbert-hoover; David E. Hamilton, "Herbert Hoover: Domestic Affairs," The Miller Center, University of Virginia, accessed October 9, 2022, https://millercenter.org/president/coolidge/domestic-affairs.

14. Alfred E. Cornebise, *The United States 15th Infantry Regiment in China, 1912–1938*, (McFarland & Company, Inc., 2004) 91, 167; *Passenger Lists 1910–1939*, US Army Transport Service. K. L. Berry, July 6, 1933, National Archives at College Park; College Park, Maryland; Records of the Office of the Quartermaster General, 1774–1985; record group 92; roll or box 605. Arrival Chinwangtao July 6, 1933. Ancestry.com website, accessed February 20, 2019.

15. Cornebise, *The United States 15th Infantry Regiment in China*, 14.

16. Cornebise, *The United States 15th Infantry Regiment in China*, 26–27; Bill Federer, "The Life and Times of Herbert Hoover; Warned against an Expanding Federal Government," World Tribune: Window on the Real World website, accessed August 10, 2019, https://worldtribune.com/life/the-life-and-times-of-herbert-hoover-warned-against-an-expanding-federal-government. Interestingly, future President Herbert Hoover and his wife were in Tientsin during the rebellion. He was a mining engineer in Tientsin at the time. During the rebellion, both Hoovers, who had learned Chinese, helped protect and care for people within the concessions as well as Chinese citizens escaping the violence.

17. Cornebise, *The United States 15th Infantry Regiment in China*, 26–27, 29, 55.

18. Adam Bisno, "The US Navy and the Sino-Japanese War of 1894–95," Naval History and Heritage Command, US Navy, accessed January 24, 2023, https://www.history.navy.mil/browse-by-topic/wars-conflicts-and-operations/steam-steel-navy/sino-japanese-war.html; "The Treaty of Portsmouth and the Russo-Japanese War, 1904–1905," Office of the Historian, Foreign Service Institute, US Department of State, accessed January 24, 2023, https://history.state.gov/milestones/1899-1913/portsmouth-treaty; Kenneth S. Latourette, *The Chinese: Their History and Culture* (MacMillan Company, 1934) 444; Cornebise, *The United States 15th Infantry Regiment in China*, 29, 49.

19. Cornebise, *The United States 15th Infantry Regiment in China*, 55, 60, 210; Latourette, *The Chinese*, 467.

20. Cornebise, *The United States 15th Infantry Regiment in China*, 211.

21. Jean Speece Kimber to the grandchildren of Helen DePass Dahlin, March 20, 1995, in author's private collection. Jean Speece, Helen DePass, and the Berry children were classmates at the Tientsin Grammar School in 1934–36. Jean wrote

Helen's grandchildren in 1995 describing the life she and other teenagers at Tientsin Grammar School experienced in Tientsin and Chinwangtao.

22. Desmond Power, *Little Foreign Devil* (Pangli Imprint, 1996), 92.

23. Tom Berry to Speece-Kimber Family, January 16, 2002, in author's private collection.

24. "Things Chinese," *The Sentinel* (Tientsin) XXII. no. 33, August 18, 1934.

25. *The Sentinel*, "Things Chinese."

26. Kimber to the grandchildren of DePass Dahlin, March 20, 1995.

27. Kimber to the grandchildren of DePass Dahlin, March 20, 1995.

28. "Col. Burt letter to Captain Kearie L. Berry, May 12, 1935," *The Sentinel* (Tientsin) XXIII, no. 20, May 18, 1935.

29. "Colonel Truesdell Radios Commendation on Work Of track & Field Squad," *The Sentinel* (Tientsin) XXIII, no. 21, May 25, 1935.

30. "Hunters Enjoy Successful Trip in Shansi," *The Sentinel* (Tientsin) XXIV, no. 25, December 21, 1935.

31. "US Troops in North China to Stay," *The Sentinel* (Tientsin) XXV, no. 17, April 25, 1936.

32. "Sailing of the Grant Next Week Takes Many Old-Times Home," *The Sentinel* (Tientsin) XXVI, no. 4, July 25, 1936; *Passenger Lists 1910–1939*, US Army Transport Service. K. L. Berry, July 28, 1936, National Archives at College Park; College Park, Maryland; Records of the Office of the Quartermaster General, 1774–1985; record group 92; roll or box 343. Departure Date: July 28, 1936. Ancestry.com website, accessed February 20, 2019.

33. Power, *Little Foreign Devil*, 96.

CHAPTER 5

1. "How Did Hitler Happen?" The National World War II Museum Organization website, accessed December 10, 2020, https://www.nationalww2museum.org/war/articles/how-did-hitler-happen; Liam Dale, *World War 2, The Call of Duty: A Complete Timeline* (The History Journals, 2019).

2. "Texas National Guard," *Historical and Pictorial Review: National Guard of the State of Texas, 1940*, XLI; Jason Dawsey, "A Shared Enmity: Germany, Japan, and the Creation of the Tripartite Pact," The National World War II Museum Organization website, accessed December 12, 2021, https://www.nationalww2museum.org/war/articles/germany-japan-tripartite-pact; Harry M. Henderson, *History of the 141st Infantry, 36th Infantry Division Texas National Guard* (The Naylor Company, 1950), 69.

3. "141st Infantry Regiment Win Hulen Trophy Match Second T," *San Antonio Evening News*, August 19, 1937; "Texas National Guard," *Historical and Pictorial Review: National Guard of the State of Texas, 1940*, 462.

4. William I. Hitchcock, *The Age of Eisenhower: America and the World in the 1950* (Simon and Schuster, 2018), 16; Bruce A. Olson, "Texas National Guard," *Handbook of Texas Online* (Texas State Historical Association), November 25, 1940, https://www.tshaonline.org/handbook/entries/texas-national-guard.

5. "Camp Bowie Army Base in Brownwood, TX.," Military Bases.com, accessed January 29, 2020, https://militarybases.com/texas/camp-bowie/; Lorene Bishop, "History of Camp Bowie," City of Brownwood, Texas website, accessed September 23, 2019, http://www.brownwoodtexas.gov/323/History-of-Camp-Bowie.

6. Bishop, "History of Camp Bowie." An interesting item about Camp Bowie during this time period: Construction at Camp Bowie was underway during most of Berry's time at Bowie and rains fell from October 1940 to June 1941. The official rain totaled 19.5 inches. With the sixty miles of dirt roads built and the laying of utility lines along these roads, the soil became very soft. The slow rain that fell over a period of days resulted in the grounds being very muddy. Workers began calling the camp "Camp Gooie" instead of Camp Bowie.

7. William Manchester, *American Caesar: Douglas MacArthur 1880–1964* (Little Brown & Company, 1978), 175.

8. James MacGregor Burns, *Roosevelt: The Lion and the Fox* (Harcourt Brace, 1956), 440–441, 457, 109.

9. William Bartsch, *December 8, 1941: MacArthur's Pearl Harbor* (Texas A&M University Press, 2003), 94.

10. Louis Morton, *US Army in World War II: The War in the Pacific: The Fall of the Philippines* (Center of Military History United States Army, 1989), 71.

11. Sam Kinch, "Texas Adjutant General One of Eisenhower's Greatest Fans," *Fort Worth Star-Telegram*, June 15, 1952; "General Walter Krueger," Military Hall of Honor website, https://militaryhallofhonor.com/honoree-record.php?id=268.

12. William I. Hitchcock, *The Age of Eisenhower: America and the World in the 1950s* (Simon and Schuster, 2018), 16; Manchester, *American Caesar*, 166 "studied dramatics."

13. Manchester, *American Caesar*, 183; Kinch, "Texas Adjutant General One of Eisenhower's Greatest Fans," 8.

14. K. L. Berry to Alice Berry, October 30–November 30, 1941, in author's private collection.

15. K. L. Berry to Alice Berry.

16. K. L. Berry to Alice Berry.

17. K. L. Berry to Alice Berry.

18. K. L. Berry to Alice Berry.

19. K. L. Berry to Alice Berry.

20. K. L. Berry to Alice Berry.

21. K. L. Berry to Alice Berry.

22. K. L. Berry to Alice Berry.
23. K. L. Berry to Alice Berry.
24. K. L. Berry to Alice Berry.
25. K. L. Berry to Alice Berry.
26. K. L. Berry to Alice Berry.
27. John Whitman, *Bataan: Our Last Ditch* (Hippocrene Books, 1990), 33.
28. Morton, *US Army in World War II*, 71.
29. "Story of the 57th Infantry," Col. Edmund J. Lilly Jr. P.O.W. Papers, notebook 19, Center for Research: Allied POWS Under the Japanese website, accessed February 19, 2019, http://www.mansell.com/pow_resources/camplists/china_hk/mukden/edmund_lilly/Lilly_Diary_Vol19_TEXT_Story_of_the_57th_Infantry.html.

CHAPTER 6

1. Duane Schultz, *Hero of Bataan: The Story of General Jonathan M. Wainwright* (St. Marten's Press, 1981), 32.
2. Louis Morton, *US Army in World War II, The War in the Pacific, The Fall of the Philippines* (Center of Military History United States Army, 1989), 62–63, "prolonged defense"; Schultz, *Hero of Bataan*, 52, "after the outbreak of war."
3. Schultz, *Hero of Bataan*, 55, "well-equipped enemy."
4. Morton, *US Army in World War II*, 64, "fighting its way."
5. Morton, *US Army in World War II*, 14.
6. John W. Whitman, *Bataan: Our Last Ditch* (Hippocrene Press, 1990), 15; Morton, *US Army in World War II*, 67, 161.
7. Morton, *US Army in World War II*, 157.
8. Morton, *US Army in World War II*, 161.
9. Col. K. L. Berry, "History of the 3rd Regiment,1st Regular Division (PA), December 19, 1941–April 9, 1942," pg. 1, unpublished manuscript in the author's private collection, "mustered into United States service on December 17."
10. Eliseo D. Rio, *Rays of a Setting Sun: Recollections of World War II*, editor Elisea Vitriolo (C & E Publishing, 2016), 29–30.
11. Berry, "History of the 3rd Regiment," 1.
12. Rio, *Rays of a Setting Sun*, 34.
13. Berry, "History of the 3rd Regiment," 1; Col. Edmund Lilly's papers vol. 19, 25, Center for Research Allied POWs Under the Japanese website, accessed February 19, 2019, http://www.mansell.com/pow_resources/camplists/china_hk/mukden/edmund_lilly/Lilly_Diary_Vol19_TEXT_Story_of_the_57th_Infantry.html; Whitman, *Bataan*, 109.
14. Rio, *Rays of a Setting Sun*, 32.

15. Berry, "History of the 3rd Regiment," 1; Clayton Chun, *The Fall of the Philippines 1941–1942* (Osprey Publishing, Ltd., 2012), 57; Morton, *US Army in World War II*, 256.

16. Berry, "History of the 3rd Regiment," 1–2.

17. Rio, *Rays of a Setting Sun*, 44.

18. Whitman, *Bataan*, 126; Morton, *US Army in World War II*, 261.

19. Whitman, *Bataan*, 107.

20. Whitman, *Bataan*, 89.

21. Morton, *US Army in World War II*, 256–57; Whitman, *Bataan*, 49.

22. Morton, *US Army in World War II*, 257.

23. Schultz, *Hero of Bataan*, 132, "body and soul."

24. Richard C. Mallonee II, *The Naked Flagpole: Battle for Bataan from the Diary of Richard C. Mallonee*, editor Richard C. Mallonee (Presidio Press, 1980), 85.

25. Berry, "History of the 3rd Regiment," 2.

26. Morton, *US Army in World War II*, 249; Berry, "History of the 3rd Regiment," 2; Whitman, *Bataan*, 211.

27. Rio, *Rays of a Setting Sun*, 46.

28. Chun, *The Fall of the Philippines*, 63; Schultz, *Hero of Bataan*, 136–137.

29. Whitman, *Bataan*, 213.

30. Rio, *Rays of a Setting Sun*, 47.

31. Rio, *Rays of a Setting Sun*, 47.

32. Mallonee, *The Naked Flagpole*, 87.

33. Morton, *US Army in World War II*, 252.

34. Rio, *Rays of a Setting Sun*, 47.

35. Whitman, *Bataan*, 213.

36. Chun, *The Fall of the Philippines*, 63, "found a seam."

37. Berry, "History of the 3rd Regiment," 2.

38. Rio, *Rays of a Setting Sun*, 51–53; Whitman, *Bataan*, 214. After the war, Ledda eventually became a colonel in US Army.

39. Berry, "History of the 3rd Regiment," 3.

40. Berry, "History of the 3rd Regiment," 3.

41. Rio, *Rays of a Setting Sun*, 55, "back to Tokyo."

42. Whitman, *Bataan*, 216.

43. Rio, *Rays of a Setting Sun*, 56, "sneaked up on his command post.".

44. Jonathan M. Wainwright, *General Wainwright's Story: The Account of four years of humiliating defeat, surrender, and captivity by, who paid the price of his country's unpreparedness*, editor Robert Considine, (Doubleday & Company Inc., 1946), 50–51.

45. Col. K. L. Berry to General Jones, November 1942, author's private collection; Berry, "History of the 3rd Regiment," 3.

46. Whitman, *Bataan*, 216.
47. Berry General Jones, November 1942.
48. Berry, "History of the 3rd Regiment," 3.
49. Whitman, *Bataan*, 218–219.
50. Gerald Astor, *Crisis in the Pacific: The Battles for the Philippine Islands by the Men Who Fought Them*, (Random House, 1996), 145.
51. Astor, *Crisis in the Pacific*, 146. Rio, *Rays of a Setting Sun*, 57–58.
52. Morton, *US Army in World War II*, 283.
53. Morton, *US Army in World War II*, 283, "no automatic weapons at all."
54. Berry, "History of the 3rd Regiment," 4, "attacked from three sides."
55. Astor, *Crisis in the Pacific*, 145–146, 148.
56. Rio, *Rays of a Setting Sun*, 59.
57. Astor, *Crisis in the Pacific*, 148–149, "hold positions."
58. Whitman, *Bataan*, 222.
59. Whitman, *Bataan*, 223.
60. Astor, *Crisis in the Pacific*, 148–149; Whitman, *Bataan*, 223.
61. Morton, *US Army in World War II*, 284.
62. John Toland, *But Not in Shame: The Six Months After Pearl Harbor* (Random House, 1961), 182.
63. Berry to Alice Berry, February 22, 1942, author's private collection.

CHAPTER 7

1. Louis Morton, *US Army in World War II, The War in the Pacific, The Fall of the Philippines* (Center of Military History US Army, 1989), 295.
2. Col. K. L. Berry, "History of the 3rd Regiment,1st Regular Division (PA), December 19, 1941-April 9, 1942," pg. 5, unpublished manuscript in author's private collection.
3. Duane Schultz, *Hero of Bataan: The Story of General Jonathan M. Wainwright* (St. Marten's Press, 1981), 150–151.
4. Schultz, *Hero of Bataan*, 148–149, 159.
5. Schultz, *Hero of Bataan*, 151.
6. Jonathan M. Wainwright, *General Wainwright's Story: The Account of four years of humiliating defeat, surrender, and captivity by Jonathan M. Wainwright, who paid the price of his country's unpreparedness*, edited by Robert Considine (Doubleday & Company Inc., 1946), 53.
7. Schultz, *Hero of Bataan*, 157.
8. Morton, *US Army in World War II*, 324.
9. John W. Whitman, *Bataan: Our Last Ditch* (Hippocrene Books, 1990), 348.
10. Whitman, *Bataan*, 348.

11. Eliseo D. Rio, *Rays of a Setting Sun: Recollections of World War II*, edited by Elisela R. Vitriolo (C&E Publishing, Inc., 2016), 80.

12. Schultz, *Hero of Bataan*, 156.

13. Schultz, *Hero of Bataan*, 156.

14. Col. K. L. Berry to Alice Berry, February 2, 1942, in author's private collection.

15. K. L. Berry to Alice Berry, February 2, 1942.

16. K. L. Berry to Alice Berry, February 2, 1942.

17. Whitman, *Bataan*, 61, 359–360; Wainwright, *General Wainwright's Story*, 62.

18. Col. Edwin E. Aldridge, diary, December 31, 1941–September 14, 1945, compiled and transcribed by Edwin E. Aldridge Jr., Cushing Library, Texas A&M University, 42B, 3. Each page of Aldridge's diary had a page number at the top, such as 3, and a designation, such as 42B, at the bottom of the page. Aldridge was General Jones's chief of staff.

19. Whitman, *Bataan*, 364.

20. Morton, *US Army in World War II*, 342, the plan for attack; Whitman, *Bataan*, 364–365.

21. Whitman, *Bataan*, 365; Morton, *US Army in World War II*, 343; Col. K. L. Berry to Gen. Albert Jones, November 1942, in author's private collection.

22. Aldridge, diary, 42B, 3.

23. Aldridge, diary, 42B, 3.

24. Berry to Jones, at Karenko POW camp, November 1942.

25. Morton, *US Army in World War II*, 342.

26. Whitman, *Bataan*, 365; Morton, *US Army in World War II*, 343.

27. Whitman, *Bataan*, 369, "barrage of incredible intensity."

28. Morton, *US Army in World War II*, 343–346.

29. John Toland, *But Not in Shame: The Six Months After Pearl Harbor* (Random House, 1961), 206–207.

30. Schultz, *Hero of Bataan*, 180–181.

31. K. L. Berry to Alice Berry, February 22, 1942, in author's private collection.

32. Morton, *US Army in World War II*, 350; Gerald Astor, *Crisis in the Pacific: The Battles for the Philippine Islands by the Men Who Fought Them* (Random House, 1996), 172; Schultz, *Hero of Bataan*, 188, "brutal and bloody.".

33. Whitman, *Bataan*, 394.

34. Astor, *Crisis in the Pacific*, 172.

35. K. L. Berry to Alice Berry, February 2, 1942.

36. K. L. Berry to Alice Berry, February 2, 1942.

37. K. L. Berry to Alice Berry, February 2, 1942.

38. K. L. Berry to Alice Berry, February 2, 1942.

39. K. C. Emerson, *Guest of the Emperor* (Self-published, fourth printing, 1987), 5. Emerson wrote: "Most of the men had not seen one another since early December,

so we spent a lot of time exchanging information on our individual actions of the past few months, and a little bit of time wondering what the Japanese would do with us. Colonel Harrison C. Browne, Chief of Staff, Philippine Division, asked me, "Do you remember the maneuvers near Balmorhea, Texas, in the fall of 1939?" I answered, "Yes." "Did you ever think about what happened to the participants?" "No." "Well, Skinny Wainwright commanded the losing side as a new brigadier general and he's here and a three-star general; Clint Pierce, as a lieutenant colonel, was the umpire and he's here as a brigadier general; and I commanded the winning side—the 9th Infantry as a colonel—and I'm here with no promotion. You know Emerson, I would have given anything to have had the 9th Infantry here; it would have made a difference."

40. K. L. Berry to Alice Berry, February 2, 1942.

41. K. L. Berry to Alice Berry, February 2, 1942.

42. Schultz, *Hero of Bataan*, 189–190; Morton, *US Army in World War II*, 353.

43. Rio, *Rays of a Setting Sun*, 97.

44. William E. Brougher, *South to Bataan, North to Mukden: The Prison Diary of Brigadier General W. E. Brougher*, edited by D. Clayton James (University of Georgia Press, 2010), 32. While at POW Camp O'Donnell, General Brougher wrote about his angry feelings regarding their sacrifice as soldiers. He believed they were "victims of a crime of neglect on the part of responsible authorities" of the government. He believed "a foul trick of deception has been played on a large group of Americans by the Commander in Chief and a small staff who are now eating steak and eggs in Australia. God Damn them!"

45. Schultz, *Hero of Bataan*, 203.

46. Schultz, *Hero of Bataan*, 192; K. L. Berry, "Wainwright summary" private notebook in author's private collection. Berry used "(?)" after a word to show sarcasm.

47. Morton, *US Army in World War II*, 365.

48. Aldridge Diary, 42B, 5.

49. Schultz, *Hero of Bataan*, 210; Whitman, *Bataan*, 421.

50. Schultz, *Hero of Bataan*, 194, "three-eights."

51. Whitman, *Bataan*, 413.

52. Whitman, *Bataan*, 420; Schultz, *Hero of Bataan*, 222.

53. Astor, *Crisis in the Pacific*, 191; Rio, *Rays of a Setting Sun*, 97.

54. Berry, "History of the 3rd Regiment," 7.

55. Whitman, *Bataan*, 469; Morton, *US Army in World War II*, 401

56. Schultz, *Hero of Bataan*, 223–224.

57. Whitman, *Bataan*, 397; Morton, *US Army in World War II*, 402.

58. Whitman, *Bataan*, 454; Morton, *US Army in World War II*, 404, 425; Astor, *Crisis in the Pacific*, 192.

59. Whitman, *Bataan*, 475–477.
60. Whitman, *Bataan*, 477–483; Astor, *Crisis in the Pacific*, 192.
61. Whitman, *Bataan*, 519–521.
62. Morton, *US Army in World War II*, 431; Whitman, *Bataan*, 516.
63. Whitman, *Bataan*, 564.
64. Morton, *US Army in World War II*, 442–447; Whitman, *Bataan*, 543.
65. Wainwright, *General Wainwright's Story*, 80; Whitman, *Bataan*, 544.
66. Whitman, *Bataan*, 544, 572, 585, "destroy their installations."
67. Schultz, *Hero of Bataan*, 240, "largest force ever lost"; Morton, 447, 452–454; Astor *Crisis in the Pacific*, 199.

CHAPTER 8

1. Col. K. L. Berry, "History of the 3rd Regiment, 1st Regular Division (PA), December 19, 1941–April 9, 1942," 7, unpublished manuscript in the author's private collection; John W. Whitman, *Bataan: Our Last Ditch*, (Hippocrene Books, 1990), 601.
2. Berry, "History of the 3rd Regiment," 8. PA stands for Philippine Army.
3. Berry, "History of the 3rd Regiment," 8.
4. Col. K. L. Berry, "Brougher Summary," private notebook 1942–1945, in author's private collection and Texas Military Forces Museum, Austin, Texas.
5. William E. Brougher, *South to Bataan North to Mukden: the Prison Diary of Brigadier General W. E. Brougher*, edited by D. Clayton James (University of Georgia Press, 2010), 37; Whitman, *Bataan*, 600–01.
6. Col. Edwin E. Aldridge, diary, December 31, 1941–September 14, 1945, compiled and transcribed by Edwin E. Aldridge Jr., Cushing Library, Texas A&M University, 42B, 3.
7. Aldridge, diary, 42B, pg. 7.
8. Whitman, *Bataan*, 600.
9. Berry, "History of the 3rd Regiment," 8.
10. Jonathan M. Wainwright, *General Wainwright's Story: The Account of four years of humiliating defeat, surrender, and captivity by Jonathan M. Wainwright, who paid the price of his country's unpreparedness*, edited by Robert Considine (Doubleday & Company Inc., 1946), 161; Maj. Gen. Albert Jones to Col. K. L. Berry, December 4, 1945, in author's collection. Jones is complaining to Berry that Wainwright has not recognized with citations any of Jones' service on Bataan; Maj. Gen. Albert Jones to Gen. K. L. Berry, September 22, 1946, in author's collection. Jones has seen Wainwright's autobiography in which Wainwright extolls General Brougher's virtues as a soldier and commander. Jones is disgusted because Wainwright still has not given Jones any citations for medals for his part in any of the battles he commanded; Maj. Gen. Albert Jones to Maj. Gen. K. L. Berry, January 15, 1952, in

author's collection. Jones wrote Berry that he told military historian Louis Morton that the escapes were through Brougher's lines.

11. Louis Morton, *US Army in World War II, The War in the Pacific, The Fall of the Philippines* (Center of Military History US Army, 1989), 466.

12. Eliseo D. Rio, *Rays of a Setting Sun: Recollections of World War II*, edited by Elisea Vitriolo (C & E Publishing 2016), 98.

13. Rio, *Rays of a Setting Sun*, 98.

14. Rio, *Rays of a Setting Sun*, 99.

15. Rio, *Rays of a Setting Sun*, 99.

16. Rio, *Rays of a Setting Sun*, 100.

17. Rio, *Rays of a Setting Sun*, 100.

18. Rio, *Rays of a Setting Sun*, 100.

19. Rio, *Rays of a Setting Sun*, 100.

20. Rio, *Rays of a Setting Sun*, 101.

21. Rio, *Rays of a Setting Sun*, 101.

22. Rio, *Rays of a Setting Sun*, 103.

23. Rio, *Rays of a Setting Sun*, 103, 104, 112.

24. Michael Norman and Elizabeth M. Norman, *Tears in the Darkness: The Story of the Bataan Death March and Its Aftermath* (Farrar, Straus and Giroux, 2009), 202–205.

25. Lester I. Tenney, *My Hitch in Hell: The Bataan Death March* (Brassey's, 1995,) 46.

26. Clayton Chunn, *The Fall of the Philippines 1941–1942* (Osprey Publishing, Ltd., 2012), 77; Whitman, *General Wainwright's Story*, 605; John Toland, *But Not in Shame: The Six Months After Pearl Harbor* (Random House, 1961), 335–336.

27. John E. Olson, *O'Donnell: Andersonville of the Pacific* (Self-published, 1985). The author bought this signed book from AbeBooks many years ago. It appeared to have been typed and then mimeograph; Doroty Cave, *Beyond Courage: One Regiment Against Japan, 1941–1945* (Sunstone Press, 2006), 177.

28. Berry, "Berry Summary," private notebook.

29. Norman and Norman, *Tears in the Darkness*, 168.

30. Donald Knox, *Death March: The Survivors of Bataan*, (A Harvest Book Harcourt Inc., 1981) 127, Wohlfeld quote.

31. Tenney, *My Hitch in Hell*, 50.

32. Tenney, *My Hitch in Hell*, 59.

33. Tenney, *My Hitch in Hell*, 59.

34. Berry, "Berry Summary," private notebook; William Dyess, *The Dyess Story: The Eye-Witness Account of the Death March on Bataan* (Pickle Partner Publishing, 2013), 56, "crushing blows."

35. Tenney, *My Hitch in Hell*, 46.

36. Lt. Col. Victor Z. Gomez Affidavit, "Humanity to a fellow POW on the Death March," October 3, 1947, in author's collection; Berry Affidavit "Unusual Aid Rendered to Prisoners of War," June 2, 1947, in author's collection; Ken Towery, "Berry Proud of His Country and Country is Proud of Him," *Austin American*, May 16, 1958, 4. Francis Crow to K. L. Berry, March 8, 1946, in author's private collection. In the letter, Francis Crow thanks Berry for the aid he gave her son Judson Crow "during his last hour on earth."

37. Dyess, *The Dyess Story*, chapter 6, "Artesian wells lined the road."

38. Norman and Norman, *Tears in the Darkness*, 178, "they stopped sweating."

39. Knox, *Death March*, 139, "right through Capt. Miller."

40. Dyess, *The Dyess Story*, chapter 5, "should we falter."

41. Aldridge, diary, 3.

42. Toland, *But Not in Shame*, 342, "rank made little difference"; John A. Adams Jr., *The Fightin' Texas Aggie Defenders of Bataan & Corregidor* (Texas A&M University Press, 2016), 68.

43. Berry, "Berry Summary," private notebook; Tenney, *My Hitch in Hell*, 60.

44. Berry, "Berry Summary," private notebook.

45. Toland, *But Not in Shame*, 344, 348.

46. Berry, "Berry Summary", private notebook.

47. Toland, *But Not in Shame*, 351–352.

48. Col. E. B. Miller, *Bataan Uncensored* (The Hart Publications Inc., 1949), 228–229, accessed on Archive.org, https://archive.org/details/BataanUncensored-nsia, "compelled to live in their own filth."

49. Tenney, *My Hitch in Hell*, 62.

50. Whitman, *Bataan*, 605; Toland, *But Not in Shame*, 329.

CHAPTER 9

1. Richard C. Mallonee, *The Naked Flagpole: Battle for Bataan from the Diary of Richard C. Mallonee*, edited by Richard C. Mallonee II, (Presidio Press, 1980), 154; Col. K. L. Berry, "Corkill Summary," private notebook 1942–1945, in author's private collection and Texas Forces Military Museum, Camp Mabry, Texas.

2. Lester I. Tenney, *My Hitch in Hell: The Bataan Death March*, (Brassey's, 1995), 66.

3. Tenney, *My Hitch in Hell*, 66; John E Olson, *The Guerilla and the Hostage*, (Burke Publishing Co., 1994), 166; K. C. Emerson, *Guest of the Emperor* (Self-published, 1977), 21, https://kipdf.com/guest-of-the-emperor-k-c-emerson_5aac75c01723dd5d283c504d.html, accessed May 30, 2025; Col. William C. Braly, *The Hard Way Home* (Infantry Journal Press, 1947) 31.

4. Mallonee, *The Naked Flagpole*, 154; Donald Knox, *Death March: The Survivors of Bataan*, (A Harvest Book Harcourt Inc., 1981), 153.

5. Knox, *Death March*, 159–160, "we lost count."

6. Col. Ed Aldridge diary, Cushman Library, Texas A&M University Library, O'Donnell, 42B,10.

7. Mallonee, *The Naked Flagpole*, 156.

8. Ken Towery, "Berry Proud of His Country and His Country Proud of Him," *Austin American*, May 16, 1958, 4, "counting 299 in one day."

9. Knox, *Death March*, 153, 168.

10. Knox, *Death March*, 166.

11. Knox, *Death March*, 166.

12. Mallonee, *The Naked Flagpole*, 155.

13. Mallonee, *The Naked Flagpole*, 160; William E. Brougher, *South to Bataan North to Mukden: the Prison Diary of Brigadier General W. E. Brougher*, edited by Clayton D. James (University of Georgia Press, 2010), 43, "two story wooden barracks."

14. Braly, 28, seventy-seven officers and fifty-two enlisted men.

15. Mallonee, *The Naked Flagpole*, 160.

16. Malcolm Fortier, *The Life of A POW Under the Japanese in Caricature as Sketched by Col. Malcolm Vaughn Fortier*, (C. W. Hill Printing Co., 1946), 8–9.

17. Brougher, *South to Bataan*, 45.

18. Braly, *The Hard Way Home*, 31.

19. Dorothy Cave, *Beyond Courage: One Regiment against Japan, 1941–1945* (Sunstone Press, 2006), 130.

"We're the battling bastards of Bataan;
No mama, no pap, no Uncle Sam;
No aunts, no uncles, no cousins, no nieces,
No pills, no planes, no artillery pieces.
And nobody gives a damn.
Nobody gives a damn."

20. Braly, *The Hard Way Home*, 267–268, 33; Fortier, *The Life of A POW*, 10; Lewis Beebe, *Prisoner of the Rising Sun: The Lost Diary of Brig. Gen. Lewis Beebe*, edited by John M. Beebe (Texas A&M University Press, 2006), 93.

21. Jonathan M. Wainwright, *General Wainwright's Story: The Account of four years of humiliating defeat, surrender, and captivity by Jonathan M. Wainwright, who paid the price of his country's unpreparedness*, edited by Robert Considine, (Doubleday & Company, Inc., 1946, 159; Berry, "Berry Summary," private notebook.

22. Braly, *The Hard Way Home*, 38.

23. Mallonee, *The Naked Flagpole*, 161.

24. Berry, "Berry Summary," private notebook; Beebe, *Prisoner of the Rising Sun*, 98.

25. Colonel Braly, General Beebe, Colonel Mallonee, General Wainwright, and General Brougher all wrote some details about Tarlac in their diaries or books.

26. Berry, "Hoffman Summary," private notebook.

27. Only Camp Tarlac had a mess hall. At the other POW Camps, prisoners ate in their rooms.

28. Braly, *The Hard Way Home*, 36–37.

29. Brougher, *South to Bataan*, 45; Wainwright, *General Wainwright's Story*, 169; Mallonee, *The Naked Flagpole*, 171, "No Filipino uttered a sound."

30. Braly, *The Hard Way Home*, 40.

31. Joris Nieuwint, "When the Allies Killed Over 20,000 of Their Own Countrymen As They Sank Japanese Hell Ships That Transported Them," War History Online: The Place for Military History News and Views, January 17, 2016, https://www.warhistoryonline.com/featured/japanese-hellships.html?edg-c=1.

32. Braly, *The Hard Way Home*, 42.

33. Braly, *The Hard Way Home*, 42.

34. Mallonee, *The Naked Flagpole*, 173.

35. Braly, *The Hard Way Home*, 46.

CHAPTER 10

1. William Braly, *The Hard Way Home* (Infantry Journal Press, 1947), 46–47.

2. Richard C. Mallonee, *The Naked Flagpole: Battle for Bataan*, edited by Richard C. Mallonee II (Presidio Press, 1980), 174; Braly, *The Hard Way Home*, 50.

3. Mallonee, *The Naked Flagpole*, 174–175; Braly, *The Hard Way Home*, 47–49.

4. William E. Brougher, *South to Bataan North to Mukden: the Prison Diary of Brigadier General W. E. Brougher*, edited by Clayton D. James (University of Georgia Press, 2010), 47, "mean and vindictive." Braly, *The Hard Way Home*, 53. Col. Paul D. Bunker, *Bunker's War: The World War II Diary of Col. Paul D. Bunker*, edited by Keith Barlow (Presidio Press, 1996), 260, "best Jap around."

5. Malcolm Fortier, *The Life of A P.O.W. Under the Japanese in Caricature as Sketched by Col. Malcolm Vaughn Fortier* (C. W. Hill Printing Co., 1946), 45.

6. Braly, *The Hard Way Home*, 50; Bunker, *Bunker's War*, 223.

7. Bunker, *Bunker's War*, 221–224.

8. Braly, *The Hard Way Home*, 50–51.

9. Col. K. L. Berry, "Braly Summary," private notebook 1942–1945, in author's private collection and the Texas Forces Military Museum at Camp Mabry, Austin, Texas; "Alphonse and Gaston," Don Markstein's Toonopedia website, accessed January 6, 2023, http://toonopedia.com/alphgast.htm; Berry, "Bowler Summary," private notebook.

10. Berry, "Bunker Summary," private notebook.

11. Berry, "Bunker Summary," private notebook.

12. Bunker, *Bunker's War*, 224.

13. Braly, *The Hard Way Home*, 266–272; Bunker, *Bunker's War*, 267.

14. Braly, *The Hard Way Home*, 51; Jonathan M. Wainwright, *General Wainwright's Story: The Account of four years of humiliating defeat, surrender, and captivity by Jonathan M. Wainwright, who paid the price of his country's unpreparedness*, edited by Robert Considine (Doubleday & Company, Inc., 1946), 200.

15. Wainwright, *General Wainwright's Story*, 201.

16. Braly, *The Hard Way Home*, 52; Bunker, *Bunker's War*, 248, 269, 273; Wainwright, *General Wainwright's Story*, 187.

17. Bunker, *Bunker's War*, 240

18. Braly, *The Hard Way Home*, 72.

19. Bunker, *Bunker's War*, 273.

20. Braly, *The Hard Way Home*, 72, "several garden-minded officers."

21. Col. K. L. Berry, POW Camp Karenko diary, August 17, 1942–June 6, 1943, in author's collection and in the Texas Military Forces Museum, Camp Mabry.

22. Berry, POW Camp Karenko diary.

23. Braly, *The Hard Way Home*, 58; Bunker, *Bunker's War*, 271.

24. Bunker, *Bunker's War*, 227, "Col. Berry tried to pass the dope to us."

25. Wainwright, *General Wainwright's Story*, 182.

26. Col. K. L. Berry, Radiogram, November 10, 1942, in author's private collection.

27. Col. K. L. Berry to General Jones, November 26, 1942, in author's collection, which the author reprinted in Col. K. L. Berry's POW Camp Karenko diary, August 17, 1942–June 6, 1943.

28. Berry, POW Camp Karenko diary.

29. Berry, POW Camp Karenko diary.

30. Berry, POW Camp Karenko diary.

31. Braly, *The Hard Way Home*, 71, 107.

32. Berry, POW Camp Karenko diary. Anzac Day is a day of remembrance in Australia and New Zealand that broadly commemorates all Australians and New Zealanders who served and died in wars, conflicts, and peacekeeping operations.

33. Bunker, *Bunker's War*, 278; Braly, *The Hard Way Home*, 104.

34. Braly, *The Hard Way Home*, 103; Berry, POW Camp Karenko diary.

35. Braly, *The Hard Way Home*, 121, 124.

36. Berry, POW Camp Karenko diary.

CHAPTER 11

1. The Japanese preferred the word Nipponese; thus, the POWs called them Nips instead of Japs.

2. Col. K. L. Berry, POW Camp Karenko diary, August 17, 1942–June 6, 1943, in author's collection and in the Texas Military Forces Museum, Camp Mabry, Texas.

3. William Braly, *The Hard Way Home* (Infantry Journal Press, 1947), 72.

4. Braly, *The Hard Way Home*, 52.

5. Col. Paul D. Bunker, *Bunker's War: The World War II Diary of Col. Paul D. Bunker*, edited by Keith Barlow (Presidio Press, 1996), 258.

6. Berry, POW Camp Karenko diary.

7. Malcolm Fortier, *The Life of A P.O.W. Under the Japanese in Caricature as Sketched by Col. Malcolm Vaughn Fortier* (C. W. Hill Printing Co. 1946), 26; Berry, Diary.

8. Berry, POW Camp Karenko diary.

9. Braly, *The Hard Way Home*, 122.

10. Berry, POW Camp Karenko diary. Berry used (?) after a word to indicate sarcasm.

11. Berry, POW Camp Karenko diary.

12. Berry, POW Camp Karenko diary.

13. Berry, POW Camp Karenko diary.

14. Berry, POW Camp Karenko diary.

15. Berry, POW Camp Karenko diary.

16. Berry, POW Camp Karenko diary.

17. Berry, POW Camp Karenko diary

18. Bunker, *Bunker's War*, 245.

19. Berry, POW Camp Karenko diary.

20. Berry, POW Camp Karenko diary.

21. Berry, POW Camp Karenko diary.

22. Berry, POW Camp Karenko diary.

23. Braly, *The Hard Way Home*, 106.

24. Braly, *The Hard Way Home*, 106.

25. Richard C. Mallonee, *The Naked Flagpole: Battle for Bataan*, edited by Richard C. Mallonee II (Presidio Press, 1980), 183.

26. Braly, *The Hard Way Home*, 106.

27. Berry, POW Camp Karenko diary.

28. Berry, POW Camp Karenko diary.

29. Berry, POW Camp Karenko diary. The Allies, including British and Americans, fought to gain control of northern Africa, with the German and Italian troops finally surrendering there on May 13, 1943. Berry and others used the term "G2" to mean "we intuit or guess."

30. Berry, POW Camp Karenko diary.

31. Berry, POW Camp Karenko diary.

32. Phil Zimmer, "Death of Admiral Isoroku Yamamoto," Warfare History Network, November 2017, https://warfarehistorynetwork.com/article/death-of-admiral-isoroku-yamamoto/.

33. Berry, POW Camp Karenko diary.
34. Berry, POW Camp Karenko diary.
35. Berry, POW Camp Karenko diary.
36. Berry, POW Camp Karenko diary.
37. William E. Brougher, *South to Bataan North to Mukden: the Prison Diary of Brigadier General W. E. Brougher*, edited by Clayton D. James (University of Georgia Press, 2010) 64–65; Braly, *The Hard Way Home*, 123.
38. Berry, POW Camp Karenko diary.
39. Berry, POW Camp Karenko diary. General Parker was left behind when the other generals left Karenko. He was the POWs' commanding officer.
40. Berry, POW Camp Karenko diary.
41. Berry, POW Camp Karenko diary.
42. Berry, POW Camp Karenko diary.
43. Braly, *The Hard Way Home*, 127; Jonathan M. Wainwright, *General Wainwright's Story: The Account of four years of humiliating defeat, surrender, and captivity by Jonathan M. Wainwright, who paid the price of his country's unpreparedness*, edited by Robert Considine (Doubleday & Company, Inc., 1946), 213–215.
44. Berry, POW Camp Karenko diary; Braly, *The Hard Way Home*, 132.
45. Brougher, *South to Bataan*, 76; Braly, *The Hard Way Home*, 133.

CHAPTER 12

1. Richard C. Mallonee, *The Naked Flagpole: Battle for Bataan from the Diary of Richard C. Mallonee*, edited by Richard C. Mallonee II (Presidio Press, 1980), 187.
2. Mallonee, *The Naked Flagpole*, 187, "quite a shock."
3. William C. Braly, *The Hard Way Home* (Infantry Journal Press, 1947), 134; Col. K.L. Berry, POW Camp Shirakawa diary, June 8, 1943–October 9, 1944, in author's collection and in the Texas Military Forces Museum at Camp Mabry, Austin, Texas.
4. Berry, POW Camp Shirakawa diary; Braly, *The Hard Way Home*, 135.
5. William E. Brougher, *South to Bataan North to Mukden: the Prison Diary of Brigadier General W. E. Brougher*, edited by D. Clayton James, (University of Georgia Press, 2010), 77, "good reading light."
6. Braly, *The Hard Way Home*, 134.
7. Berry, POW Camp Shirakawa diary. Berry is referring to Roosevelt's New Deal Civilian Conservation Corps camps, which were run by the military. Enrollees worked for six months at a time on public land projects. The camps were fairly primitive.
8. Col. K. L. Berry, "Galbraith Summary," private notebook 1942–1945, in author's private collection and the Texas Forces Military Museum at Camp Mabry,

Austin, Texas; Col. Nicoll Galbraith, POW diary and papers. Copies of Galbraith's diary and papers were given to the author by Galbraith's son, Whitney Galbraith.

9. Berry, POW Camp Shirakawa diary.
10. Berry, Diary; Brougher, *South to Bataan*, 91, "sitting in each other's laps."
11. Berry, POW Camp Shirakawa diary.
12. Berry, POW Camp Shirakawa diary.
13. Berry, POW Camp Shirakawa diary. Berry became a Mason as a POW. After the war, Berry became the highest level of Mason and was invited to be a Shriner.
14. Berry, POW Camp Shirakawa diary.
15. Berry, POW Camp Shirakawa diary.
16. Berry, POW Camp Shirakawa diary.
17. Brougher, *South to Bataan*, 106.
18. Galbraith, diary, "K grumbles"; Berry, POW Camp Shirakawa diary.
19. Berry, POW Camp Shirakawa diary.
20. Berry, POW Camp Shirakawa diary.
21. Berry, POW Camp Shirakawa diary.
22. Col. Edmund Lilly, Notebook 1, Diary, Part 1, Center for Research: Allied POWS Under the Japanese website, accessed February 19, 2019, http://www.mansell.com/pow_resources/camplists/china_hk/mukden/edmund_lilly/Lilly_Diary_Vol01_TEXT.htm, "Eyes of Texas"; Berry, POW Camp Shirakawa diary.
23. Berry, POW Camp Shirakawa diary.
24. Berry, POW Camp Shirakawa diary.
25. Berry, POW Camp Shirakawa diary.
26. Berry, POW Camp Shirakawa diary.
27. Berry, POW Camp Shirakawa diary. That Christmas a water pipeline had just been finished by the POWs. The commander and guards benefited from the pipeline, as did the POWs, so perhaps the Japanese were appreciative of the work completed.
28. Berry, POW Camp Shirakawa diary.
29. Braly, *The Hard Way Home*, 179, "thrillers and poetry."
30. Berry, POW Camp Shirakawa diary.
31. Berry, POW Camp Shirakawa diary.
32. Brougher, *South to Bataan*, 87, "thirty-two officers"; Berry, POW Camp Shirakawa diary.
33. Berry, POW Camp Shirakawa diary. Radiogram news: Berry's daughter Celeste was married to Bill Reilly, a West Point graduate, who was fighting the war in Europe and became a POW of the Germans. Tom, Berry's youngest son, had hoped to get an appointment to West Point from Mr. South, and K. L. Jr. had graduated from West Point and married Phyllis Boarman.

34. Berry, POW Camp Shirakawa diary.
35. Berry, POW Camp Shirakawa diary.
36. Berry, POW Camp Shirakawa diary.
37. Berry, POW Camp Shirakawa diary.
38. Berry, POW Camp Shirakawa diary. Berry did not know that Alice had moved to San Antonio from Brownsville. He was angry that he did not receive more letters from her that had been written over time like others had received. For instance, Aldridge received eighteen letters and Brougher received twenty-one letters—letters that had been written over several months and then appeared all at once when the Japanese allowed mail to come through. Alice's letters (in the author's collection) were very short and had practically no content in them compared with other POWs' letters received. However, Alice did write that she had traveled to Washington, DC, to learn about what happened to Berry's promotion.
39. Berry, POW Camp Shirakawa diary.
40. Berry, POW Camp Shirakawa diary.
41. Berry, POW Camp Shirakawa diary.
42. Berry, POW Camp Shirakawa diary. K. L. Jr. piloted B-17s over France and Germany. In April 1944, his plane was hit and all on board had to bail out over the English Channel. K. L. Jr., the last man off the plane, was one of only two survivors.
43. Berry, POW Camp Shirakawa diary.
44. Berry, POW Camp Shirakawa diary.

CHAPTER 13

1. Col. K. L. Berry, POW Camp Shirakawa diary, June 8, 1943–October 9, 1944, in author's collection and in the Texas Military Forces Museum at Camp Mabry, Austin, Texas.
2. Berry, POW Camp Shirakawa diary.
3. Berry, POW Camp Shirakawa diary.
4. Berry, POW Camp Shirakawa diary; William C. Braly, *The Hard Way Home* (Infantry Journal Press, 1947), 136.
5. Berry, POW Camp Shirakawa diary.
6. Berry, POW Camp Shirakawa diary.
7. Berry, POW Camp Shirakawa diary.
8. Berry, POW Camp Shirakawa diary.
9. Berry, POW Camp Shirakawa diary.
10. Berry, POW Camp Shirakawa diary.
11. Berry, POW Camp Shirakawa diary.
12. Berry, POW Camp Shirakawa diary.

13. Berry, POW Camp Shirakawa diary; William E. Brougher, *South to Bataan, North to Mukden: the Prison Diary of Brigadier General W. E. Brougher*, edited by D. Clayton James (University of Georgia Press, 2010), 114.

14. Berry, POW Camp Shirakawa diary; Malcolm Fortier, *The Life of A P.O.W. Under the Japanese in Caricature as Sketched by Col. Malcolm Vaughn Fortier* (C. W. Hill Printing CO., 1946), 77.

15. Braly, *The Hard Way Home*, 163.

16. Braly, *The Hard Way Home*, 165–166.

17. Berry, POW Camp Shirakawa diary.

18. Berry, POW Camp Shirakawa diary.

19. Berry, POW Camp Shirakawa diary; Col. Edmund Lilly, notebook 1, diary, part 1, Center for Research: Allied POWS Under the Japanese website, accessed February 19, 2019, http://www.mansell.com/pow_resources/camplists/china_hk/mukden/edmund_lilly/Lilly_Diary_Vol01_TEXT.htm; Col. Nicoll Galbraith, POW diary and papers, August 10, 1944, "Texas mustache." Copies of Galbraith's diary and papers were given to the author by Galbraith's son, Whitney Galbraith.

20. Berry, POW Camp Shirakawa diary. Berry's dream about having a ranch someday recurred in his diary, but he expressed concern that a ranch and house would be beyond his means. He did in fact buy a ranch northwest of Austin with his sons after the war. It became a place to hunt with family and friends.

21. Berry, POW Camp Shirakawa diary.

22. Berry, POW Camp Shirakawa diary.

23. Berry, POW Camp Shirakawa diary.

24. Berry, POW Camp Shirakawa diary.

25. Berry, POW Camp Shirakawa diary; Brougher, *South to Bataan*, 125-127.

26. Berry, POW Camp Shirakawa diary.

27. Berry, POW Camp Shirakawa diary.

28. Braly, *The Hard Way Home*, 175.

29. Braly, *The Hard Way Home*, 175–176.

30. Berry, POW Camp Shirakawa diary; John M. Beebe, *Prisoner of the Rising Sun: The Lost Diary of Brig. Gen. Lewis Beebe* (Texas A&M University Press, 2006), 211.

31. Berry, POW Camp Shirakawa diary.

32. Berry, POW Camp Shirakawa diary; Brougher, *South to Bataan*, 135, 137–138.

33. Berry, POW Camp Shirakawa diary.

34. Berry, POW Camp Shirakawa diary.

35. Berry, POW Camp Shirakawa diary; Brougher, *South to Bataan*, 137.

36. Braly, *The Hard Way Home*, 186; Brougher, *South to Bataan*, 141.

37. Berry, POW Camp Shirakawa diary.

CHAPTER 14

1. Col. William E. Corkill, diary, in author's private collection.
2. Corkill, diary; William Braly, *The Hard Way Home* (Infantry Journal Press, 1947), 189.
3. Braly, *The Hard Way Home*, 190.
4. Braly, *The Hard Way Home*, 190–192; Col. Edmund Lilly, Notebook 1: Diary, part 1, December 26, 1943–December 7, 1944, Center for Research Allied POWS Under the Japanese website, accessed February 19, 2019, http://www.mansell.com/pow_resources/camplists/china_hk/mukden/edmund_lilly/Lilly_Diary_Vol01_TEXT.htm.
5. Braly, *The Hard Way Home*, 192; Lilly, Notebook 1.
6. Corkill, Diary; Braly, *The Hard Way Home*, 192–193.
7. Braly, *The Hard Way Home*, 193; Corkill, Diary; Lilly, Notebook 1.
8. Braly, *The Hard Way Home*, 194.
9. Lilly, Notebook 1.
10. Corkill, diary; Lilly, Notebook 1.
11. Lilly, Notebook 1.
12. Adam Bisno, PhD, "The Japanese 'Hell Ships' of World War II," Naval History and Heritage Command, NHHC Communication and Outreach Division, November 2019, accessed June 13, 2021, https://www.history.navy.mil/browse-by-topic/wars-conflicts-and-operations/world-war-ii/1944/oryoku-maru.html.
13. Lt. Col. Ovid O. Wilson had served with II Corps under General Parker in the Battle of Bataan. After the Death March and Camp O'Donnell, he ended up in Camp Cabanatuan on Luzon instead of Camp Tarlac where Colonel Berry and other colonels were imprisoned; Rob Clark, "The Life of Ovid O. Wilson," *Bryan-College Station Eagle*, November 11, 2018, accessed June 12, 2021, https://theeagle.com/news/local/the-life-of-ovid-o-wilson/article_4e860ec7-6a58-50bc-b8e7-435faa5ab7ed.html.www.theeagle.com.
14. Bisno, "The Japanese 'Hell Ships' of World War II."
15. Clark, "The Life of Ovid O. Wilson"; Scottie Kersta-Wilson, "My Granddaddy," *The Quan*, Spring 2021, 90:4–5.
16. John M. Gibbs, "Beppu," July 31, 1946, American Ex-Prisoners of War website, accessed June 13, 2021, https://www.axpow.org/medsearch/beppu.htm.
17. Corkill, diary; Gibbs, "Beppu."
18. Corkill, diary.
19. Gibbs, "Beppu."
20. Gibbs, "Beppu;" Corkill, diary.
21. Corkill, diary.

22. Braly, *The Hard Way Home*, 197–198.
23. Corkill, diary; Gibbs, "Beppu."
24. Corkill, diary; Lilly, Notebook 1.

CHAPTER 15

1. William Braly, *The Hard Way Home* (Infantry Journal Press, 1947), 200, 202.
2. Richard C. Mallonee, *The Naked Flagpole Battle for Bataan from the Diary of Richard C. Mallonee*, edited by Richard C. Mallonee II (Presidio Press, 1980), 194.
3. Braly, *The Hard Way Home*, 203.
4. Braly, *The Hard Way Home*, 201–203.
5. Col. William E. Corkill, diary, in author's private collection.
6. Col. K. L. Berry, diary, POW Camp Cheng Chia Tun, December 9, 1944–May 19, 1945, in author's private collection and the Texas Military Forces Museum at Camp Mabry, Austin, Texas.
7. William E. Brougher, *South to Bataan North to Mukden: The Prison Diary of Brigadier General W. E. Brougher*, edited by D. Clayton James (University of Georgia Press, 2010), 147.
8. Berry, POW Camp Cheng Chia Tun diary. Berry used (?) after a word to convey sarcasm.
9. Berry, POW Camp Cheng Chia Tun diary.
10. Berry, POW Camp Cheng Chia Tun diary.
11. Berry, POW Camp Cheng Chia Tun diary; Braly, *The Hard Way Home*, 218.
12. Berry, POW Camp Cheng Chia Tun diary; Mallonee, *The Naked Flagpole*, 195.
13. Col. Edmund Lilly, diary, December 8, 1944–April 26, 1945, Center for Research: Allied POWs Under the Japanese website, accessed February 19, 2019, http://www.mansell.com/pow_resources/camplists/china_hk/mukden/edmund_lilly/Lilly_Diary_Vol02_1944-12-08to1945-04-26_3318.pdf.
14. Corkill, diary.
15. Berry, POW Camp Cheng Chia Tun diary.
16. Corkill, diary.
17. Berry, POW Camp Cheng Chia Tun diary; Lilly, diary.
18. Berry, POW Camp Cheng Chia Tun diary.
19. Corkill, diary.
20. Berry, POW Camp Cheng Chia Tun diary.
21. Lilly, diary.
22. Berry, POW Camp Cheng Chia Tun diary.
23. Jonathan M. Wainwright, *General Wainwright's Story: The Account of four years of humiliating defeat, surrender, and captivity by Jonathan M. Wainwright, who paid the price of his country's unpreparedness*, edited by Robert Considine (Doubleday & Company Inc., 1946), 244–245.

24. Berry, POW Camp Cheng Chia Tun diary.
25. Berry, POW Camp Cheng Chia Tun diary.
26. Berry, POW Camp Cheng Chia Tun diary.
27. Lilly, diary.
28. Berry, POW Camp Cheng Chia Tun diary.
29. Berry, POW Camp Cheng Chia Tun diary.
30. Berry, POW Camp Cheng Chia Tun diary.
31. Berry, POW Camp Cheng Chia Tun diary.
32. Berry, POW Camp Cheng Chia Tun diary.
33. Berry, POW Camp Cheng Chia Tun diary.
34. Col. K. L. Berry diary, POW Camp Mukden, May 21–1945-August 13, 1945, in author's private collection and the Texas Military Forces Museum at Camp Mabry, Austin, Texas.
35. Berry, POW Camp Mukden diary.
36. Berry, POW Camp Mukden diary.
37. Berry, POW Camp Mukden diary.
38. Berry, POW Camp Mukden diary.
39. Berry, POW Camp Mukden diary.
40. Berry, POW Camp Mukden diary.
41. Corkill, diary.
42. Berry, POW Camp Mukden diary.
43. Berry, POW Camp Mukden diary.
44. Berry's friends were Gen. Claudius "Speck" Easley and Lt. Gen. Simon Bolivar Buckner. US 10th Army Commander Lt. Gen. Simon B. Buckner was killed June 18, 1945, by fragments of Japanese artillery fire. Easley was killed on June 19, 1945, by Japanese machine gun fire. Both were killed during the last days of the Okinawa campaign; "Battle of Okinawa," The National WWII Museum, https://www.nationalww2museum.org/war/topics/battle-of-okinawa, accessed February 24, 2022.
45. "'Downfall,' The Plan for The Invasion of Japan: Evolution of Downfall" Campaigns of MacArthur in the Pacific website, vol I, chapter XIII, accessed on October 15, 2022, https://www.ibiblio.org/hyperwar/USA/RptsMacA/I/RptsI-13.html; Richard B. Frank, *Downfall: The End of the Imperial Japanese Empire*, (Random House, 1999), 138, 340.
46. Lee Lacy, "Harry Truman and The Bomb," National Archives, *Pieces of History* (blog), August 5, 2014, https://prologue.blogs.archives.gov/2014/08/05/harry-truman-and-the-bomb/
47. Berry, POW Camp Mukden, diary.
48. Lacy, "Harry Truman and The Bomb."
49. Berry, POW Camp Mukden, diary.

50. Brougher, *South to Bataan*, 181.

51. Berry, POW Camp Mukden diary.

52. "Japanese Order Posted in POW Camps [1944]" Library of Congress, John L. Stensby Sr. Collection, accessed June 7, 2025, https://www.loc.gov/resource/afc2001001.00454.pm0003001/?r=-0.845,-0.062,2.689,1.421.

CHAPTER 16

1. William E. Brougher, *South to Bataan North to Mukden: The Prison Diary of Brigadier General W. E. Brougher*, edited by D. Clayton James (University of Georgia Press, 2010,) 183.

2. Col. Nicoll F. Galbraith, POW diary, August 16, 1945. A copy of the diary was given to the author by Whitney Galbraith, Colonel Galbraith's son.

3. "Office of Strategic Services: America's First Intelligence Agency," Central Intelligence Agency, accessed December 26, 2022, https://www.cia.gov/legacy/museum/exhibit/the-office-of-strategic-services-n-americas-first-intelligence-agency/. The OSS was established on June 13, 1942. President Franklin D. Roosevelt appointed William J. Donovan, a highly decorated World War I officer, as OSS director. Donovan organized the OSS to reflect his vision of a national intelligence center, uniquely combining research and analysis, covert operations, counterintelligence, espionage, and technical development—the core missions of today's Central Intelligence Agency; Hal Leith, *POWs of Japanese Rescued!* (Trafford Publishing, 2003), 8.

4. Leith, *POWs of Japanese Rescued!*, 14.

5. Galbraith, POW diary.

6. William C. Braly, *The Hard Way Home* (Infantry Journal Press, 1947), 248.

7. Richard C. Mallonee, *The Naked Flagpole: Battle for Bataan from the Diary of Richard C. Mallonee*, edited by Richard C. Mallonee II (Presidio Press, 1980), 201.

8. Col. William E. Corkill diary, in author's private collection.

9. Leith, *POWs of Japanese Rescued!*, 25.

10. Galbraith, POW diary.

11. Galbraith, POW diary.

12. Braly, *The Hard Way Home*, 254.

13. Braly, *The Hard Way Home*, 268.

14. V. N. Cordero, *My Experiences During the War With Japan* (Zerreiss & Co., 1950), 59; Virgilio N. Cordero, *Bataan Y La Marcha De La Muerte*, (Afrodisio Aguado, 1957), 176.

15. This story was told to K. L. Berry's son Tom Berry by Gerta herself at a Tientsin Grammar School reunion in 1982. Tom Berry told this story to his nephew William King Reilly. Reilly could not remember the date.

16. "Mukden Rescue and Evacuation: Evacuation Air and Train Rosters," Center for Research: Allied POWs Under the Japanese website, edited by Roger Mansell, accessed on April 9, 2019, http://mansell.com/pow_resources/camplists/china_hk/mukden/mukden_train_evac_2.html.

17. "War Diary of USS *RELIEF*," September 1945, Center for Research: Allied POWs Under the Japanese website, edited by Roger Mansell, accessed January 28, 2021,. http://www.mansell.com/Resources/special_files//FOLD3/USS%20Relief%20Diary%201945-09.pdf.

18. "Mukden Rescue and Evacuation."
19. Braly, *The Hard Way Home*, 261.
20. "War Diary of USS *RELIEF*."
21. Braly, *The Hard Way Home*, 261.
22. "War Diary of USS *RELIEF*."
23. "War Diary of USS *RELIEF*."
24. "War Diary of USS *RELIEF*."
25. *American and Allied Personnel Recovered from Japanese Prisons: A Pictorial History*, (US Army, Replacement Command, AFWESPAC, 1945).

CHAPTER 17

1. Alice Berry to Colonel Berry, September 15, 1945, in author's private collection. Only one letter that Alice wrote to him was returned to her and that letter was in May 1942.

2. "Prisoner of War Covers: Background," Manchuko Stamps website, accessed December 5, 2022, http://www.manchukuostamps.com/POWcovers.htm. The American POW recovery team discovered storerooms containing Red Cross parcels and thousands of items of mail withheld from the prisoners. Some of the mail had been checked by the censor and had a censors mark; other mail had yet to be checked. This mail was then made available to surviving prisoners within the camp and sent on to those already evacuated.

3. Alice Berry to K. L. Berry, March 18, 1945, and August 6, 1945, in author's private collection.

4. Alice Berry telegram to Colonel Berry, October 10, 1945, in author's private collection.

5. "General Berry, Pacific War Hero Born on a Farm Near Bolivar," *Denton Record-Chronicle*, August 7, 1953. Berry wrote most of what appeared in this article and submitted it at the newspaper's request. The paper took almost verbatim what Berry wrote about himself, including this quote about his wife.

6. Col. Harrison Browne, interview by King Reilly, February 28, 1984, in author's private collection. Browne and Berry remained close friends when they got

back home to Texas, and they enjoyed many hunting trips with other hunting buddies until Berry died. Years later, Browne told King Reilly: "K. L. had everything. A great athlete, a great shot, and a very fine man in every respect. I had never known a man like him. And the greatest compliment I ever had was that he liked me better than anyone he had ever known, too. He said Alice used to get a little bit jealous because he liked me so much."

7. Gen. Jonathan Wainwright to Colonel Berry, November 7, 1945, in author's private collection.

8. Col. Harvey Matthews to Colonel Berry, October 27, 1945, in author's private collection. Matthews was stationed in Washington, DC, after the war and had associates who could keep him informed about the situation with Berry's case.

9. Gen. George C. Marshall letter to General Wainwright, Nov 16, 1945, copy of letter in author's private collection.

10. Tom Connally to Charles Clark, November 26, 1945; Paul J. Kilday to Maj. Gen. Claude Birkhead, January 24, 1946; Lyndon B. Johnson to General Dwight D. Eisenhower, June 6, 1946, Paul J. Kilday telegram to Winchester Kelso, June 27, 1946, in author's private collection; "Colonels to be brigadier generals," *The Evening Star*, Washington D.C., January 22, 1946.

11. "Kearie Lee Berry," Hall of Valor, Military Medals Database, accessed December 20, 2018, https://valor.militarytimes.com/hero/6024.

12. "Kearie Lee Berry," Hall of Valor.

13. Maj. Gen. Edward Whitsell to General Berry, February 26, 1946, in author's private collection. The letter informs Berry that he was awarded the Silver Star; "Kearie Lee Berry," Hall of Valor; Berry's family donated his medals to the Texas Military Forces Museum at Camp Mabry, Austin, Texas.

14. Colonel Matthews to Colonel Berry, January 30, 1945, in author's private collection.

15. Colonel Matthews to General Berry, March 8, 1946, in author's private collection; Berry wrote Eisenhower on June 4, 1946; Lyndon Johnson wrote Eisenhower on June 6, 1946; Lt. Col. James Stack to Col. K. L. Berry, June 7, 1946, in author's private collection. Stack was aide to General Eisenhower, chief of staff of the War Department.

16. General Eisenhower to General Berry, July 9, 1946, in author's private collection.

17. Paul J. Kilday to General Berry, August 28, 1946, in author's private collection.

18. General K. L. Berry to Claude Birkhead, May 26, 1947, in author's private collection.

19. Berry to Birkhead, May 26, 1947.

20. *Recommended Lists for Promotion to General Officer Grades*, pamphlet (Department of the Army, December 5, 1947), in author's private collection.

CHAPTER 18

1. Beauford Jester to Gen. K. L. Berry, May 10, 1947, in author's private collection.
2. "Generals Berry and Martin Promoted," *The National Guardsman*, July 1, 1947, 6.
3. Dick Smith and Laurie E. Jasinski, "History and Role of the Texas Adjutant General's Office,"
in *Handbook of Texas Online* (Texas State Historical Association), updated November 1, 1994, https://tshaonline.org/handbook/online/articles/mba01.
4. "Gen. Berry Named Terrell Trustee," *The National Guardsman*, July 1, 1947, 4.
5. "Berry Sworn In As Draft Head," *Austin American*, July 23, 1948, 1.
6. "Berry to Spark Campaign," *Austin American*, February 17, 1953, 11.
7. "Texas Adjutant to Head March of Dimes in State," Extension of Remarks of Hon. Lyndon B. Johnson, 100 Cong. Rec. (Bound) vol.100, part 20 (June 7, 1954–December 2, 1954), June 29, 1954; https://www.govinfo.gov/app/details/GPO-CRECB-1954-pt20/context.
8. "Gen. Berry Reappointed To Head of March of Dimes," *Houston Post*, September 3, 1955.
9. Berry's family members who attended the funeral in Austin remembered that the plane circled the Capitol building twice. Governor Jester would have been pleased to know that Berry's son Tom and Jester's daughter, Joan, found each other, fell in love, and married in 1951. "Plane Brings Governor's Body Home To Corsicana for Funeral Today," *Fort Worth Star-Telegram*, Morning Edition, July 13, 1949, 2; William McGill to General Berry July 18, 1949, in author's private collection.
10. Archie P. McDonald, "Allan Shivers," TexasEscapes.com, accessed March 16, 2022, http://www.texasescapes.com/DEPARTMENTS/Guest_Columnists/East_Texas_all_things_historical/AllanShivers112600McDonald.htm.
11. "Texas Adjutant General Urges Americans to Demand Action," *Abilene Reporter-News*, September 7, 1950, 2.
12. K. L. Berry to Lyndon B. Johnson, July 25, 1950, in author's private collection.
13. Berry to Johnson, July 25, 1950.
14. "Kennady to Attend Guard Group Session," *Fort Worth Star-Telegram*, October 4, 1950, 7; Bill Boehm, "Six Guard divisions play key role in Korean War," *National Guard News*, June 25, 2020, https://www.nationalguard.mil/News/Article-View/Article/580794/six-guard-divisions-play-key-role-in-korean-war/.
15. Jimmy Banks, "Gen. Berry Denies Camp Hulen Misused as Recreation Center," *Dallas Morning News*, August 17, 1950.
16. Banks, "Gen. Berry Denies Camp Hulen Misused as Recreation Center."
17. Banks, "Gen. Berry Denies Camp Hulen Misused as Recreation Center."

18. Ray O'Day, "Chit Chat," *The Quan: Official Publication of the American Defenders of Bataan and Corregidor*, November 3, 1948.

19. "Prisoner of War Club Opens Meeting," *Valley Morning Star*, July 22, 1950, 1.

20. Chester J. Pach Jr., "Dwight D. Eisenhower: Campaigns and Elections," The Miller Center, University of Virginia, accessed February 26, 2021, https://millercenter.org/president/eisenhower/campaigns-and-elections.

21. Sam Kinch, "Texas Adjutant General One of Eisenhower's Greatest Fans," *Fort Worth Star-Telegram*, June 15, 1952, 28.

22. Pach, "Dwight D. Eisenhower: Campaigns and Elections."

23. Kinch, "Texas Adjutant General One of Eisenhower's Greatest Fans."

24. "At Inauguration: Cow Bell 'Silent' in 1948 to Ring Out for Ike Now," *Austin American*, January 15, 1953, 1; "Ike Gets Texas Cow Bell for 'Fog'-Time Use," *Houston Chronicle*, January 30, 1953.

25. Col. K. L. Berry, POW Camp Shirakawa diary, June 9, 1943–October 8, 1944, in author's collection and at the Texas Military Forces Museum at Camp Mabry, Austin, Texas.

26. Author visited the ranch and small house many times in the 1980s.

27. K. L. Berry III, conversation with the author, 2022.

28. Bruce Reilly and William King Reilly conversations with the author, 2022.

29. Reilly and Reilly, conversation with author.

30. Reilly and Reilly, conversation with author.

31. Reilly and Reilly, conversation with author.

32. Reilly and Reilly, conversation with author.

33. "POW Memory Brings General Back: Berry is Guest of Old Friend in Hunt," *Brownwood Bulletin*, September 5, 1948.

34. Harold Scherwitz, *San Antonio Sunday Light Sports*, November 4, 1945.

35. Bill Harding, "Berry Suggests Army Borrow Strategy from Football Coaches," *Austin American-Statesman*, July 16, 1950.

36. Abe Asis to General Berry, May 15, 1955, in author's private collection.

37. Abe Asis to Colonel Berry, Nov 13, 1946, in author's private collection; Abe Asis to General Berry, May 15, 1955, in author's private collection.

38. "Gen. Berry Sees Bataan Commanders," *Manila Bulletin*, May 21, 1955.

39. Abe Asis to General Berry, May 15, 1955, in author's private collection.

40. "Honored U.S. General Weeps," *Saturday Mirror* (Manila), May 21, 1955.

41. "Berry Recalls Bataan Days with his Former Officers," *The Sunday Times* (Manila), May 22, 1955, "seasoned combat men."

42. General Berry made special mention of Santos after his release from prison camps when he wrote the "History of the 3rd Infantry Regiment, 1st Regular Division (PA)." Berry referred to the Battle of the Pockets when he wrote, "Major Howard Hinman and Captain Alfredo M. Santos, both later promoted one grade, distinguished themselves by their valiant conduct under fire and were awarded the

Distinguished Service Cross." In 1962, Santos became a four-star general and chief of staff of the Armed Forces of the Philippines.

43. General Berry to Lieutenant General Vargas, June 22, 1955, in author's private collection.

44. "Austinites Honor Adjutant General," *Austin American*, June 2, 1955.

45. Vivian Elizabeth Smyrl, "Camp Mabry," *Handbook of Texas Online* (Texas State Historical Association), accessed April 3, 2019, https://tshaonline.org/handbook/online/articles/qbc18.

46. "Honors Go to Phinney and Berry," *Austin American*, June 15, 1958.

47. "Honors Go to Phinney and Berry," *Austin American*; Berry's mother, Viola, would have been especially proud of her soldier son being compared to Travis, Crockett, and Houston, who were heroes in her mind and in her book of poetry.

48. Nat Henderson, "Ike Arrives Here for Night's Stay," *Austin American*, February 19, 1959.

49. Henderson, "Ike Arrives Here for Night's Stay."

50. "Adjutant General Gets Top NG Award," *Austin American*, October 8, 1959.

51. "Distinguished Service Medal," National Guard Association of the United States, accessed February 18, 2021, https://www.ngaus.org/about-ngaus/awards.

52. "Four UT Stars to Enter Hall," *Austin American*, October 25, 1959.

53. John Roehm to General Berry, January 23, 1961, in author's private collection; "Adjutant General Appointment Hangs: Governor Reluctant to Stir Boiling Pot," *Austin Times Herald*, February 16, 1961.

54. K. L. Berry to Lloyd Gregory, January 25, 1961, in author's private collection.

55. Price Daniel to Eldon Young, February 9, 1961, copy in author's private collection.

56. Hundreds of letters from First Officers Training Camp members and other prominent citizens of Texas, including Meade F. Griffin, Supreme Court of Texas (January 13, 1961); William H. Simpson, retired US Army general (January 17, 1961); Bob Bullock (January 25, 1961); Lloyd Bentsen (January 13, 1961); Clyde Littlefield (January 14, 1961); A. Garland Adair (January 27, 1961), in author's private collection; "Adjutant General Appointment Hangs: Governor Reluctant to Stir Boiling Pot," *Austin Times Herald*, Feb 16, 1961; Col. John Roehm to Governor Daniel, January 18, 1961, in author's private collection.

57. "Taylor is Adjutant General," *Houston Post*, May 23, 1961, "the boss man has spoken." K. L. Berry to Sam Riley, June 16, 1961; "Gen. Bishop In Favor of Lease," *Austin American*, August 6, 1964; "Mabry Bill Approved," *Austin American*, June 6, 1963. To replace Berry, Governor Daniel appointed Gen. James E. Taylor, former state senator and House member, to be the new adjutant general. Berry thought highly of General Taylor and thought he was a good choice for the job: "The governor picked a good man in Gen. Taylor, and he has my 100 percent support." Taylor took a leave of absence from his position with the Texas Motor Association to

become adjutant general, so perhaps it was not surprising that Taylor resigned in December, just six months into the job, to return to the Texas Motor Association. Governor Daniel immediately appointed General Bishop to the adjutant general position effective January 1, 1962. One wonders if the whole business with the adjutant general position was just to remove Berry, replace him with a prominent person for six months, and then appoint Bishop, because Bishop supported the golf course idea. The golf course was approved by the legislature in 1963, and Governor John Connolly signed the bill. However, in the end, after several years of talking about building the golf course, it never happened. Lack of resources for such a project scuttled it.

58. Gen. K. L. Berry, "Farewell Message," General Orders No. 66, State of Texas Adjutant General's Department, Austin, June 30, 1961, copy in Author's private collection.

59. Berry, "Farewell Message."

60. "Taylor Named To Command Texas Guard," *Vernon Daily Record*, May 22, 1961.

CHAPTER 19

1. K.L. Berry to Sam Riley, June 16, 1961, copy of letter in author's private collection.

2. K.L. Berry to Gen. Holt Atherton, June 21, 1961, copy of letter in author's private collection.

3. "General Berry to Retire, Head Forty Acres Club," *Austin American*, June 25, 1961, 24.

4. "The 9th Column," *Austin American*, October 17, 1961, 17.

5. "Gen. Berry Ailing in SA Hospital," *Austin American*, October 20, 1961, 1; *Corpus Christi Times*, October 21, 1961, 12.

6. King Reilly, memories of K. L. Berry, conversation with author in 2022.

7. "Gen. K. L. Berry, 72, Dies in San Antonio," *Austin American*, April 28, 1965, 1, 6.

EPILOGUE

1. Texas House Concurrent Resolution 110, May 4, 1965, in author's private collection.

2. "Heroism Award: General K. L. Berry Award," National Guard Association of Texas, NGAT Awards Program, accessed February 13, 2021, https://ngat.org/awards-program/.

3. Bill Little, "Standing Sentinel," The University of Texas at Austin Athletics, November 10, 2017, https://texassports.com/news/2017/11/10/football-standing-sentinel.

BIBLIOGRAPHY

PRIMARY SOURCE BOOKS

Beebe, Lewis. *Prisoner of the Rising Sun: The Lost Diary of Brig. Gen. Lewis Beebe.* Edited by John M. Beebe. Texas A&M University Press, 2006.

Berry, Viola R. *The Alamo and Other Poems.* News Publishing Company, 1906.

Braly, Col. William C. *The Hard Way Home.* Infantry Journal Press, 1947.

Brougher, William E. *South to Bataan, North to Mukden: The Prison Diary of Brigadier General W. E. Brougher.* Edited by Clayton D James. University of Georgia Press, 2010.

Bunker, Paul D. *Bunker's War: The World War II Diary of Col. Paul D. Bunker.* Edited by Keith Barlow. Presidio Press, 1996.

Cordero, V. N. *My Experiences During the War With Japan.* Zerreiss & Co., 1950.

Cordero, Virgilio N. *Bataan Y La Marcha De La Muerte.* Afrodisio Aguado, 1957.

Fortier, Malcolm. *The Life of A P.O.W. Under the Japanese in Caricature as Sketched by Col. Malcolm Vaughn Fortier.* C. W. Hill Printing CO, 1946.

Dittmar, Gus C. *They Were First, Recollections of The First Officers Training Camp of Leon Springs, Texas, May 8 to August 15, 1917.* The Steck-Warlick Company, 1969.

Dyess, William. *The Dyess Story: The Eye-Witness Account of the Death March on Bataan.* Pickle Partner Publishing, 2013.

Emerson, K. C. *Guest Of The Emperor.* Published by the author, fourth printing, 1987.

Leith, Hal. *POWs of Japanese Rescued!* Trafford Publishing, 2003.

Mallonee, Richard C. *The Naked Flagpole: Battle for Bataan from the Diary of Richard C. Mallonee.* Edited by Richard C. Mallonee II. Presidio Press, 1980.

Olson, John, *O'Donnell: Andersonville of the Pacific.* Published by the author, undated.

Olson, John E. *The Guerilla and the Hostage.* Burke Publishing Co., 1994.

Power, Desmond. *Little Foreign Devil*. Pangli Imprint, 1996.
Rio, Eliseo D. *Rays of a Setting Sun: Recollections of World War II*. Edited by Elisela R. Vitriolo. C & E Publishing, 2016.
Tenney, Lester I. *My Hitch in Hell: The Bataan Death March*. Brassey, 1995.
Wainwright, Jonathan M. *General Wainwright's Story: The Account of four years of humiliating defeat, surrender, and captivity by Jonathan M. Wainwright, who paid the price of his country's unpreparedness*. Edited by Robert Considine. Doubleday & Company Inc., 1946.

DIARIES

Aldridge, Ed. Diary and papers. Cushing Library, Texas A&M University Archives. Author visited archives in October 2021.
Berry, K. L. Diary, February 19, 1943–August 13, 1945. Author's private collection; Texas Forces Military Museum, Camp Mabry, Austin, Texas.
Lilly, Edmund. Diary vol. 19 and papers, Center for Research: Allied POWS Under the Japanese, Accessed February 19, 2019, http://www.mansell.com/pow_resources/camplists/china_hk/mukden/hoten_main.htm.
Galbraith, Nicoll F. Diary. Copy of diary given to author by Whitney Galbraith.
Corkill, Ed. Diary. Author's private collection.
"War Diary of USS RELIEF." Center for Research: Allied POWS Under the Japanese, edited by Roger Mansell. September 1945. http://www.mansell.com/Resources/special_files/FOLD3/USS%20Relief%20War%20Diary%201945-09.pdf.

DOCUMENTS

Berry, Alice. Telegram to Col. K. L. Berry. October 10, 1945. Author's private collection.
Berry, Col. K. L. Radiogram home. November 10, 1942. Author's private collection.
Berry, Col. K. L. Private Notebook,1942–1945. Author's private collection and transcribed copy given to Texas Military Forces Museum Camp Mabry, Austin, Texas.
Berry, Col. K. L. *History of the 3rd Regiment, 1st Regular Division (PA), December 19, 1941–April 9, 1942*. Author's private collection.
Berry, Gen. K. L. Short biography. Author's private collection.
Browne, Col. Harrison, interview by William K. Reilly, February 28, 1984. Transcript in author's private collection.
Gomez, Lt. Col. Victor Z. *Affidavit on Col. Berry's humanity to a fellow POW on the Death March, Oct. 3, 1947*. Author's private collection.
Invitation to the ball at Malacañan Palace. Author's private collection.

"Recommended Lists for Promotion to General Officer Grades." Department of the Army, Washington, DC, December 5, 1947.

Record of Lieutenant Berry's posting to California. US Military Forces Museum, Camp Mabry, Texas. Sent to the author by Lisa Sharik, deputy director of the museum. Author's private collection.

Texas House Concurrent Resolution, HCR No. 110. Introduced by Slack. May 4, 1965. Author's private collection.

UT certificate. Box contents of K. L. Berry. H. J. Lutcher Stark Museum, University of Texas at Austin.

LETTERS

Asis, Abe to Col. K. L. Berry, Nov 13, 1946. Author's private collection.

Asis, Abe to Gen. K.L. Berry, May 15, 1955. Author's private collection.

Berry, Alice to Col. K.L. Berry, September 15, 1945. Author's private collection.

Berry, K. L. to Lyndon B. Johnson, July 25. 1950. Author's private collection.

Berry, K. L. to Gen. Albert Jones, November 26, 1942. Author's private collection.

Berry, K. L. to Howard V. Ratliff, August 15, 1955. Author's private collection.

Berry, K. L. Series of letters to Alice Berry, October 30, 1942–November 30, 1942. Author's private collection.

Berry, K .L. to Alice Berry, February 2, 1942. Author's private collection.

Berry, K. L. to Alice Berry, February 22, 1942. Author's private collection.

Berry, K. L. to Sam Riley, June 16, 1961. Author's private collection.

Berry, Gen. K. L. to Gen. Claude Birkhead, May 26, 1947. Author's private collection.

Berry, K. L. to Gen. Holt Atherton, June 21, 1961. Copy in author's private collection.

Berry, Gen. K. L. to Lieutenant General Vargas, June 22, 1955. Author's private collection.

Berry, Gen. K. L. to Gen. Albert Jones. Letters, 1945–1947. Author's private collection.

Berry, Gen. K. L. to Lloyd Gregory, January 25, 1961. Author's private collection.

Berry, Gen. K. L. to Associate Justice Meade Griffin, January 30, 1961. Author's private collection.

Berry, Gen. K. L. Thank you letters to hundreds of First Officers Training Camp members and other Texas citizens, 1961. Author's private collection.

Berry, Tom to Speece-Kimber family, January 16, 2002. Author's private collection.

Daniel, Price to Eldon Young, February 9, 1961. Author's private collection.

Eisenhower, Gen. Dwight to Gen. K. L. Berry, July 9, 1946. Author's private collection.

Jester, Beauford to Gen. K. L. Berry, May 10, 1947. Author's private collection.
Kilday, Paul J. to Gen. K. L. Berry, August 28, 1946. Author's private collection.
Kimber, Jean Speece to the grandchildren of Helen DePass Dahlin, March 20, 1995. Author's private collection.
Marshall, Gen. George C. to General Wainwright, Nov 16, 1945. Author's private collection.
Matthews, Col. H. to Gen. K. L. Berry. Author's private collection.
Matthews, Col. H. to Col. K. L. Berry, October 27, 1945. Author's private collection.
Matthews, Col. H. to Col. K. L. Berry, January 30, 1945. Author's private collection.
McGill, William to Gen. K. L. Berry, July 18, 1949. Author's private collection.
Roehm, Col. John to Governor Daniel, January 18, 1961. Author's private collection.
Roehm, John to Gen. K. L. Berry, January 23, 1961. Author's private collection.
Stack, Lt. Col. James to Col. K. L. Berry, June 7, 1946. Author's private collection.
Texas Congressmen Connolly, Johnson, and Kilday letters, 1946–47. Author's private collection.
Wainwright, Gen. Jonathan to Col. K. L. Berry, November 7, 1945. Author's private collection.
Whitsell, Maj. Gen. Edward to Gen. K. L. Berry, February 26, 1946. Author's private collection.

REPORTS

1850 US Federal Census. Ancestry.com website. Census Place: Summerfield, Dallas, Alabama, roll M432_4, pg. 322A, image 656. Accessed February 20, 2019. https://www.ancestry.com/imageviewer/collections/8054/images/4187294-00656?pId=18139460.

1860 US Federal Census. Ancestry.com website. Census Place: Division 1, Montgomery, Alabama, enumerated August 4, 1860l, roll M653_19, pg. 121, image 249; Family History Library Film 803019. Accessed February 20, 2019. https://www.ancestry.com/imageviewer/collections/7667/images/4211192_00249?pId=.

1870 US Federal Census. Ancestry.com website. Census Place: Township 20, Elmore, Alabama, roll M593_15, pg. 139A, image 470; Family History Library Film 545514. Accessed February 20, 2019. https://www.ancestry.com/imageviewer/collections/7163/images/4257535_00470?pId=.

Barnes, Alexander F. and Cassandra J. Rhodes. "Logistics in Reverse: The U.S. Intervention in Siberia, 1918–1920," in *US Intervention in Siberia and Northern Russia 1918–1920*. Progressive Management Publications, 2019.

"Campaigns of MacArthur in the Pacific. Vol. I." US Army Center of Military History website. Accessed October 15, 2022. https://history.army.mil/Publications/Publications-Catalog-Sub/Publications-By-Title/Reports-of-General-MacArthur-Collection/.

"Chronology Timeline, December 7, 1944." Cynthia B. Caples, Public Affairs Officer, US Consulate, Shenyang Hoten POW Camp (Mukden, China). Center for Research Allied POWS Under the Japanese website. Edited by Roger Mansell. Accessed February 20, 2019. http://mansell.com/pow_resources/camplists/china_hk/mukden/mukden_timeline.html.

"Directory of Deceased American Physicians, 1804–1929." Ancestry.com website. Accessed February 20, 2019. Cincinnati Medical College, 1836.

"Japanese Order Posted in POW Camps [1944]." Stensby, John L. Sr. Collection. Library of Congress. Accessed June 7, 2025, https://www.loc.gov/resource/afc2001001.00454.pm0003001/?r=-0.845,-0.062,2.689,1.421,0.

"Mukden Rescue and Evacuation: Evacuation Air and Train Rosters." Center for Research Allied POWS Under the Japanese website. Edited by Roger Mansell. Accessed February 20, 2019. http://www.mansell.com/pow_resources/camplists/china_hk/mukden/mukden_train_evac_2.html.

Passenger Lists 1910–1939: US Army Transport Service. The National Archives at College Park, College Park, Maryland, Records of the Office of the Quartermaster General, 1774–1985. Record group 92, roll or box 571, June 5, 1919. Departure Place: San Francisco. Ancestry.com website.

Passenger Lists 1910–1939: US Army Transport Service. The National Archives at College Park, College Park, Maryland, Records of the Office of the Quartermaster General, 1774–1985. Record group 92, roll or box 605. Arrival Chinwangtao July 6, 1933. Ancestry.com website.

Passenger Lists 1910–1939: US Army Transport Service. The National Archives at College Park; College Park, Maryland; Records of the Office of the Quartermaster General, 1774–1985. Record group 92, roll or box 343. Departure Date: July 28, 1936. Ancestry.com website.

Smalser, Robert L. "The Siberia Expedition 1918–1920: An Early Operation Other than War," in *US Intervention in Siberia and Northern Russia 1918–1920.* Progressive Management Publications, 2019.

"Texas Adjutant to Head March of Dimes in State," Extension of Remarks of Hon. Lyndon B. Johnson, 100 Cong. Rec. (Bound) vol.100, part 20 (June 7, 1954–December 2, 1954), June 29, 1954; https://www.govinfo.gov/app/details/GPO-CRECB-1954-pt20/context.

US Select Military Registers, 1862–1985: K.L. Berry. Ancestry.com website. This collection was indexed by Ancestry World Archives Project contributors from the original data at US Military Registers, 1902–1985. Oregon State Library, 2019.

"World War I Draft: Topics in Chronicling America." Library of Congress Research Guides website. Accessed May 23, 2025. https://guides.loc.gov/chronicling-america-wwi-draft.

SECONDARY SOURCE BOOKS

Adams, John A. Jr. *The Fightin' Texas Aggie Defenders of Bataan and Corregidor*. Texas A&M University Press. 2016.

American and Allied Personnel Recovered From Japanese Prisons: A Pictorial History. Replacement Command. AFWESPAC, November 11, 1945.

Astor, Gerald. *Crisis in the Pacific: The Battles for the Philippine Islands by the Men Who Fought Them*. A Dell Book, a Division of Random House. 1996.

Bartsch, William H. *December 8, 1941: MacArthur's Pearl Harbor*. Texas A&M University Press, 2003.

Burns, James MacGregor. *Roosevelt: The Lion and the Fox*. Harcourt Brace, 1956.

Cactus Yearbook. University of Texas, 1913. http://hdl.handle.net/2152/61703

Cactus Yearbook. University of Texas, 1915. http://hdl.handle.net/2152/2428.

Cactus Yearbook. University of Texas, 1916. http://hdl.handle.net/2152/23763

Cave, Dorothy. *Beyond Courage: One Regiment against Japan, 1941–1945*. Sunstone Press, 2006.

Chun, Clayton. *The Fall of the Philippines 1941–1942*. Oxford: Osprey Publishing Ltd., 2012.

Clubb, Edmond O., *Twentieth Century China*, Columbia University Press, 1964.

Cornebise, Alfred E. *The United States 15th Infantry Regiment in China, 1912–1938*. McFarland & Company, Inc., 2004.

Dale, Liam. "The Road to War." In *World War 2, The Call of Duty: A Complete Timeline*. The History Journals, 2019.

The Bronco, Yearbook of Denton High School 1911. Denton High School, 1911.The Portal to Texas History website. https://texashistory.unt.edu/ark:/67531/metapth743014/?q=The%20Bronco%20Yearbook%201911.

Frank, Richard B. *Downfall: The End of the Imperial Japanese Empire*. Random House, 1999.

Henderson, Harry M. *History of the 141st Infantry, 36th Infantry Division Texas National Guard*. The Naylor Company, 1950.

Hitchcock, William I. *The Age of Eisenhower: America and the World in the 1950s*. Simon and Schuster, 2018.

Knox, Donald. *Death March: The Survivors of Bataan*. Harcourt Inc., 1981.

Latourette, Kenneth S. *The Chinese: Their History and Culture*. The MacMillan Company, 1934.

Manchester, William. *American Caesar: Douglas MacArthur 1880–1964*. Little Brown & Company, 1978.

Miller, Ernest B. "Bataan Uncensored." The Hart Publications, 1949. Accessed January 10, 2021, via the Internet Archive, https://archive.org/details/BataanUncensored-nsia.

Morton, Louis. *The War in the Pacific: The Fall of the Philippines*. Center of Military History United States Army, 1989.

Norman, Michael, and Elizabeth M. Norman. *Tears in the Darkness: The Story of the Bataan Death March and Its Aftermath*. Farrar, Straus and Giroux, 2009.

Schultz, Duane. *Hero of Bataan: The Story of General Jonathan M. Wainwright*. St. Marten's Press, 1981.

Sides, Hampton. *Ghost Soldiers: The Epic Account of World War II's Greatest Rescue Mission*. Anchor Books, 2002.

Stewart, Richard W., ed. "The United States Army in a Global Era, 1917–2008" in *American Military History Vol. II*, Center of Military History US Army website. https://www.armyupress.army.mil/Portals/7/educational-services/military-history/american-military-history-volume-2.pdf.

Texas National Guard. *Historical and Pictorial Review: National Guard of the State of Texas*. 1940. University of North Texas Libraries, The Portal to Texas History; crediting Boyce Ditto Public Library website. https://texashistory.unt.edu/ark:/67531/metapth833790/.

Toland, John. *But Not in Shame: The Six Months After Pearl Harbor*. Random House, 1961.

Trask, David. "The Entry of the USA into the War and Its Effects." In *The Oxford Illustrated History of the First World War*. Edited by Hew Strachan. Oxford University Press, 2014.

Tuchman, Barbara W. *The Zimmermann Telegram: American Enters the War, 1917–1918*. Random House, 2014.

Whitman, John. *Bataan: Our Last Ditch*. Hippocrene Books, 1990.

ARTICLES

"A Brief History of the 27th Regiment." The Wolfhound Pack: US 27th Infantry Regimental Historical Society. Accessed September 27, 2019. https://wolfhound-pack.org/history-of-the-regiment/#brief_history.

"Alphonse and Gaston." Don Markstein's Toonopedia. Accessed January 6, 2023. http://toonopedia.com/alphgast.htm.

Barnes, Alexander F. "On the Border: The National Guard Mobilizes for War in 1916." *Army Sustainment Magazine*, March–April 2016. https://www.army.mil/article/162413/on_the_border_the_national_guard_mobilizes_for_war_in_1916.

"Bataan Death March." Map. National Museum of the United State Air Force. Accessed January 24, 2023. https://www.nationalmuseum.af.mil/Visit/Museum-Exhibits/Fact-Sheets/Display/Article/196797/bataan-death-march/.

"Battle of Okinawa." The National World War II Museum. Accessed February 24, 2022. https://www.nationalww2museum.org/war/topics/battle-of-okinawa.

Bisher, Jaime. "American Expeditionary Force-Siberia (AEFS)." Cossack Warlords of the Trans-Siberian, 2016. https://cossackwarlords.weebly.com/american-expeditionary-force-siberia.html.

Bishop, Lorene. "History of Camp Bowie." City of Brownwood Texas website. Accessed September 23, 2019. http://www.brownwoodtexas.gov/323/History-of-Camp-Bowie.

Bisno, Adam. "The Japanese 'Hell Ships' of World War II." Naval History and Heritage Command, November 2019. https: //www.history.navy.mil/browse-by-topic/wars-conflicts-and-operations/world-war-ii/1944/oryoku-maru.html.

Bisno, Adam. "The U.S. Navy and the Sino-Japanese War of 1894–95." Naval History and Heritage Command. Accessed January 24, 2023. https://www.history.navy.mil/browse-by-topic/wars-conflicts-and-operations/steam-steel-navy/sino-japanese-war.html.

Boehm, Bill. "Six Guard Divisions Play Key Role in Korean War." *National Guard News*. June 25, 2020. https://www.nationalguard.mil/News/Article-View/Article/580794/six-guard-divisions-play-key-role-in-korean-war/.

"Camp Bowie Army Base in Brownwood, TX." Texas Military Bases website. Accessed January 29, 2020. https://militarybases.com/texas/camp-bowie/.

Clark, Rob. "The Life of Ovid O. Wilson." *Bryan-College Station Eagle*, November 11, 2018. https://theeagle.com/news/local/the-life-of-ovid-o-wilson/article_4e860ec7-6a58-50bc-b8e7-435faa5ab7ed.html.

Cochran, Mike. "A Brief History of Denton County." The History of Denton, Texas, Online Since 1996. Accessed April 13, 2019. http://dentonhistory.net/.

Dawsey, Jason. "A Shared Enmity: Germany, Japan, and the Creation of the Tripartite Pact." National World War II Museum. Accessed December 12, 2021. https://www.nationalww2museum.org/war/articles/germany-japan-tripartite-pact.

De León, Arnoldo and Robert A. Calvert. "Segregation." *Handbook of Texas Online* (Texas State Historical Association). Updated January 27, 2021. https://www.tshaonline.org/handbook/entries/segregation.

"Distinguished Service Medal." National Guard Association of the United States. Accessed February 18, 2021. www.ngaus.org.

"Exhibits: The Office of Strategic Services: America's First Intelligence Agency." Central Intelligence Agency. Accessed April 6, 2022. https://www.cia.gov/legacy/museum/exhibit/the-office-of-strategic-services-n-americas-first-intelligence-agency/.

Federer, Bill. "The Life and times of Herbert Hoover; Warned against an Expanding Federal Government." *World Tribune: Window on the Real World*. August 10, 2019. https://worldtribune.com/life/the-life-and-times-of-herbert-hoover-warned-against-an-expanding-federal-government/.

"First Officers Training Camp" Texas Historic Markers website. Marker 11744. 1999. https://texashistoricalmarkers.weebly.com/first-officers-training-camp.html.

"Generals Berry and Martin Promoted," *The National Guardsman*, vol. 5, no. 2, July 1, 1947.

"Gen. Berry Named Terrell Trustee," *The National Guardsman*, vol. 5, no. 2 July 1, 1947.

"General Walter Krueger." Military Hall of Honor. Accessed January 20, 2020. https://militaryhallofhonor.com/honoree-record.php?id=268.

Gibbs, John M. "Beppu." American Ex-Prisoners of War. July 31, 1946. https://www.axpow.org/medsearch/beppu.htm.

Graff, Henry F. "Grover Cleveland: Domestic Affairs." The Miller Center, University of Virginia. Accessed October 11, 2022. https://millercenter.org/president.

Graff, Henry F. "Grover Cleveland: Impact and Legacy." The Miller Center, University of Virginia. Accessed October 11, 2022. https://millercenter.org/president.

Greenberg, David. "Calvin Coolidge: Domestic Affairs." The Miller Center, University of Virginia. Accessed October 9, 2022. https://millercenter.org/president/coolidge/domestic-affairs.

Grieco, Michael C. "Making Sense of the Unknown: The AEF in Siberia." US Army School of Advanced Military Studies, US Army Command and General Staff College, Ft. Leavenworth, Kansas website. 2018 (Monograph). https://apps.dtic.mil/sti/pdfs/AD1071803.pdf

Hamilton, David E. "Herbert Hoover: Domestic Affairs." The Miller Center, University of Virginia. Accessed October 9, 2022. https://millercenter.org/president/hoover/domestic-affairs.

"Heroism Award: General K. L. Berry Award." National Guard Association of Texas. NGAT Awards Program. Accessed February 13, 2021. http://www.ngat.org/awards.htm.

"How Did Hitler Happen?" National World War II Museum. Accessed December 10, 2020. https://www.nationalww2museum.org/war/articles/how-did-hitler-happen.

Hoffmann, Fritz L. "Villa, Francisco [Pancho]." *Handbook of Texas Online* (Texas State Historical Association). Updated July 2, 2016. https://www.tshaonline.org/handbook/entries/villa-francisco-pancho.

Johnson, Erik. "An Army With No Country: the Czechoslovak Legion in Europe." Czech Center Museum Houston. November 11, 2021. https://www.czechcenter.org/blog/2021/11/11/an-army-with-no-country-the-czechoslovak-legion-in-europe.

"Kearie Lee Berry." Hall of Valor, Military Medals Database. Accessed February 20, 2021. https://valor.militarytimes.com/hero/6024.

Kolb, Richard K. "Walking on Eggs Loaded with Dynamite: America's Adventure in Siberia." *Veterans of Foreign Wars*, February 1991, 14–17.
Lacy, Lee. "Harry Truman and The Bomb." *Pieces of History* blog, US National Archives, August 5, 2014. https://prologue.blogs.archives.gov/2014/08/05/harry-truman-and-the-bomb.
Larson, Vaughn R. "Wisconsin Guard Protected Mexican Border a Century Ago." Wisconsin National Guard, June 24, 2016. https://ng.wi.gov/news/16063.
Leatherwood, Art. "Leon Springs Military Reservation." *Handbook of Texas Online* (Texas State Historical Association). Accessed March 29, 2019. https://tshaonline.org/handbook/online/articles/qbl06—leatherwood.
Little, Bill. "Standing Sentinel." University of Texas at Austin Athletics. Accessed March 6, 2021. https://texassports.com/news/2017/11/10/football-standing-sentinel.
"Prisoner of War Covers" Manchuko Stamps website. Accessed October 25, 2021. http://www.manchukuostamps.com/POWcovers.htm.
"Map of the Philippines." National Museum of the United States Air Force website. https://www.nationalmuseum.af.mil/Upcoming/Photos/igphoto/2000510499/.
McDonald, Archie P. "Allan Shivers." *Texas Escapes Magazine*. Accessed March 16, 2022. http://www.texasescapes.com/DEPARTMENTS/Guest_Columnists/East_Texas_all_things_historical/AllanShivers112600McDonald.htm.
Metz, Leon C. "Fort Bliss." *Handbook of Texas Online* (Texas State Historical Association) Updated October 3, 2019. https://www.tshaonline.org/handbook/entries/fort-bliss.
Milkis, Sidney, "Theodore Roosevelt: Foreign Affairs." The Miller Center, University of Virginia. Accessed November 24, 2022. https://millercenter.org/president/roosevelt/foreign-affairs.
"Mobilizing for War: The Selective Service Act in World War I." National Archives Foundation. Accessed June 1, 2025. https://www.archivesfoundation.org/documents/mobilizing-war-selective-service-act-world-war/#:~:text=Mobilizing%20for%20War%3A%20The%20Selective%20Service%20Act%20in,21%20to%2045%20to%20register%20for%20military%20service.
Nieuwint, Joris. "When the Allies Killed Over 20,000 of Their Own Countrymen As They Sank Japanese Hell Ships That Transported Them." War History Online. January 17, 2016. https://www.warhistoryonline.com/featured/japanese-hellships.html.
"1914: A Perfect Season." H. J. Lutcher Stark Center for Physical Culture and Sports website. University of Texas. Accessed May 9, 2019. https://www.starkcenter.org/1914-a-perfect-season/.
Odom, E. Dale. "Denton, TX (Denton County)." *Handbook of Texas Online* (Texas State Historical Association). Updated September 9, 2020. https://www.tshaonline.org/handbook/entries/denton-tx-denton-county.

"Office of Strategic Services: America's First Intelligence Agency." Central Intelligence Agency, exhibits. Accessed December 26, 2022. https://www.cia.gov/legacy/museum/exhibit/the-office-of-strategic-services-n-americas-first-intelligence-agency/.

Oko, Dan. "Teddy Roosevelt, San Antonio, and the Birth of the Rough Riders: Texas Cowboys in Cuba." *Texas Highways*. October 29, 2019. https://texashighways.com/culture/history/teddy-roosevelt-san-antonio-and-the-birth-of-the-rough-riders/.

Olson, Bruce A. "Texas National Guard." *Handbook of Texas Online* (Texas State Historical Association). Accessed November 5, 2021. https://www.tshaonline.org/handbook/entries/texas-national-guard.

Overfelt, Robert C. "Mexican Revolution." *Handbook of Texas Online* (Texas State Historical Association). Updated December 2, 2020. https://www.tshaonline.org/handbook/entries/mexican-revolution.

Pach, Chester J. Jr. "Dwight D. Eisenhower: Campaigns and Elections." The Miller Center, University of Virginia. Accessed February 26, 2021. https://millercenter.org/president/eisenhower/campaigns-and-elections.

"President Herbert Hoover." National Archives, Herbert Hoover Presidential Library and Museum. Accessed December 10, 2022. https://hoover.archives.gov/hoovers/president-herbert-hoover.

Pope, S. W. "An Army of Athletes: Playing Fields, Battlefields, and the American Military Sporting Experience, 1890–1920." *The Journal of Military History* 59(3), 1995. https://doi.org/10.2307/2944617.

Putnam, Christine L. "AEF Siberia." The Doughboy Center: The Story of the American Expeditionary Forces. Accessed October 20, 2021. http://www.worldwar1.com/dbc/ghq2arm.htm.

"Schofield Barracks Army Base Guide." Military.com Network. Accessed October 19, 2019. https://www.military.com/base-guide/schofield-barracksfort-shafter.

Scottie, Kersta-Wilson. "My Granddaddy." *The Quan: The Official Newsletter of the American Defenders of Bataan and Corregidor Memorial Society*, Spring 2021.

Smith, Dick and Laurie E. Jasinski. "History and Role of the Texas Adjutant General's Office." *Handbook of Texas Online* (Texas State Historical Association). Accessed April 3, 2019. https://tshaonline.org/handbook/online/articles/mba01.

Smith, Gibson Bell. "Guarding the Railroad, Taming the Cossacks: The U.S. Army in Russia 1918–1920." *Prologue Magazine* (National Archives and Records Administration). Accessed September 9, 2019. https://www.archives.gov/publications/prologue/2002/winter/us-army-in-russia-1.html.

Smyrl, Vivian E. "Camp Mabry." *Handbook of Texas Online* (Texas State Historical Association). August 11, 2020. https://tshaonline.org/handbook/online/articles/qbc18.

"The Camps." Never Forgotten website. Accessed June 1, 2025. http://taiwanpow
.org/The%20Camps/index.php.

"The Defining Role of the National Guard in WWI." National Guard Bureau
Historical Services. August 7, 2017. https://www.army.mil/article/191849.

"The Early Years at Fort Benning," in *Fort Benning: Home of the Infantry Booklet*.
Chattahoochee Valley Libraries website. Accessed June 5, 2025. https://dlg
.galileo.usg.edu/chattahoochee/do:cvl159.

"The Formative Period: Formation of the Division and Training at Camp Travis."
90th Infantry Division Association. Accessed May 21, 2019. http://www
.90thdivisionassoc.org/90thDivisionFolders/mervinbooks/WWI90/WWI9002
/WWI9002.htm.

"The Treaty of Portsmouth and the Russo-Japanese War, 1904–1905." Office of the
Historian, US Department of State. Accessed January 24, 2023. https://history.
state.gov/milestones/1899–1913/portsmouth-treaty.

"The 24th Infantry and Fort Benning." Chattahoochee Valley Libraries. Accessed
May 20, 2019. https://georgialibraries.omeka.net/s/CVL-Columbus-Georgia-
Hear-Us-Talking/page/24thinfantry.

"The 24th Infantry Regiment." 25th Infantry Division Association. Accessed
December 20, 2022. https://www.25thida.org/units/infantry/24th-infantry
-regiment/.

Trickey, Erick. "Forgotten Doughboys Who Died Fighting in the Russian Civil
War." *Smithsonian Magazine*. February 12, 2019. https://www.smithsonianmag.
com/history/forgotten-doughboys-who-died-fighting-russian-civil-war
-180971470/.

"The 21st Infantry Regiment." 25th Infantry Division Association. Accessed
November 14, 2021. https://www.25thida.org/units/infantry/21st-infantry-
regiment/.

White, Lonnie J. "The 36th Division in World War I." Texas Military Forces
Museum. Accessed May 21, 2019. http://www.texasmilitaryforcesmuseum
.org/36division/archives.

White, Lonnie J. "141st Infantry Regiment." Texas Military Forces Museum.
Accessed May 21, 2019. http://www.texasmilitaryforcesmuseum.org/36division
/archives.

"Wilson, Woodrow." The White House. Accessed December 10, 2022. https://
www.whitehouse.gov/about-the-white-house/presidents/woodrow-wilson/.

"Winter Texan Wednesday: The Story of the 2nd Infantry Football Team." Museum of South Texas History. February 6, 2019. https://mosthistory.org/events
/winter-texan-wednesday-the-story-of-the-2nd-infantry-football-team/.

"World War I: Short History of the 90th Division at Camp Travis." The 90th
Division Association. Accessed May 21, 2019. http://www.90thdivisionassoc

.org/90thDivisionFolders/WWONE/a_short_history_of_the_90th_divi.htm

Yockelson, Mitchell. "The United States Armed Forces and the Mexican Punitive Expedition: Part 1." *Prologue Magazine.* Winter 1997, vol. 29, no. 4. Accessed September 2, 2021. https://www.archives.gov/publications/prologue/1997/winter/mexican-punitive-expedition-2. html.

Zimmer, Phil. "Death of Admiral Isoroku Yamamoto." Warfare History Network website. Accessed January 15, 2023. https://warfarehistorynetwork.com/article/death-of-admiral-isoroku-yamamoto/.

PERIODICALS

Abilene Reporter-News
Austin American
Austin American-Statesman
Austin Times Herald
Brownwood Bulletin
Dallas Morning News
Denton Record Chronicle
Evening News (Manila)
Evening Star (Washington, DC)
Fort Worth Record Sports
Fort Worth Star-Telegram
Galveston Daily News
Houston Chronicle
Houston Post
Longhorn T (University of Texas at Austin)
Manila Bulletin
National Guardsman
Philippines Herald (Manila)
Quan: Official Publication of the American Defenders of Bataan and Corregidor, Inc
Reveille (Corpus Christi)
San Antonio Evening News
San Antonio Light Sports
Saturday Mirror (Manila)
Sentinel (Tienstin, China)
Sunday Times
Valley Morning Star
Vernon Daily Record

INDEX

Italicized page numbers indicate photos, illustrations, or maps.

Aldridge, Edwin E. (Colonel): Battle of Bataan, 82–84, 86–88, 95, 96; and Bell, Gilmer M., 155; birthday gift for Berry, K. L., at Mukden camp, 205; Camp Karenko, 151; Camp O'Donnell, 110; Camp Shirakawa, 155, 156, 159, 164, 166, 170–71, 175; Cheng Chia Tun camp, 198, 199; Fort McKinley, 54; letters from home while POW, 166; post-Battle of Bataan, 103, 104; in *The Quan* "Chit Chat" column, 233; Wainwright Travelers 1950 gathering, 235
American Expeditionary Force (AEF), 26
Amis, William M. (Colonel), 151
Anderson County, Texas: Berry, Thomas Eugene, in Tennessee Colony of, 35
Anyasan Point, 77
Asis, Abe, 241, 243
Atherton, Holt, 250–51
athletic training: and military programs, 33, 34, 37, 44
Atkinson, Edward C. (Colonel), 151
atomic bombs: "Fat Boy" dropped on Nagasaki, 207; "Little Boy" dropped on Hiroshima, 207; POWs with no knowledge of, 207; successful test, 206; and Truman, Harry S., 206

Ausmus, Delbert (Colonel), 151
Austin, Texas, 16, 23, 35, 251; Driskill Hotel, 16; Eisenhower, Dwight D., meeting in, 245
Austin American, 246
Austin American-Statesman, 36
Australia: Battle of Bataan and stalled Japanese invasion of, 99, 243; Japanese attack on Darwin, 85
avitaminosis, 77
awards and honors for Berry, K. L.: Combat Infantryman's Badge, 223; Distinguished Service Cross, 223; Distinguished Service Medal, 223, 224; Distinguished Service Medal, National Guard Association of the United States, 245–46; Longhorn Hall of Honor, 246, *246*; Pershing Trophy, 48; Philippine Legion of Honor, 242, 243–44; Purple Heart, 223; Silver Star, 223; Texas Distinguished Service Medal, 244; Texas Heritage Foundation recognition, 244

Bagac, Philippines, 64, 65, 75, 76, 78, 79, 98; as Bataan Death March starting point, 98, 99, 100, 101–3, 105; Berry, K. L., interaction with Japanese soldier in, 102

"Baggy Pants." *See* Wakasugi, Jiro (2nd Lieutenant)
Baker, Newton D.: in Siberia, 26; and soldier athletic programs, 14–15
Balanga, Philippines, 64, 98; Bataan Death March POWs, arrival of first, 103; Bataan Death March POWs, food for, 104; Bataan Death March stop of, 100, 104
Bali: Japanese invasion of, 85
Balsam, Alfred S. (Colonel): birthday gift for Berry, K. L., at Mukden camp, 205; Camp Karenko, 125, 151; Camp Shirakawa garden, 159
Bardowski, Zenon, 82
Barrell, Len. *See* Longhorns
Barry, Edwin F. (Colonel): Bataan Death March, 99–105, 114, 125, 153; death in Camp Tarlac, 117–18
Bataan Death March, 14, 189; Bagac to Balanga, 98, 99, 100, 101–3, 105; Balanga to Orani, 100, 103, 104; bayoneting of POWs by Japanese soldiers, 103; beatings of POWs by Japanese soldiers, 103; beheadings of POWs by Japanese soldiers, 103; and Camp O'Donnell, 105, 109; Capas to Camp O'Donnell, 100, 105; deaths, 105; food, 104; health of POWs on, 99; Lubao to San Fernando, 100, 104–5; map, *100*; Mariveles to Balanga, 99, 101, 103, 105; Orani to Lubao, 100, 104; road conditions, 102; San Fernando to Capas train ride, 99, 100, 101, 103, 104–5; shootings of POWs by Japanese soldiers, 103, 104; survivors, 114; water, lack of on, 102, 104, 105. *See also* Bataan Death March, POWs on; Berry, K. L., and Bataan Death March
Bataan Death March, POWs on: Barry, Edwin F., 99–105, 114, 125, 153; Browne, E. Harrison, 103; Cordero, Virgil N., 104; Crow, Judson, 102; Dyess, William, 102, 103; Gomez, Victor, 102; Lay, Kermit, 103; Lough, Maxon S., 103; MacDonald, Stuart D., 103; Miller (Captain), 103; Tenney, Lester, 99, 101–2, 104, 105; Wetherby, Loren A., 104; Wohlfeld, Mark, 101. *See also* Berry, K. L., and Bataan Death March
Bataan Peninsula, 58, 59, 62, 64; January 8, 1942, situation on, 65; US Sixth Army recapture of, 200; US withdrawal to, 62. *See also* Bataan Death March; Battle of Bataan
Batalan River, 70
Battle of Bataan, 85–87, 113, 189, 243; April 1 to 10, 1942, 90–93; Battle of the Pockets, 79–85, *80*, 85, 95, 132; Battle of the Points, 79, 85; December 1941 to March 1942, 87–90; 11th Division, 77, 79, 80, 82, 92, 93, 94, 95, 96; 51st Division, 69, 92; 1st Regular Philippine Division, 64, 66, 67, 69, 70, 71, 73, 74, 75, 76, 77, 79, 80, 81, 83, 84–85, 94, 96; food rations, cuts to and lack of, 66, 73, 77, 86, 89, 90, 92, 93, 99; 41st Division, 92; 42nd Infantry, 92; health problems, soldiers, 77, 87, 89, 92; I Corps, 64, 65, 67, 68, 69, 77, 81, 86, 87; Igorot soldiers, 82; II Corps, 64, 65, 67, 68, 69, 86, 87, 96; January 1942, 64–75; Ledda, Daniel, injury to in, 69; 91st Division, 64, 77, 79, 98; 92nd Infantry, 73; 192nd Tank Battalion, 99; Philippine scouts (I Corp), 86; planning and preparations, 59–64, *61*; 2nd Philippine Constabulary, 77; 3rd Battalion, 63; 3rd Infantry, 63, 70, 71, 72, 74, 81; 3rd Regiment, Philippine Army, 62, 63, 64, 66; 31st Division, 64, 92; 21st Division, 92; 26th Cavalry, 64, 70, 73, 77, 89; view as successful, 99. *See also* Battle of Bataan, surrender of Americans and Filipinos after; Battle of Bataan, US and Filipino soldiers in
Battle of Bataan, surrender of Americans and Filipinos after, 93, 94–99; Berry, K. L., and 1st Division, 96–99, 101; and King, Edward P., 93, 96, 103; and Rio, Eliseo, 96–98
Battle of Bataan, US and Filipino soldiers in, 125, 132; Aldridge, Edwin E., 82–84, 86–88; Bardowski, Zenon, 82; Berry,

K. L. (*see* Berry, K. L., and Battle of Bataan); Bluemel, Clifford, 73, 92; Brougher, William E., 80, 82, 84, 88, 92, 93, 95, 96; Fitch, Alva, 71–74; Halstead, Earl T., 74; Houser, Houston, 71; Jones, Albert M., 68, 69, 82–84, 87–89, 93, 95; King, Edward P., 88, 89, 92, 93; Laird (Captain), 71, 74; Ledda, Daniel, 69; Lilly, Edmund J., 92; Lim, Vicente, 90, 92; Lough, Maxon S., 91; Mallonee, Richard C., 66, 68; Maury (Major), 68; McCollum, O. S., 69; Parker, George M., 60, 64, 65, 67, 70, 75, 77, 89, 90, 92; Rio, Eliseo, 62–64, 67, 68, 74, 79, 87; Rodman, John H., 73; Segundo, Fidel, 63, 67, 69–74, 82, 83; Sharp, William F., 60; Stevens, Luther R., 95, 96; Townsend, Glen R., 86, 94; Volckmann, Russell W., 95; Wainwright, Jonathan M., 64–71, 73–77, 79, 82, 83, 86–90, 92, 93; Wetherby, Loren A., 92. *See also* Battle of Bataan; Battle of Bataan, surrender of Americans and Filipinos after

Battle of the Bismarck Sea, 147

Battle of the Bulge, 200

Battle of the Coral Sea, 117

Battle of Okinawa, 205, 206; civilian deaths, 205; Japanese army deaths, 205; US victory, 205

Battle of the Pockets, 79–85, *80*, 85, 95, 132. *See also* Battle of Bataan

Battle of the Points, 79, 85. *See also* Battle of Bataan

Battling Bastards of Bataan, 114

Beckwith-Smith, Merton (Major General), 141

Beebe, Lewis C. (Brigadier General): birthday gift for Berry, K. L., at Mukden camp, 205; brigadier general status, awarded permanent, 227; Camp Tarlac, 116, 134; Wainwright Travelers 1950 gathering, 234

Bell, Gilmer M. (Colonel): and Aldridge, Edwin E., 155; and Berry, K. L., 163; birthday gift for Berry, K. L., at Mukden camp, 205; Camp Karenko, 143, 151; Camp Shirakawa, 155

Bennett, John (Colonel), 142

Beppu, Kyushu (Japan), 188, 191. *See also* Nitchi Man hotel (Beppu)

beriberi: and Battle of Bataan, 77; and Camp Karenko POWs, 141, 142, 143

Berlin, Germany, 200–201, 202

Berlin Wall: construction of, 250

Berry, Ada: letter to brother Berry, K. L., 166

Berry, Alice Fleming, 221, 223; births of children, 24, 34–35; Camp Mabry, 239; China, 43; death of husband, Berry, K. L., 254; death of son, Berry, K. L., Jr., 252; Hawaii, 34; wedding photo, *17*; Wright-Patterson Army Air Corps Base stay with son's family, 220

Berry, Anne: letter to brother Berry, K. L., 166. *See also* Matthews, Anne Berry

Berry, Celeste Viola, 162; birth of, 34–35; China, 43; letter to father, Berry, K. L., 166

Berry, Eugene "Gene," 6; football, 8, 9; and letter-writing campaign about Berry, K. L., brigadier general promotion, 222; letters to brother, Berry, K. L., 165; track, University of Texas, 10; and University of Texas, 7, 8, 9, 10

Berry, Jim, 46

Berry, John, 6; and letter-writing campaign about Berry, K. L., brigadier general promotion, 222

Berry, K. L., 7, 9, 36, 177, 215, 221; and Bell, Gilmer M., 163; Berry, Alice, not receiving mail from while POW, 52, 81, 82, 163–64, 165, 166, 219; and Birkhead, Claude, 226; birth of, 3; and Bolte, Gerta, 213; books read while POW, 135, 161–62, 198, 261–63; Brooke General Hospital stays, 220, 222, 251, 254; and Browne, E. Harrison, 205, 220, 259; Camp Bowie, 49, 50, 51; Camp Bullis, 48; Camp Kearney, 23, 48; Camp Mabry, move from, 25; Camp Perry, 39; China, 40–46, 48, 213; and Connally,

John, 254; death of, 254; education, college, 16, 35–37; education, high school, 6–7, 8; and Eisenhower, Dwight D., 15, 37, 51, 147, 225, 226, 236, 237, 245, 247, 251; First Officers Training Camp at Leon Springs, 18–19, 248; football, 8, 9, 9, 10, 15–16, 33, 35–36, 36, 37, 38, 241; Fort Benning, 37–40, 38, 40, 48, 245; Fort D. A. Russell, 24; Fort McKinley, 54; Fort Sam Houston, 34–35, 37, 47; as Forty Acres Club president, 250–51; health during captivity, 157; health during retirement, 251–52; hearing issues, 130–31; and Hoffman, Robert G., 117; hunting, 6, 44–45, 237, 238–39; and Jester, Beauford, death of, 231; and Johnson, Lyndon B., 222, 225, 230, 252–53; and Jones, Albert M., 132, 133, 165; and Kolchak, Alexander, 29; and Krueger, Walter, 50–51; Lazy K Ranch, 237–39; and Littlefield, Clyde, 11, 36–37, 245; and Magsaysay, Ramon, 243; military athletic training programs, directing, 33, 34, 37, 44; military career, early, 23–32; and Mitchell, Eugene H., 51; nicknames, high school, 7, 8; Officers Reserve Corps (ORC), 19; 141st Infantry Regiment Rifle Team, 48, 48; Philippines, postwar visit to, 242–43; presidential politics, involvement in, 236–37; promotions and promotion issues, military, 24, 44, 49, 63, 82, 86, 131–33, 222, 225, 228; in *The Quan* "Chit Chat" column, 233; retirement, military, 227; Schofield Barracks, 34, 34; as Selective Service Director for Texas, 229; and Semenov, Grigori, 30; sense of humor, 158, 240; 7th Company, 19; Siberia, 24, 27–30, 32, 34, 117, 148, 230; Sigma Delta Psi Honorary Athletic Fraternity, induction into, 11; smoking addiction, 158; as State Guard Advisory Board chairman, 251; as Terrell Military College trustee, 229; as Texas Military District executive officer, 226, 228; in Texas National Guard, 11, 13; in Texas 2nd Infantry, 12, 13–14, 16; and Townsend, Glen R., 94; track, 8, 10, 36, 44; transport home to United States, postwar, 214–15; University of Texas, 7, 8, 9, 9, 10, 16, 33, 35–37, 36; University of Vermont, 39, 40; and Vargas, Jesus, 242, 243, 244; and Wainwright, Jonathan M., 131, 132, 133, 222, 224, 226; Wainwright Travelers 1950 gathering, 233, 234, 235; will, writing, 82; and Wilson, Ovid "Zero," 53, 191, 204, 205; as Wolfhound, 27, 28, 32; wrestling, 8, 10. *See also* awards and honors for Berry, K. L.; Battle of Bataan, surrender of Americans and Filipinos after; Berry, K. L., as adjutant general of Texas; Berry, K. L., and Bataan Death March; Berry, K. L., and Battle of Bataan; Berry, K. L., and Camp Karenko; Berry, K. L., and Camp O'Donnell; Berry, K. L., and Camp Shirakawa; Berry, K. L., and Camp Tarlac; Berry, K. L., and Cheng Chia Tun POW camp; Berry, K. L., and Mukden POW camp; Berry, K. L., and Nitchi Man hotel (Beppu); Berry, K. L., and WWII military service in Philippines; family of Berry, K. L.; Longhorns

Berry, K. L., as adjutant general of Texas, 227, 228–49, 229; as advocate for strong military, 231; Beauford, Jester, reappointment by, 231; Camp Mabry, 239–40; Daniels, Price, reappointment by, 245; Eisenhower, Dwight D., meeting in Austin, 245; end of final term, 248, 250; farewell message, 249; gardening, 239; and Herring, Charles, Camp Mabry golf course bill, 247–48; Johnson, Lyndon B., letter to on building up military faster, 232; leisure pursuits, 237–40; POW friends, 233–35; responsibilities, 228–30, 233; seven-point plan for US preparedness, 231–32; Shivers, Robert Allan, reappointments by, 231; swearing in, 228; and Texas National Guard Officer Candidate Academy

School, 244; and Texas State Guard Reserve Corps, 232; and Texas state auditor, 232–33

Berry, K. L., and Bataan Death March, 14, 100, 103, 105, 223, 240, 243; interaction with Japanese soldier in Bagac, 102; stop in Orani, 104

Berry, K. L., and Battle of Bataan, 37, 60, 62–64, 66, 69–76, 79–81, 83–86, 89, 90, 94–96, 223, 243; surrender of 1st Division, 96–99

Berry, K. L., and Camp Karenko, 126, 132, 151; appreciation letters to generals, 133; arrival, 122; attending talks, 134; benjo guard duty, 140; books and reading, 135; and Bunker, Paul D., 126–27; calisthenics, 125; dates/time in, 122, 136; deafness, 131; diary entries, 136–38, 140–41, 142–43, 144, 145, 146, 148, 150, 151, 196; on food, 114–15; garden, personal, 127, 130; guardhouse punishment, 144; leisure activities, 133; long-hair and mustache protest, 145–47, 176; no-escape pledge, 124; optimism about war ending, 149; postcard to wife Alice, 132; punishment by guards, 144, 145–46; radiogram to wife Alice, 131; Red Cross supplies, 150; snail hunting and cooking, 128–29

Berry, K. L., and Camp O'Donnell, 109, 110, 111

Berry, K. L., and Camp Shirakawa, 154–56, 159–67, 173, 181, 184, 237; act of defiance, 182; anger about food given to camp animals, 172; anger at malingerers, 174; birthday celebrations, 160; books and reading, 161–62, 198; building bamboo pipeline, 170–71; bull fest host, 160; dates/time in, 153, 168; diary entries, 154–55, 157, 161, 180, 183, 185, 196; essay answers as protest, 178–79; exercising, 159; fly catching punishment, 175; on food, 157, 158–59; garden, camp, 172; garden, personal, 159; health problems, 157; lecture, 160; library, 161; long hair and mustache protest, 176–78;

New Year's in Yasume Park, 161; peanut planting and gathering, 175–76; on POWs fighting each other, 155–56; radiograms and letters from wife Alice, 162, 165, 166; Red Cross package, 157; separating fighting POWs, 155; stinkweed cutting and fetching, 171–72; talk on shooting leopard in China, 160; walks, 159–60, 174; water detail, 168–70; and Yasume Park, 174–75

Berry, K. L., and Camp Tarlac, 113, 116, 117, 125; dates/time in, 111; on food, 114–15; as personnel adjutant, 115

Berry, K. L., and Cheng Chia Tun POW camp, 196, 200, 201; awareness of upcoming move, 203; bird watching, 198, 240; books and reading, 198; bridge playing, 197–98; calisthenics leader, 197; Corkill, William E., throwing surprise birthday celebration for, 198–99; dates/time in, 194, 203; disappointment in not receiving radiogram, 199; golfing, 198; leisure time activities, 197–98; postcard sent from, 199; walking, 198

Berry, K. L., and Mukden POW camp, 191, 204–5, 219; birthday celebration, 205; bridge playing, 204; and brigadier general pay, 204; citations for 1st Division officers and enlisted men, 205; dates/time in, 203; food, 206; knowledge of Russia invading Manchuria while in, 207; mail received, 204; return of old diary, 196; sunbathing, 204; swimming, 240; tailor shop, POW, 204; thoughts about wife Alice, while in, 202; and Walker, Jack, 204, 265; walking, 204; and Wilson, Ovid "Zero", 204

Berry, K. L., and Nitchi Man hotel (Beppu), 192; dates/time in, 191

Berry, K. L., and WWII military service in Philippines, 32, 34, 37, 42, 54, 55, 241; and Corkill, William E., 51; and Fort McKinley, 54; and USS *Coolidge*, 52. *See also specific POW camps*; Berry,

K. L., and Bataan Death March; Berry, K. L., and Battle of Bataan

Berry, K. L., Jr., 34, 163, 165; birth of, 24; burial at Fort Sam Houston, 252; China, 43; death of in plane crash, 252; Lazy K Ranch, 237; letter to father, Berry, K. L., 165; marriage to Boarman, Phyllis, 162, 163; promotion to Paine Field base commander, 252

Berry, Kearie Albert, 3

Berry, Kearie Lee. *See* Berry, K. L.

Berry, Kearie Lee, Jr. *See* Berry, K. L., Jr.

Berry, Kearie Lee, III, 220, 235

Berry, Thomas Eugene (father of Berry, K. L.), 3, 4; death of, 35; as farmer, 6

Berry, Thomas Eugene (son of Berry, K. L.), 19, 162, 163; Army track events, participation in as teenager, 44; birth of, 35; China, 43; Chinwangtao summers, 43; Lazy K Ranch, 237; in Marines, 204

Berry, Viola, 35; "Manila, Our War with Spain" poem, 5

Berry, William David, 4

Bevan, Gerald (Captain): Camp Shirakawa, 176; caricature of Berry, K. L., 177

Binuangan River, 93

birds: Berry, K. L., shooting on ranch, 240; Berry, K. L., watching at Cheng Chia Tun POW camp, 198, 240

Birge, Bill: Texas 2nd Infantry football team, 15; University of Texas football team, 15. *See also* Longhorns

Birkhead, Claude (Lieutenant General), 226

birthday celebrations, POW camp: Camp Shirakawa for Berry, K. L., 160; Cheng Chia Tun POW camp for Corkill, William E., 198–99; Mukden POW camp for Berry, K. L., 205

Bittner, Robert D. (Staff Sargeant), 161

Bluemel, Clifford (Brigadier General): Battle of Bataan, 73, 92; birthday gift for Berry, K. L., at Mukden camp, 205

Boarman, Phyllis: marriage to Berry, K. L., Jr., 162, 163

Boatwright, John R. (Colonel), 215; Bunker, Paul D., fight with, 127; Camp Karenko, 125, 127, 151; Cordero, Virgil N., fight with, 127; postwar transport home, 214–15

Bolsheviks, 24–25, 26, 27, 28, 30, 32

Bolte, Gerta, 213

Bonham, Roscoe (Colonel), 215; birthday gift for Berry, K. L., at Mukden camp, 205; Camp Karenko, 125, 151; Camp Shirakawa, 160, 170–71, 180; Cheng Chia Tun camp, 198; postwar transport home, 214–15

"Boots." *See* Nakashima (2nd Lieutenant)

Borneo, North: Japanese capture of, 69

Bougainville (Solomon Islands), 148; Japanese invasion of, 69

Bowler, Louis J. (Colonel): Camp Karenko, 125, 126, 127, 130, 134, 142, 144, 150, 151; Camp Shirakawa, 155, 156, 161; postwar transport home, 214–15

Boxer Rebellion, 41

Braddock, William H. (Colonel): Camp Shirakawa, 176; Cheng Chia Tun camp, 196

Braly, William C. (Colonel): birthday gift for Berry, K. L., at Mukden camp, 205; Camp Karenko, 125, 126, 127, 128, 129, 130, 131, 134, 140, 145–46, 151; Camp Shirakawa, 154, 155, 156, 160, 161, 168, 171–72, 175; Camp Tarlac, 114, 115, 117; Cheng Chia Tun camp, 197; Cheng Chia Tun camp, chronicling journey to, 185; Mukden camp, rescue from, 210–11; postwar transport home, 214–15; violin, 117, 126, 134, 214; Wainwright Travelers 1950 gathering, 234

Brawner, Pembroke A. (Colonel): birthday gift for Berry, K. L., at Mukden camp, 205; Camp Karenko, 125, 143, 144, 151; Camp Shirakawa, 160

Brazil Maru, 191

Brezina, Frank (Colonel): Camp Karenko, 125, 128, 151; death in Camp Shirakawa, 157

Bridge, Robert P. (Colonel), 171–72

INDEX 323

Brokaw, Frank (Lieutenant Colonel), 52
Brooke General Hospital, 220, 222, 251, 254
Brougher, William E. (Brigadier General): Battle of Bataan, 80, 82, 84, 88, 92, 93, 95; and Battle of the Pockets/Bataan, 80; and Berry, K. L., 94, 95; birthday gift for Berry, K. L., at Mukden camp, 205; Camp Karenko, 130; Camp Shirakawa, 154, 155, 156, 168, 180, 183; Camp Tarlac, 115; Cheng Chia Tun camp, 196; and Jones, Albert M., 95; low morale during Battle of Bataan, 88; military medals received, 96; Mukden camp, 208; Mukden camp, rescue from, 208. *See also* 11th Division
Brown, Ernest (Captain), 203
Brown, Robert (Private), 111
Browne, E. Harrison (Colonel), 87; Bataan Death March, 103; and Berry, K. L., 220; birthday gift for Berry, K. L., at Mukden camp, 205; birthday poems for Berry, K. L., 259; Camp Karenko, 125, 126, 127, 128, 133, 146, 151; Camp O'Donnell, 110; Camp Shirakawa, 156, 159–60, 172, 174–75; Cheng Chia Tun camp, 196, 197, 199, 201; Fort McKinley, 54; Lazy K Ranch, 239; and Lilly, Edmund J., 55; postwar transport home, 213; in *The Quan* "Chit Chat" column, 233; Wainwright Travelers 1950 gathering, 234, 235; Walter Reed Hospital, 220
Brownwood, Texas: Berry family move to, 49; relocation of Camp Bowie to, 49
Bunker, Paul D. (Colonel): and Berry, K. L., 126–27, 130–31, 141; Boatwright, John R., fight with, 127; Camp Karenko, 125, 126–27, 128–29, 134, 139–40, 144; death of in Camp Karenko, 141; and MacArthur, Douglas, 126
Burleson, Texas, 4
Burma: Allies' gains in, 180; British liberation of Mandalay, 201; Japanese invasion of, 60, 69; Japanese losses in, 145
Burt, R. J. (Colonel), 44
Bushido code, 101

Cabanatuan POW camp. *See* Camp Cabanatuan
Callahan, James W. (Colonel): birthday gift for Berry, K. L., at Mukden camp, 205; Camp Karenko, 125, 134
Camp Bowie: Berry, K. L., at, 49, 50, 51; expanded training facility, 49; relocation to Brownwood, 49
Camp Bullis, 48
Camp Cabanatuan, 111, 203, 204
Camp Funston, 18, 19. *See also* Camp Stanley
Camp Heito, 182
Camp Hulen, 47, 232
Camp Karenko, 153; beriberi treatment, 142, 143; books and reading, 134–35; calisthenics, 125; camp commanders, 123–24, 135; Camp O'Donnell, compared with, 122; Camp Shirakawa, compared with, 153, 171; church services, 134; deaths, 141; diaries, POW, 126; fire drills, 140; food, 114–15, 122, 126, 127–30, 137, 138; Fortier, Malcolm V., caricatures of, 123, 124, 129, 139, 142, 143; garden, camp, 138; gardens and gardening, individual, 126, 130; generals and high-ranking civilians leaving, 149–50, 151; goat tending, 139; guard violence, 122, 142–44, 146, 148, 154; health of POWs, 141–42; and Imamura, Yayohachi (Captain), 123, 125, 142; and Kojima, Toshio (Lieutenant), 135; lecture series, 160; leisure activities, 133–35; library, 135; location, 153; and Matsumura, Yoshio (Private 1st Class/Corporal), 142, 156; and Nakano, Junichi (Lieutenant Colonel), 123, 140–41; and Nakashima (2nd Lieutenant), 124; no-escape pledge, 124–25; paying respect to Japanese emperor, 125; "policing up," 140, 174; post exchange (PX), 134, 138, 150; POW march to, 121; radiograms, 131; Red Cross supplies, 149–50, 157–58; Red Cross visitor to, 151; room assignments, 125, 151; rules and regulations, 123, 124–25, 145; singing

program, 134; snails as food, 128–29; vaccinations, POW, 142; vigilant guard duty, 139–40; and Wakasugi, Jiro (2nd Lieutenant), 124, 127, 130, 145; war news and bad treatment of POWs, 142–45, 148; work, coerced and forced, 138–41, 171; workers' rice, 138. *See also* Camp Karenko, Allied POWs in

Camp Karenko, Allied POWs in, 120, 121, 157, 176; Aldridge, Edwin E., 151; Amis, William M., 151; Atkinson, Edward C., 151; Ausmus, Delbert, 151; Balsam, Alfred S., 125, 151; Beckwith-Smith, Merton, 141; Beebe, Lewis C., 134; Bell, Gilmer M., 143, 151; Bennett, John, 142; Berry, K. L. (*see* Berry, K. L., and Camp Karenko); Boatwright, John R., 125, 127, 151; Bonham, Roscoe, 125, 151; Bowler, Louis J., 125, 126, 127, 130, 134, 142, 144, 150, 151; Braly, William C., 125, 126, 127, 128, 129, 130, 131, 134, 140, 151; Brawner, Pembroke A., 125, 143, 144, 151; Brezina, Frank, 125, 128, 151; Brougher, William, 130; Browne, E. Harrison, 125, 126, 127, 128, 133, 146, 151; Bunker, Paul D., 125, 126–27, 128–29, 134, 139–40, 144; Bunker, Paul D., death of, 141; Callahan, James W., 125, 134; Campbell, Alex H., 125, 151; Carter, James D., 125, 151; Cavanaugh, James, death of, 141; Chase, Theo M., 125, 151; Churchill, Lawrence S., 143; Collier, James V., 125; Cooper, Wibb E., 151; Cordero, Virgil N., 125, 151; Corkill, William E., 151; Cottrell, Joseph F., 151; Dougherty, Louis R., 143; Duke, C. L. B., 144; Galbraith, Nicoll F., 155; Glattly, Harold W., 141; Heath, Lewis, 144; Ives, Albert R., 143; Jones, Albert M., 132; King, Edward P., 131; Lathrop, Leslie T., 146; Lucas (Brigadier), 143; Mallonee, Richard C., 146; Quesenberry, Marshall H., 143; Richards, Harrison H. C., 130; Simson, Ivan, 134; Steel, Charles L., 130; Wainwright, Jonathan M., 127–28, 131, 139, 140, 144; Wood, Stuart, 146

Camp Kearney, 23, 48

Camp Mabry, and Berry, K. L., 232–33; Herring, Charles, golf course legislative bill, 247–48; retirement and forced move from, 251; Reilly grandsons living with at, 239

Camp Murphy, 62

Camp O'Donnell, 109–11; as Bataan Death March final destination, 109, 105; compared with Camp Karenko, 122; deaths, 110–11; duties of enlisted men, 111; food, 111, 114; health of POWs, 110–11; living quarters, 110; medical barracks, 111; POWs trucked to, 99; rules, 110; and Tsuneyoshi, Yochio, 109; water, lack of, 110, 111. *See also* Camp O'Donnell, Allied POWs in

Camp O'Donnell, Allied POWs in, 99, 100, 105, 109–11, 113, 114, 122, 125; Aldridge, Edwin E., 110; Berry, K. L. (*see* Berry, K. L., and Camp O'Donnell); Brown, Burton R., 111; Browne, E. Harrison, 110; Corkill, William E., 109; Falconer, John, 110; Glattly, Harold W., 141; Hoffman, Robert G., 110; Keltner, Ed H., 110; Mallonee, Richard C., 111; MacDonald, Stuart C., 110; Mills, Loyd, 111; Quesenberry, Marshall H., 110; Stevens, Luther R., 110; Stoudt, Daniel, 111; Tenney, Lester, 109

Camp Perry: Berry, K. L., as infantry rifle team assistant coach at, 39

Camp Shirakawa, 189, 194; altercations between POWs, 155; amateur shows, 161; bamboo pipeline construction, 170–71; bathing in rain, 168; camp commander and staff, 154; Camp Karenko, compared with, 153, 171; Christmas celebration, 160–61; deaths, 157; defiant acts, POW, 182; exercise, 159; fly-catching punishment, 175; food, 155, 157–59, 183; Fortier, Malcom V., caricatures of, 155, 169, 170, 171, 173, 174; garden, camp, 172; generals' housing, 154; health of POWs, 156–57; and Hioki, Shiro (Lieutenant), 154, 172, 176, 180; hospital, 156; and Imamura, Yayohachi (Captain),

154, 172; influenza, 156; lecture series, 160; leisure activities, 159–62; library, 161; location, 153; and Matsumura, Yoshio (Private 1st Class/Corporal), 154, 156; mosquito problem, 153; move out of, 184, 185; and Nagatomo, Y. (Sergeant), 154; news photographers at, 175; park construction, 174; peanut planting and gathering, 175–76; "policing up," 174; post exchange (PX), 158; POWs hiking to, 152; quarters, 154–55; radiograms, 162; Red Cross supplies, 157–58; room assignments, 154–55; rules, 181–82; and Sazawa, Hideo (Colonel), 179, 181; stinkweed cutting and fetching, 171–72; and Wakasugi, Jiro (2nd Lieutenant), 154; water detail, 169–70; water problems, 153, 154, 168–71; work, coerced and forced, 171–76; and Yamanaka, Bob, 154; Yasume Park, 174. *See also* Camp Shirakawa, Allied POWs in

Camp Shirakawa, Allied POWs in: Aldridge, Edwin E., 155, 156, 164, 166, 170, 175; Bell, Gilmer M., 155; Berry, K. L. (*see* Berry, K. L., and Camp Shirakawa), Bevin (Captain), 176; Bittner, Robert D., 161; Bonham, Roscoe, 160, 170, 180; Bowler, Louis J., 155, 156; Braddock, William H., 176; Braly, William C., 154, 155, 156, 160, 168, 175; Brawner, Pembroke A., 160; Brezina, Frank, 157; Bridge, R. P., 172; Brougher, William, 154, 155, 168, 180, 183; Browne, E. Harrison, 156, 159, 160, 172, 174–75; Carter, James D., 170; Collier, James V., 160; Cordero, Virgil N., 155, 156, 159, 160, 176; Corkill, William E., 155, 159–60, 171, 175; Fortier, Malcolm V., 176; Foster, Valentine D., 172; Frissell, Howard N., 160, 176; Galbraith, Nicoll F., 155, 156, 158, 160, 161, 178; Hilton, Roy C., 176; Hughes, James C., 164, 180, 181; Ives, Albert R., 180; Johnson, Edwin H., 180, 181; Jones, Albert M., 159–60; Keltner, Ed H., 160; Kohn, Joe P., 180; Lilly, Edmund J., 160, 178; Mallonee, Richard C., 168; McBride, Allan C., 157; McLeod, Torquil, 180; Parker, George M., 181; Pechek, Frank, 161, 176; Quesenberry, Marshall H., 159–60; Rice (Captain), 157; Rutherford, Dorsey J., 164, 180; Selby (Commodore), 157; Sheppard (Private), 157; Sherry, Dean, 164; Sledge, Theodore J., 170, 175; Ward, Frederick A., 161; Weaver, James R. N., 154, 158; Whitehurst, Matthew S., 161; Worthington, Josiah W., 176

Camp Stanley, 18

Camp Tamazato, 149–50, 151

Camp Tarlac, 111–18, 125; daily schedule, 116; deaths, 117; food, 114, 115, 118, 127; Fortier, Malcolm V., caricatures of, *112, 113, 116*; gardens, personal, 115, 130; Ito (Colonel), visits to, 114, 117, 118; latrines, 113; laundry, 113–14; leisure activities, 117; living quarters, 113; mess hall, 117; and Nishiyama (Corporal), 114, 117; post exchange (PX), 116–117; reading material, 117; and Ugi, (Lieutenant), 114; and Ura (Lieutenant), 114, 118. *See also* Camp Tarlac, Allied POWs in

Camp Tarlac, Allied POWs in, 111–18; Barry, Edwin F., death in, 117–18; Beebe, Lewis C., 116; Berry, K. L. (*see* Berry, K. L., and Camp Tarlac); Bowler, Louis J., 125; Braly, William C., 114, 115, 117, 125; Brougher, William, 115; Bunker, Paul D., 126; Cordero, Virgil N., 117; Elmes, Chester H., 117; Fortier, Malcolm V., 112, 115; Galbraith, Nicoll F., 155; Glattly, Harold W., 141; Hoffman, Robert G., 117; King, Edward P., 114; Mallonee, Richard C., 115, 118; Monihan, James G., 117; Stowell, Allen L., 117; Wainwright, Jonathan M., 114; Wood, Stuart, 117

Camp Travis, 23

Camp Wilson, 15

Campbell, Alex H. (Colonel): birthday gift for Berry, K. L., at Mukden camp, 205; Camp Karenko, 125, 151

"Can Do" regiment, 41, 46. *See also* 15th Infantry, and China

Capas: Bataan Death March stop of, 100, 105
caricatures by Fortier, Malcom V. (Colonel): Camp Karenko, *123, 124, 129, 139, 142, 143*; Camp Shirakawa, *155, 169, 170, 171, 173, 174*; Camp Tarlac, *112, 113, 116*; Cheng Chia Tun camp, *195*; Mukden camp rescue, *209, 211, 212*; *Nagaru Maru, 119*; *Ōryoku Maru, 186*. See also Fortier, Malcom V. (Colonel)
Carlton, Alva. *See* Longhorns
Carranza, Venustiano, 13
Carter, James D. (Colonel): Camp Karenko, 125, 151; Camp Shirakawa, 159, 170–71
Cavanaugh, James (Master Sargeant), 141
Cavite naval base (Philippines): Japanese capture of, 64
Cavness, C. H., 232–33
Chase, Theo M. (Colonel), 125, 151
Chastaine, Ben Hur (Colonel), 52, 53
Cheng Chia Tun POW camp, 195, 204; bomb shelters, digging, 201; Camp Shirakawa clothing brought to, 194; check-in, 194; clothing issued to POWs, 196; Corkill, William E., birthday celebration, 198–99; food, 196, 197; Fortier, Malcom V., caricature of, *195*; gardens, non-US POW, 197; health of POWs, 200; and Ikeda (Lieutenant), 197; leisure time, 197–98; and Matsumiya (Lieutenant), 194–95, 197; medical clinic, 200; POW march from train station to, 194; POW train ride to Mukden camp from, 203; POWs learning of Germany's surrender, 203; POWs learning of President Roosevelt's death, 203; radiograms, 199; ration cuts, 197; Red Cross food and supplies, 196–97; rumors, 200–201; Russian stove "pachikas," 195; supplies given to POWs as war was ending, 202; Texas Independence Day celebration, 198; war news and guard treatment of POWs, 201. *See also* Cheng Chia Tun POW camp, Allied POWs in

Cheng Chia Tun POW camp, Allied POWs in, 185, 194–203; Aldridge, Edwin E., 198, 199; Berry, K. L. (*see* Berry, K. L., and Cheng Chia Tun POW camp); Bonham, Roscoe, 198; Braddock, William H., 196; Braly, William C., 197; Brougher, William, 196; Browne, E. Harrison, 196, 197, 199, 201; Corkill, William E., 196, 198–99; Dougherty, Louis R., 198; Evans, Leonard, 198; Fortier, Malcom V., 197; Gillespie, James O., 196; Hersee, Philip, 198; Hilsman, Roger, 196; Hoffman, Robert G., 196, 200; Keltner, Ed H., 198, 199; King, Edward P., 195; Lilly, Edmund J., 198, 199, 201; Mallonee, Richard C., 194, 197; Marshall, Floyd, death of in, 200; Moore, George F., 195; Parker, George M., 197; Pilet, Nunez C., 196; Rogers, Richard G., 198; Selleck, Clyde A., 196, 198; Quesenberry, Marshall H., 196, 199; Traywick, Jesse T., 196; Wainwright, Jonathan M., 195; Wood, Stuart, 196, 200
China: Berry, K. L., military service in, 40–46, 48; Boxer Rebellion, 41; declaration of war against Japan, 55; and 15th Infantry, 40–46; Great Wall, 42, 44; Japanese occupation, 46; Japanese territorial gains, 42, 49; map of 1930s Japanese aggression in, *45*; Twenty-One Demands agreement with Japan, 42. *See also specific provinces, towns, and villages*
China Sea, 41
Chinwangtao, China, 41, 43–44; Berry children and summers in, 43
Churchill, Lawrence S. (Colonel): birthday gift for Berry, K. L., at Mukden camp, 205; Camp Karenko, 143
Churchill, Winston: and Roosevelt, Franklin D., 49–50; and Yalta Conference, 200
Chynoweth, Bradford G. (Brigadier General), 51, 52, 53
Civil Rights Act, 252

Civil War, 3
Clark Air Force Base, 243
Clark Field: Japanese attack on, 55, 60
Cleveland, Grover, 3
Clipper, 53
Coastal Artillery Command, 125. *See also* Moore, George F. (Major General)
Collier, James V. (Colonel): Camp Karenko, 125; Camp Shirakawa, 160
Collin County, Texas, 4
Columbus, New Mexico, 13
Combat Infantryman's Badge, 223
communism: fear of spread of, 231, 232, 252
Connally, John, 254
Connolly, Tom: letter to about Berry, K. L., brigadier general promotion, 222
Cook, John D. (Colonel), 215
Cooke County, Texas, 4
Coolidge, Calvin, 35, 39
Cooper, Wibb E. (Colonel), 151
Cordero, Virgil N. (Colonel): Bataan Death March, 104; birthday gift for Berry, K. L., at Mukden camp, 205; and Boatwright, John R., 127; Camp Karenko, 125, 151; Camp Shirakawa, 155, 156, 159, 160, 176; Camp Tarlac, 117; and Corkill, William E., 155; post-rescue Mukden town exploration, 212–13; postwar transport home, 214–15; in *The Quan* "Chit Chat" column, 233
Corkill, William E. (Colonel), 52; and Berry, K. L., 51, 52; birthday gift for Berry, K. L., at Mukden camp, 205; Camp Karenko, 151; Camp O'Donnell, 109; Camp Shirakawa, 155, 159–60, 171, 175; Cheng Chia Tun camp 196, 198–99; Cheng Chia Tun camp, chronicling journey to, 185, 188; Cheng Chia Tun camp, surprise birthday celebration, 198–99; and Cordero, Virgil N., 155; on end of war, 210; Mukden camp, 205, 210; on Nitchi Man hotel, 191–92; post-rescue Mukden town exploration, 212–13; postwar transport home, 213; in *The Quan* "Chit Chat" column, 233;

USS *Coolidge* adjutant, 52; Wainwright Travelers 1950 gathering, 235
Corpus Christi, Texas, 14, 16; protest in streets, 14, 15
Corregidor (island), 53, 59, 64, 67, 77, 78, 79, 89, 90, 92, 114, 117, 125, 126, 134, 153; Allies' recapture of, 201; Japanese capture of, 117; Japanese preparing to attack, 99; US surrender of, 133
Corsicana, Texas, 231
Cossacks, 24, 25, 26, 28, 30
Cottrell, Joseph F. (Colonel): birthday gift for Berry, K. L., at Mukden camp, 205; Camp Karenko, 151
Craig, William H., 215
Crow, Judson (Major), 102
Cuba: 1960s troubles in, 250
Curtis, Donald (Colonel): birthday gift for Berry, K. L., at Mukden camp, 205
Czechoslovak Legion, 25, 26, 30

Dairen, Manchuria, 213, 214
Dallas, Texas, 5, 229
Daniels, Price: Berry, K. L., reappointment as Texas adjutant general, 245; Berry, K. L., and Texas Distinguished Service Medal, 244; decision to not reappoint Berry, K. L., as Texas adjutant general, 248; Eisenhower, Dwight D., meeting in Austin, 245
dehydration: and Bataan Death March POWs, 102, 104, 105; and Camp O'Donnell POWs, 110, 111; and *Ōryoku Maru* POWs, 187
Democratic Convention of 1948: Berry, K. L., at, 236; and cowbell, 237; Jester, Beauford, at, 236
dengue: and Bataan Death March POWs, 99; and Battle of Bataan, 77, 92
dental problems: and Camp Shirakawa POWs, 156
Denton, Texas, 3, 5; as college town, 6
Denton County, Texas, 4, 5–6; cotton farming, 5; wheat production, 6
Denton High School: and Berry, K. L., 6–7; yearbook, 7, 8

328 INDEX

Depression, Great, 39
Dewey, George (Admiral), 4–5
diarrhea: and Bataan Death March POWs, 105; and Camp Shirakawa POWs, 157
Diaz, Porfirio, 12
Distinguished Service Cross, 223
Distinguished Service Medal, 223, 224
Distinguished Service Medal (National Guard Association of the United States), 245–46
Dittmar, Gus C., 246; First Officers Training Camp at Leon Springs, 19; 2nd Company, 19. *See also* Longhorns
Doane, Irvin E. (Colonel), birthday gift for Berry, K. L., at Mukden camp, 205
Dooley, Thomas O. (Major), and Wainwright Travelers 1950 gathering, 234
Dougherty, Louis R. (Colonel): birthday gift for Berry, K. L., at Mukden camp, 205; Camp Karenko, 143; Cheng Chia Tun camp, 198
draft, military: Roosevelt, Franklin D., peacetime, 49; World War I National Guard, 19
Drake, Charles C. (Brigadier General), 66; birthday gift and song for Berry, K. L., at Mukden camp, 205
Duke, Cecil L. B. (Brigadier), 144
Dumas, Hugh A. (Colonel), 52; birthday gift for Berry, K. L., at Mukden camp, 205
Dyess, William (Captain), 102, 103
dysentery, amoebic: and Bataan Death March POWs, 99, 104, 105; and Battle of Bataan, 77, 90, 92; and Camp Karenko POWs, 141; and Camp O'Donnell POWs, 110, 111

Eisenhower, Dwight D.: and Berry, K. L., 15, 37, 51, 147, 225, 226, 247, 251; Berry, K. L. thoughts about as 1952 presidential candidate, 236; Columbia University president, 236; cowbell gift from Texas delegation, 237; end of second presidential term, 250; as football coach, 15; as Fort Benning assistant football coach, 37, 38, 245; as Forty Acres Club honorary member, 251; and Krueger, Walter, 51; and MacArthur, Douglas, 51; meeting with Berry, K. L. in Austin as president of the United States, 245; as 1952 presidential candidate, 236; 1952 presidential election victory, 237; and Normandy invasion/D-Day, 182.
El Paso, Texas, 14, 18
11th Division, 77, 79, 80, 82, 92, 93, 94, 95, 96
Elmes, Chester H. (Colonel), 117
Enos, William A. (Colonel), 215
Enoura Maru: POW deaths, 191; POW survival post-attack, 191; US bombing of, 191
Evans, Leonard (Private), 198

Falconer, John (Private), 110
family of Berry, K. L.: births of children, 24, 34–35; brothers, 6; death of son Berry, K. L., Jr., 252; and grandfather Berry, Thomas, 4, 5; hunting with sons and grandsons, 237, 238–39; letters to wife Alice while deployed, 28, 29, 51, 52, 53, 54, 75, 81, 86, 87, 131–32; letters from wife Alice, 219, 220; meeting future wife Alice, 16; mother, influence of, 5; parents, 3; and Reilly grandsons, 239. *See also* Berry, Ada; Berry, Alice Fleming; Berry, Anne; Berry, Celeste Viola; Berry, Eugene "Gene"; Berry, Jim; Berry, John; Berry, K. L., Jr.; Berry, Kearie Albert; Berry, Kearie Lee, III; Matthews, Anne Berry; Reilly, Bruce; Reilly, William King; Riley, John Sleychk; Riley, Sam H.; Riley, Viola Eugenia
15th Infantry, and China: and Berry, K. L., 40–46; and Burt (Commander), 46; and Lynch, Thomas A., 46; nicknamed "Can Do" regiment, 41, 46; pulled from, 46
Fiftieth Texas Legislature: law about rank of adjutant general, 228

INDEX

51st Division, 69, 92
57th Infantry, 63. *See also* Lilly, Edmund J.
Filipino soldiers: Battle of Bataan, surrender after, 96–99; in I Corps, 64; trained by US military, 50. *See also specific Filipino soldiers*; Battle of Bataan, surrender of Americans and Filipinos after; Battle of Bataan, US and Filipino soldiers in; 1st Regular Philippine Division, and Battle of Bataan; 3rd Regiment, Philippine Army
Finland: Russians fighting Germans in, 182
fireside chats: Franklin D. Roosevelt, 57, 87
First Army Corps. *See* I Corps
First Camp Men, 19
First Officers Training Camp Association, 19
First Officers Training Camp at Leon Springs, 23; and Berry, K. L., 18–19, 248; and Jester, Beauford, 226
1st Regular Philippine Division, and Battle of Bataan, 64, 66, 67, 69, 70, 71, 73, 74, 75, 76, 77, 79, 80, 81, 83, 84–85, 94; in I Corps, 64; surrender of, 96–99
Fitch, Alva (Major), 71–74
Fleming, Alice Celeste: meeting Berry, K. L., 16; wedding photo, *17*. *See also* Berry, Alice Fleming
food rations, cuts to/lack of: Battle of Bataan, 66, 73, 77, 86, 89, 90, 92, 93, 99. *See also specific POW camps*
football. *See* University of Texas football
football, and Berry, K. L.: Fort Benning head coach, 37, 38; high school, 8; Texas 2nd Infantry team, 15–16, 33. *See* University of Texas football
Fort Benning: Lim, Vicente, at, 91; Nara, Akira, at, 91. *See also* Fort Benning, and Berry, K. L.
Fort Benning, and Berry, K. L., 39; Army refresher course, 223; Company Officers course, 37; head football coach, 37, 38; infantry rifle team member, 37;

military service at, 37–40, 48, 91; and 24th Infantry, 39–40, *40*
Fort Bliss, 223
Fort D. A. Russell, 24
Fort Knox, 223
Fort McKinley: and Berry, K. L., 54; and Wainwright, Jonathan M., 93
Fort Riley, 223
Fort Sam Houston: Berry, K. L., at, 34–35, 37, 51; Brooke General Hospital, Berry, K. L. stays in, 220, 222, 251, 254; Eighth Army Personnel Headquarters, 226; Krueger, Walter, at, 50; Summer Training Camp, 37; Texas National Guard move to, 16; 23rd Infantry, 34; Wainwright Travelers 1950 gathering at, 233–35
Fort Sam Houston National Cemetery: Berry, K. L., buried in, 254; Berry, K. L., Jr., buried in, 252
Fort Sill, 223
Fort Worth, Texas, 5
Fort Worth Record Sports, 35
Fortier, Malcolm V. (Colonel): Camp Shirakawa, 176; Camp Tarlac, 112, 115; Cheng Chia Tun camp, 197; Cheng Chia Tun camp, chronicling journey to, 185; postwar transport home, 214–15. *See also* caricatures by Fortier, Malcom V. (Colonel)
Forty Acres Club, 250–51; integration of, 253
Forty Acres Society, 251
45th Infantry, 79
41st Division, 92
42nd Infantry, 92
Foster, Valentine D. (Colonel), 215; birthday gift for Berry, K. L., at Mukden camp, 205; Camp Shirakawa, 171–72
Freedom Riders, 250
French Indochina: and Japanese, 49
"Frisco Bob." *See* Yamanaka, Bob
Frissell, Howard N. (Colonel), 160, 176
Fry, Philip T. (Colonel), 51
Fukota No. 3 POW camp, 191

Funk, Arnold J. (Brigadier General): birthday gift for Berry, K. L., at Mukden camp, 205
Fusan, Korea, 192

Galardi, Fred (Private): birthday gift for Berry, K. L., at Mukden camp, 205
Galbraith, Nicoll F. (Colonel): birthday gift for Berry, K. L., at Mukden camp, 205; Camp Karenko, 155; Camp Shirakawa, 126, 155, 156, 158, 160, 161, 178; Camp Tarlac, 155; comments about Berry, K. L.'s hair, 178; Distinguished Service Medal ceremony, 224; Mukden camp, 208, 209; Mukden camp, rescue from, 208, 209, 210; postwar transport home, 214–15; Wainwright Travelers 1950 gathering, 234
Geneva Convention of 1929, 110
Germany, 47, 49; Allies' defeat of in North Africa, 149; British army inside of, 201; declaration of war against other European countries, 49; Hamburg, Great Britain bombing of, 180; invasion of Belgium, France, and Holland, 49; London, bombing of, 49; POWs learning of its surrender, 203; surrender, 202; US Army inside of, 201; and Versailles Treaty, 47; World War I defeat, 47. *See also* Hitler, Adolph
Gillespie, James O. (Colonel), 196
Girls Industrial College, 6
Glattly, Harold W. (Lieutenant Colonel): Camp Karenko, 141; Camp O'Donnell, 141; Camp Tarlac, 141
Glen Springs, Texas, 13
Gobi Desert, 194
Gomez, Victor (Lieutenant Colonel), 102, 243
Goodman, James H. *See* Longhorns
Graves, William S. (Brigadier General): and Hoffman, Robert G., 117; command in Siberia, 26–27, 30, 32, 117
Great Britain: bombing of Hamburg, Germany, 180; declaration of war against Japan, 55; in Tunisia, 147; US aid to during WWII, 60

Great Wall of China, 42, 44
"Grumpy." *See* Matsumura, Yoshio (Private 1st Class/Corporal)
Grunert, George (Major General), 59
Guadalcanal: Japanese losses at, 145
Guam: Japanese seizure of, 60, 62

Halsey, William (Admiral): Task Force air raids over Taiwan, 188
Halstead, Earl T. (Lieutenant Colonel), 74
Hamilton, Stuart A. (Colonel): birthday gift for Berry, K. L., at Mukden camp, 205
"Handle Bars." *See* Nagatomo, Y. (Sergeant)
"Handlebars." *See* Matsumiya (Lieutenant)
Harding, Warren G., 35
Harries, Herb, 54
Harries, Mary, 54
Harrison, Benjamin, 3
Hawaii: military service, 34; Schofield Barracks, 34
Heath, Lewis (Lieutenant General), 144
hell ships, 118. *See also Brazil Maru; Enoura Maru; Hozan Maru; Nagaru Maru; Ōryoku Maru* (December 14–15, 1944), and POWs; *Ōryoku Maru* (October 10–28, 1944), and POWs; *Rakuyo Maru; Suzuya Maru*
Herring, Charles: Berry, K. L. and Camp Mabry golf course bill, 247–48
Hersee, Philip (Private), 198
Hilsman, Roger (Colonel), 196
Hilton, Roy C. (Colonel), 176
Hioki, Shiro (Lieutenant), 154, 172, 176, 180
Hiroshima, Japan: "Little Boy" atomic bomb dropped on, 207
Hitler, Adolph, 47; and Nazi Party, 47; postponement of invasion of Great Britain, 50; suicide, 202
Hodges, Mack, 166
Hoffman, Robert G. (Colonel): and Berry, K. L., 117; birthday gift for Berry, K. L., at Mukden camp, 205; Camp

O'Donnell, 110; Camp Tarlac, 117; Cheng Chia Tun camp, 196, 200; Siberia, military service in, 117

Homma, Masaharu, 64, 70, 71, 85

Hong Kong, 134; bombing of, 200; Japanese invasion of, 60

Hoover, Herbert, 39

Horan, John P. (Colonel), 215

Hotel Beppu. *See* Nitchi Man hotel (Beppu); Nitchi Man hotel (Beppu), Allied POWs in

Hoten Camp. *See* Mukden POW camp

Hoten POW Camp system, 194; headquarters, 194, 203. *See also* Mukden POW camp

Houser, Houston (Major), 71

Hozan Maru: transporting POWs from Camp Karenko, 152

Hualien, Taiwan, 120; inlet of Karenko, 121

Huerta, Victoriano, 13

Hughes, James C. (Colonel), 215; birthday gift for Berry, K. L., at Mukden camp, 205; Camp Shirakawa, 164, 180, 181

Hulen, John A. (Brigadier General): and Texas 2nd Infantry, 13, 14

Hulen Trophy, 48

hunting, 6, 44–45, 237, 238–39

Hurley, Walter (1st Sargeant), 210

I Corps, 64, 65, 67, 68, 69, 77, 81, 86, 87, 90, 93; and Bataan Death March, 101, 103; Bataan Peninsula main lines of resistance, 65, 79; decision to surrender, 93; divisions and regiments in, 64; front lines, 76–77, 89; headquarters, 68; II Corps, inability to connect with, 68. *See also* Wainwright, Jonathan M.

Igorots, 82

II Corps, 64, 65, 67, 68, 69, 86, 87, 89, 91, 92, 96; and Bataan Death March, 101, 103; Bataan Peninsula main lines of resistance established by, 65; decision to surrender, 93; destruction of, 93; 51st Division, 68; front lines, 89, 90; inability to connect with I Corps, 68; Japanese attack on, 69; POWs, arrival of in Balanga after Bataan Death March, 103. *See also* Parker, George M.

Imamura, Yayohachi (Captain): and Camp Karenko, 123, 125, 142; and Camp Shirakawa, 154, 172

influenza: and Camp Shirakawa POWs, 156; and Cheng Chia Tun POWs, 200

Irkutsk, Siberia, 27, 28

Italy: Allies in, 180; Allies' capture of Rome, 182; Allies' defeat of in North Africa, 149

Ito (Colonel), 123; Brougher, William, remembrance of, 114; as Luzon POW camps commandant, 114; visits to POW camps, 114, 117, 118

Ives, Albert R. (Colonel): Camp Karenko, 143; Camp Shirakawa, 180

Iwai (Corporal), 154

Janin, Maurice, 30

Japan: Beppu, 188, 191; Hiroshima, atomic bomb dropped on, 207; Kyoto, 192; Kyushu, 188, 189; Moji, 188, 191; Nagasaki, atomic bomb dropped on, 207; and Operation Downfall, 205; Shirakawa, 194; Tokyo, 201; US bombing raids on, 182, 184, 193, 201, 202, 206; Yawata, 192. *See also* Japanese military, and World War II; Nitchi Man hotel (Beppu)

Japan Advertiser, 117

Japan Times, 117

Japanese military, and World War II, 47, 49; Bali, invasion of, 85; Bataan January 23–February 1942 landings map, 78; Battle of Bataan plan of attack map, 91; Battle of the Bismarck Sea, 147; Borneo, capture of North, 69; Borneo, invasion of, 60; Bougainville, invasion of, 69; Burma, capture of, 117; Burma, invasion of, 60, 69; Burma, losses in, 145; and China, territorial gains, 42, 49; Clark Field attack, 55, 60; Corregidor, capture of, 117; Corregidor, preparing to attack, 99; Darwin, Australia, attack on, 85; and French Indochina, 49; Guadalcanal, losses at, 145; Guam, seizure of, 60; Hong Kong (British), invasion of, 60; Hospital No. 1 (Bataan), bombing of, 92; Java, invasion of, 79; Korea, control of, 42; Kyushu, buildup of forces on, 206; Manchuria, control of, 42; Manila, bombing of, 60; Manila, capture of, 64; map of aggression in 1930s China, 45; Mindanao, capture of, 117; Moron, capture of, 70; Peking, occupation of, 46; Philippines, attack on December 8, 1941, 55, 60, 62; Philippines, invasion of, 60, 62; Philippines, propaganda campaign in, 89; Rabaul (Solomon Islands), capture of, 69; renaming Manchuria, 42; and Russia, 42; Russian declaration of war against, 207; Santa Barbara, California, attack on oil refinery near, 85; Singapore, invasion of, 79; Singapore, siege of, 69; South China Sea, control of, 118; Sumatra, invasion of, 79; surrender, 208; Taiwan, control of, 42; Tientsin, occupation of, 46; Tulagi (Solomon Islands), capture of, 117; USS *Houston*, sinking of, 85; USS *Langley*, sinking of, 85; Wake Island, takeover of, 60. *See also specific POW camps*; Bataan Death March; Battle of Bataan

Java: Japanese invasion of, 79

Jester, Beauford, 226, 236; Berry, K. L., and adjutant general role, 227, 228, 231; daughter Joann, 19; death of, 231; 1st Company, 19

Johnson, Edwin H. (Colonel): birthday gift for Berry, K. L., at Mukden camp, 205; Camp Shirakawa, 180, 181

Johnson, Lyndon B.: and Berry, K. L., 222, 225, 230; and Civil Rights Act, 252; and desegregation, 252–53; as vice president of the United States, 250; Vietnam War and presidency of, 252

Johnson Bill (S.1533 and H.R. 2993), 225, 227

Jones, Albert M. (Major General): appreciation letter to Berry, K. L., 133; Battle of Bataan, 68, 69, 82–84, 87–89, 93, 95; birthday gift for Berry, K. L., at Mukden camp, 205; and Brougher, William, 95, 96; Camp Karenko, 132; Camp Shirakawa, 159–60; major general status, awarded permanent, 227

Kai San, China, 46

Karenko POW camp. *See* Berry, K. L., and Camp Karenko; Camp Karenko; Camp Karenko, Allied POWs in

Keelung Harbor (Taiwan), 185, 188; US air raids on, 187

Keltner, Ed H. (Colonel), 52; and Berry, K. L., 51; birthday gift for Berry, K. L., at Mukden camp, 205; Camp O'Donnell, 110; Camp Shirakawa, 160; Cheng Chia Tun camp, 198, 199; Wainwright Travelers 1950 gathering, 235

Kennedy, John F.: assassination of, 252; and US space program, 250

Kerensky, Alexander, 24

Kilday, Paul: and Berry, K. L. brigadier general promotion, 222, 227

King, Edward P. (Major General): Battle of Bataan, surrender after, 93, 96, 98, 99, 103, 131; and Battle of Bataan, 88, 89, 92, 93, 99; Camp Karenko, 131; Camp Karenko, move from, 151; Camp Tarlac, 114; Cheng Chia Tun camp, 195; Wainwright Travelers 1950 gathering, 234

Kohn, Joe P. (Colonel), 180

Kojima, Toshio (Lieutenant), 135

Kolchak, Alexander (Admiral), 28, 29, 29, 30; Berry, K. L., meeting with, 29; execution of, 32

Korea: Fusan, 192; Japanese control of, 42

Korean War, 236; Berry, K. L., on, 232; and fear of communism spreading, 231, 232; Texas Air National Guard, federalization of during, 232

Krueger, Walter (Lieutenant General): and Berry, K. L., 50–51; and Eisenhower, Dwight D., 51; and MacArthur, Douglas, 51; military career, 50

Kyushu, Japan: Beppu, 188, 191; buildup of forces on, 206; Moji, 188, 191; and Operation Downfall, 206; US bombing parts of, 193

INDEX 333

Laird (Captain), 71, 74
Lake Baikal, Siberia, 28
Lathrop, Leslie T. (Colonel), 146
Lay, Kermit (Lieutenant), 103
Lazy K Ranch, 237–39; Berry, K. L., and sons and grandsons hunting at, 237, 238–39; and Berry, K. L., Jr., 237; and Berry, Tom, 237; and Browne, E. Harrison, 239; and Reilly, Bruce, 238; and Reilly, King, 238–39; Wainwright, Jonathan M., visit to, 237–38, 238
Ledda, Daniel: injury to during Battle of Bataan, 69
Lee, Robert E., 5; and Berry, Kearie Lee, middle name, 4
Leith, Hal (Sargeant): news from about Japan surrender, 209; as OSS rescue team member, 208
Leon Springs, Texas, 18. *See also* First Officers Training Camp at Leon Springs
Leyte, Philippines: US air raids on, 188
Lilly, Edmund J. (Colonel), 55; Battle of Bataan, 63, 92; and Browne, E. Harrison, 55; Camp Shirakawa, 160, 178; Cheng Chia Tun camp, 198, 199, 201; Cheng Chia Tun camp, chronicling journey to, 185, 188; comments about Berry, K. L.'s hair, 178; 57th Infantry, 63; Fort McKinley, 54, 55; postwar transport home, 214–15
Lim, Vicente (General): and Battle of Bataan, 90, 92; and Nara, Akira, 90–91. *See also* 41st Division
Lingayen Gulf: US Sixth Army invasion of, 200
"Little Snake Eyes." *See* Imamura, Yayohachi (Captain)
Littlefield, Clyde, 246; and Berry, K. L., 11, 36–37, 245; Sigma Delta Psi Honorary Athletic Fraternity, induction into, 11; as University of Texas track coach, 36. *See also* Longhorns
Lodge, Henry Cabot, 236
Longhorn Hall of Honor, 246, 246
Longhorns: 50th reunion of 1914 team, 253–54, 253
Longoskawayan Point, 77

Lough, Maxon S. (Brigadier General): Bataan Death March, 103; Battle of Bataan, 91; and Berry, K. L., 52–53; birthday gift for Berry, K. L., at Mukden camp, 205
Lowman, Kenneth E. (Captain): birthday gift for Berry, K. L., at Mukden camp, 205
Lubao: Bataan Death March stop of, 100, 104–5
Lucas, Hubert (Brigadier), 143
lugao, 111
Luzon, 58, 59, 95; map showing US forces in WWII, 61; North, 63, 82; POW camps on, 111; San Fernando area, 63; US divisions on, 200
Luzon Force headquarters, 99
Lynch, Thomas A. (Colonel), 46

Mabatang, Philippines, 64
MacArthur, Douglas (General): Bataan visit, 67, 88; and Battle of Bataan, 60–62, 64, 67, 77, 86; Berry, K. L., nomination of for Distinguished Service Cross, 223; and Bunker, Paul D., 126; cutting rations of Bataan troops, 66, 77, 89, 90; and Eisenhower, Dwight D., 51; and Krueger, Walter, 51; and Operation Downfall, 206; Parker, George M., meeting with in Bataan, 67; and Philippines, 50; Philippines, troops' reactions to leaving, 87–88; Philippines plan, 60, 61, 62; Roosevelt, Franklin D., pulling out of Philippines, 87; Roosevelt, Franklin D, recalling to active duty, 60; Truman, Harry S, firing of, 236; Wainwright, Jonathan M., meeting with in Bataan, 67; War Plan Orange, discarding, 60, 61; War Plan Orange, reverting to, 62
MacDonald, Stuart D. "Shorty" (Colonel): 52, 53; Bataan Death March, 103; and Camp O'Donnell, 110; birthday gift for Berry, K. L., at Mukden camp, 205
Madero, Francisco: assassination of, 13; Huerta, Victoriano, ousting of as

president of Mexico, 13; as president of Mexico, 12
Magsaysay, Ramon, 243
malaria: and Bataan Death March POWs, 99; and Battle of Bataan, 77, 89, 90, 92; and Camp Karenko POWs, 141; and Camp O'Donnell POWs, 110, 111; and Camp Shirakawa POWs, 153, 159; and quinine, 156
Mallonee, Richard C. (Colonel): Battle of Bataan, 66, 68; Camp Karenko, 146; Camp O'Donnell, 111; Camp Shirakawa, 168; on Camp Shirakawa climate versus Camp Karenko climate, 153; Camp Tarlac, 115, 118; Cheng Chia Tun camp, 194, 197; Mukden camp, 210; on not offending Japanese guards post-Japan surrender, 210
Manchukuo, 192, 194. See also Manchuria
Manchuria, 191; camels seen by Berry, K. L., in, 28; Dairen, 213, 214; Japanese control of, 42; Japanese expansion into, 26, 29; Japanese renaming of, 42, 45; Japanese retreat from, 202, 203; Russian advance toward, 202; Russian invasion of, 207. See also Manchukuo; Mukden POW camp
Mandalay: British liberation of, 201
"Manila, Our War with Spain," 5
Manila, Philippines, 54, 58, 59, 61, 90, 114, 125, 215, 219, 244; Allies' recapture of, 201; declared open city by Douglas MacArthur, 62; Japanese bombing of, 60, 62; Japanese capture of, 64; US Sixth Army attack on, 200
Manila Bay, Philippines, 53, 59, 78
maps: Bataan, Japanese landings (January 23–February 1942), 78; Bataan Death March, 100; Battle of Bataan, Japanese plan of attack, 91; China, Japanese aggression in 1930s, 45; Luzon, US forces in WWII, 61; Philippines, Japanese December 8, 1941, attack on, 58; Siberia, post-WWI, 25; Taiwan POW camps, 120
March of Dimes, 230–31

Mariana Islands: Allies' victory in, 180
Mariveles, Philippines, 77, 95; as Bataan Death March starting point, 99, 101, 103, 105
Mariveles mountain range, 76, 78
Marshall, Floyd (Colonel), 200
Marshall, George C. (General), 60; and brigadier general promotion of Berry, K. L., 222; and MacArthur, Douglas, 88; revision of Fort Benning instructional system, 39, 41; and Wainwright, Jonathan M., 90
Marshall Islands: Allies' victory in, 180
Masuda (Commander): and OSS rescue team, 209–10
Matsumiya (Lieutenant), 194–95, 197
Matsumura, Yoshio (Private 1st Class/Corporal): and Camp Karenko, 142; and Camp Shirakawa, 154, 156
Matthews, Anne Berry, 220
Matthews, Harvey, 220; and Berry, K. L., 225; and Berry, K. L., letter-writing campaign about brigadier general promotion, 222
Mauban, Philippines, 64, 67, 69, 76
Mauban line: January 18–25, 1942, 72
Maury (Major), 68
McBride, Allan C. (Brigadier General), 157
McCollum, Osa S. (Major), 69
McGill, William, 231
McKinley, William: assassination of, 6
McLeod, Torquil (Brigadier General), 180
Mensheviks, 24
Mexican American War, 12
Mexican War, 3; and Riley, John Sleychk, 4
Miller (Captain), 103
Miller, Tom: Eisenhower, Dwight D., meeting with in Austin, 245
Mills, Loyd (Captain), 111
Mindanao (island), 60, 63; Allies' recapture of, 201
Mitchell, Eugene H. (Colonel), 51
Moji, Kyushu (Japan), 188; *Brazil Maru*, arrival in, 191

Monihan, James G. (Colonel): birthday gift for Berry, K. L., at Mukden camp, 205; Camp Tarlac, 117
Monroe Doctrine: Roosevelt Corollary of, 6
Moody, Dan, 244
Moore, Arthur P. (Colonel), 208, 215; Camp Karenko, move from, 151; Cheng Chia Tun camp, 195; and OSS team, 210
Moore, George F. (Major General): permanent major general status awarded to, 227
Moron, Philippines, 64; 1st Infantry at, 69, 70, 71; Japanese capture of, 70
Mount Natib, 64, 67, 68, 76
Mount Samat, 90, 91
Mount Silanganan, 64, 67, 68, 76
Mukden, Manchuria, 194; POWs exploring town post-release, 212–13
Mukden POW camp, 191, 203–7, 219; concert, 210; food, 206; leisure activities, 204; and Masuda (Commander), 209–10; POW train ride to, 203; tailor shop, POW, 204; war news and withholding Red Cross supplies and food, 206. *See also* Mukden POW camp, Allied POWs in; Mukden POW camp, rescue from
Mukden POW camp, Allied POWs in: Aldridge, Edwin E., 204, 205; Balsam, Alfred S., 205; Beebe, Lewis C., 205; Bell, Gilmer M., 205; Berry, K. L. (*see* Berry, K. L., and Mukden POW camp); Bluemel, Clifford, 205; Boatwright, John R., 205; Bonham, Roscoe, 205; Braly, William C., 205, 210–11; Brawner, Pembroke A., 205; Brougher, William E., 205, 208; Brown, Burton R., 203; Browne, E. Harrison, 205; Callahan, James W., 205; Campbell, Alex H., 205; Churchill, Lawrence S., 205; Cordero, Virgil N., 205; Corkill, William E., 205, 210; Cottrell, Joseph F., 205; Curtis, Donald, 205; Doane, Irvin E., 205; Dougherty, Louis R., 205; Drake, Charles C., 205; Dumas, Hugh A., 205; Foster, Valentine D., 205; Fry, Phillip T., 204; Funk, Arnold J., 205; Galardi, Fred, 205; Galbraith, Nicoll F., 205, 208, 209; Hamilton, Stuart A., 205; Hoffman, Robert G., 205; Hughes, James C., 205; Hurley, Walter, 210; Johnson, Edwin H., 205; Jones, Albert M., 205; Keltner, Ed H., 205; Lough, Maxon S., 205; Lowman, Kenneth E., 205; MacDonald, Stuart D., 205; Mallonee, Richard C., 210; Monihan, James G., 205; O'Day, Ray, 205; Parker, M. (General), 209–10, 211; Pierce, Clinton A., 205; Quesenberry, Marshall H., 205; Rawitser, Emil C., 205; Selleck, Clyde A., 205; Stansell, Joshua A., 205; Stevens, Luther R., 205; Stowell, Allen L., 205; Tarpley, Tom, 203; Uhrig, Jacob E., 205; Vachon, Joseph P., 205; Walker, Jack, 203, 204, 205; Weaver, James R. N., 205; Wilson, Ovid "Zero," 191, 203, 204, 205. *See also* Mukden POW camp, rescue from
Mukden POW camp, rescue from, 208–13; and Braly, William C., 210; Braly, William C., on, 210–11; Brougher, William E., on, 208; Fortier, Malcolm V., caricatures of, 209, 211, 212; Galbraith, Nicoll F., on, 208, 209, 210; and OSS discovery of withheld mail to POWs, 219–20; and OSS rescuers, 208–10; and Parker, M. (General), 209–10, 211; and Russians, 210–12, 219; and US food drop, 212

Nagaru Maru, 118–19, 120, 121; arrival at Takao, Taiwan, 121; Berry, K. L., dates/time on, 118; food, 119; Fortier, Malcolm V., caricature of POWs aboard, *119*; quarters, POW, 118–19
Nagasaki, Japan: "Fat Boy" atomic bomb dropped on, 207
Nagatomo, Y. (Sergeant), 154
Nakano, Junichi (Lieutenant Colonel), 123, 140–41
Nakashima (2nd Lieutenant), 124

Nara, Akira: and Battle of Bataan, 67, 69, 90, 92; and Lim, Vicente, 90–91
National Defense Act (1920), 33
National Guard: change of to reserve component of Army, 47, 49
Nazi Party, 47
Neilson, H. H. *See* Longhorns
Neutrality Act of 1937, 49, 50
New Braunfels, Texas, 14
New Deal, 39
New Guinea: Allies' victory in, 180
Nitchi Man hotel (Beppu), 188, 191–93; bathing and washing facilities, 192; food, 192; latrines, 192; mess hall, 191; POWs leaving, 192; sleeping arrangements, 191. *See also* Nitchi Man hotel (Beppu), Allied POWs in
Nitchi Man hotel (Beppu), Allied POWs in: Braly, William C., 192; Browne, E. Harrison, 192; Berry, K. L., 191, 192; Corkill, William E., 191–92; Hilsman, Roger, 192; Hoffman, Robert G., 192; Mallonee, Richard C., 192; O'Day, Ray, 192; Parsons, John E., 192; Pechek, Frank, 192; Quesenberry, Marshall H., 192; Wainwright, Jonathan M., 191. *See also* Berry, K. L., and Nitchi Man hotel (Beppu)
Nicholas II, 24
91st Division: Bataan Death March, 102; Battle of Bataan, 64, 77, 79; in I Corps, 64
92nd Infantry, 73
Nishiyama (Corporal), 114, 117
North Africa, and World War II: Allies' campaign in, 153; Allies' defeat of Germans in, 149; Allies' defeat of Italians in, 149; and WWII news about, POWs, 148

O'Connor, Edwin (Colonel), at Wainwright Travelers 1950 gathering, 234, 235
O'Day, Ray (Colonel), 215; birthday gift for Berry, K. L., at Mukden camp, 205; in Nitchi Man hotel (Beppu), 192; in *The Quan* "Chit Chat" column, 233
O'Donnell POW camp. *See* Berry, K. L., and Camp O'Donnell; Camp O'Donnell; Camp O'Donnell, Allied POWs in
Office of Strategic Services (OSS): and discovery of withheld mail to POWs, 219–20; Leith, Hal, as rescue team member, 208; Masuda (Commander), speaking with rescue team, 209–10; and Moore, Arthur P., 210; Mukden camp POW rescue by, 208–10; rescuers, 208–10; and Wainwright, Jonathan M., 210
Okinawa, 214; typhoon, 215; US air raids against, 187; US fighting in, 202. *See also* Battle of Okinawa
"Old Sourpuss." *See* Nakano, Junichi (Lieutenant Colonel)
141st Infantry Regiment: Berry, K. L., and Rifle Team, 48, *48*; inducted into federal service (36th Division), 49
192nd Tank Battalion, 99; Company B and Bataan Death March, 101
Operation Downfall, 205–6; and MacArthur, Douglas, 206; planned invasion of Kyushu, 206
Orani: Bataan Death March stop of, 100, 104
Orias (Captain), 62, 66
Orion, Philippines, 65, 76, 78
Ōryoku Maru (December 14–15, 1944), and POWs, 189–91, 203; deaths, 190; survival post-attack, 190–91; under US attack, 189–90, *190*; and Wilson, Ovid "Zero," 191
Ōryoku Maru (October 10–28, 1944), and POWs, 184, 189; arrival at Kyushu, 188; bathing, 188; Berry, K. L., 187, 188; boarding, 185; Bowler, Louis J., 187; Braly, William C., 186–87; Browne, E. Harrison, 187; Corkill, William E., 187, 188; food, 186; Fortier, Malcolm V., caricature of, *186*; Galbraith, Nicoll F., 187; Lilly, Edmund J., 188; living conditions,

186–87; and US air raids on Keelung Harbor, 187; water, lack of, 187
Oswald, Lee Harvey: murder of by Ruby, Jack, 252

pachikas, 195
Panic of 1893, 3, 5
Pantingan River, 94, 98
Pararinci (Doctor): Camp Karenko Red Cross visit by, 151
Parker, George M. (Major General): Battle of Bataan, 60, 64, 65, 67, 70, 75, 77, 89, 90, 92; Camp Shirakawa, 181; Cheng Chia Tun camp, 197; and II Corps, 89, 90; and Masuda (Commander), 209; Mukden camp, 209–10, 211; and Mukden camp rescue, 209–10, 211; as Southern Luzon Commander, 53, 54; speech to POWs about not reacting to Japan surrender news, 210; and Wilson, Ovid "Zero," 53. *See also* II Corps
"Pasadena Kid." *See* Hioki, Shiro (Lieutenant)
Pearl Harbor: Japanese attack on, 55, 60, 62; and Yamamoto, Isoroku (Admiral), 148
Pechek, Frank (Staff Sargeant), 161, 176
Peking, China, 41, 43, 45; Japanese occupation of, 46
Penrose, Arthur W. (Colonel), 215
Pershing, John (General), 13, 16; failure to capture Pancho Villa, 16
Pershing Trophy, 48
Philippine Legion of Honor, 242, 243–44
Philippine Military Academy, 62, 63; Eliseo Rio graduation from, 241
Philippines: ball at presidential palace, 34; Berry, K. L., postwar visit, 241, 242–43; ceded to United States, 6; Clark Field, Japanese bombing planes at, 55, 60; Eisenhower, Dwight D., under MacArthur, Douglas, in, 51; Japanese attack on December 8, 1941, 55, 60, 62; map, 58; and Spanish-American War, 4–5; 27th Infantry in, 34; US forces landing in, 180. *See also specific cities, islands, and geographic areas;* Bataan Death March; Battle of Bataan; Berry, K. L., WWII military service in Philippines
Pierce, Clinton A. (Brigadier General): birthday gift and song for Berry, K. L., at Mukden camp, 205
Pilar, Philippines, 65, 75, 76
Pilet, Nunez C. (Colonel), 196
Plan of San Diego, 12
post exchanges: Camp Karenko, 134, 138, 150; Camp Shirakawa, 158; Camp Tarlac, 116–117; workers and extra rice, 172
presidential elections: 1912, 8; 1924, 35; 1928, 39; 1932, 39; 1940, 50; 1948, 236; 1952, 236–37
POW (prisoner of war) camps: birthday celebrations, 160, 198–99; 205; deaths, POW, 110–11, 117, 141, 157; food, 111, 114–15, 118, 122, 126, 127–30, 137, 138, 155, 157–59, 183, 192, 196, 197, 206; gardens and gardening, 115, 126, 127, 130, 138, 159, 172, 197; health problems, POW, 110–11, 141–42, 156–57, 200; leisure activities, POW, 117, 133–35, 159–62, 197–98, 204; libraries, 135, 161; "Order to Kill" documents, 207; Red Cross supplies, 149–50, 157–58, 196–97, 206, 230; rules and regulations, 110, 123, 124–25, 145, 181–82; starvation and malnutrition, 110, 141, 150, 155, 156, 176, 203, 206; treatment by guards, POW, 122, 142–46, 148, 154, 201; work, forced, 138–41, 171–76. *See also* Camp Cabanatuan; Camp Heito; Camp Karenko; Camp O'Donnell; Camp Shirakawa; Camp Tamazato; Camp Tarlac; Cheng Chia Tun POW camp; Fukota No. 3 POW camp; Mukden POW camp; Nitchi Man hotel (Beppu)
Purple Heart, 223

Quan, The: Ray O'Day "Chit Chat" column in, 233

Quanah, Texas, 3
Quesenberry, Marshall H. (Colonel): birthday gift for Berry, K. L., at Mukden camp, 205; Camp Karenko, 143; Camp O'Donnell, 110; Camp Shirakawa, 159–60; Cheng Chia Tun camp, 196, 199; Fort McKinley, 54; postwar transport home, 214–15
Quezon City, Philippines, 62
Quinauan Point, 77

Rabaul (Solomon Islands): Japanese capture of, 69
radiograms, 131, 162, 199
Rakuyo Maru: US attack on, 118
RAMPs, 213; postwar transport home, 213–15. *See also specific POWs*
Rawitser, Emil C. (Colonel): birthday gift for Berry, K. L., at Mukden camp, 205
Red Cross: Berry, K. L., as Texas State Chair of, 230; food and supply packages for POWs, 149–50, 157–58, 196–97, 206, 230; visit to Camp Karenko by Dr. Pararinci, 151
Reilly, Bruce, 237; Berry, K. L., as father figure to, 239; Camp Mabry, living at, 239; and Lazy K Ranch, 238, 239
Reilly, William King: Berry, K. L., as father figure to, 239; Camp Mabry, living at, 239–40; and Lazy K Ranch, 238–39, 240–41. *See also* Longhorns
Republic of Texas, 3, 4
Reserve Officers Training Corps, 39
Rice, E. B. (Captain), 157
Richards, Harrison H. C. (Colonel), 130
Riley, John Sleychk, 4
Riley, Sam H., 222
Riley, Viola Eugenia, 3, 4; poetry, 5
Rio, Eliseo (2nd Lieutenant): Battle of Bataan, 62–64, 67, 68, 74, 79, 87; Bronze Star Medal, 98; graduation from Philippine Military Academy, 241; as guerilla fighter against Japanese, 98; surrender to Japanese, 96–98. *See also* Battle of Bataan, surrender of Americans and Filipinos

Rodman, John H. (Colonel), 73
Roehm, John, 248
Rogers, Richard G. (Colonel), 198
Rome: Allies' capture of, 182
Roosevelt, Franklin D., 39; and Churchill, Winston, 49–50; death of, 202; death of confirmed for POWs, 203; declaration of war against Japan, 55; and draft, peacetime, 49; fireside chats, 57, 87; freezing German, Italian, and Japanese assets in United States, 50; and Lend Lease Act, 50; MacArthur, Douglas, pulling out of Philippines, 87; MacArthur, Douglas, recalling to active duty, 60; National Guard change to Army reserve by, 47, 49; National Guard, WWII mobilization of, 49; New Deal, 39; 1940 reelection of, 50; and 1932 presidential election, 39; and USAFFE creation, 50; Yalta Conference, 200
Roosevelt, Theodore "Teddy," 6, 8; Nobel Peace Prize, 6; nomination to temporary brigadier general, 222
Ruby, Jack: murder of Oswald, Lee Harvey, by, 252
Russia: declaration of war against Japan, 207; invasion of Manchuria, 207; Japanese expansion into western, 26; Japanese territorial gains, 42; and WWII news about POWs, 148. *See also* Siberia, post-World War I
Russian army, 203; Berlin, near, 200–201; Berlin, reaching, 202; liberation of Mukden POW camp, 210; Warsaw, near, 200
Russian Revolution, 30
Russo-Japanese War (1904–1906), 42
Rutherford, Dorsey J. (Colonel), 180

Saipan: US invasion of, 182
San Antonio, Texas, 14, 18, 251
San Diego, California: Berry, K. L., military service in, 23
San Fernando, Philippines: arrival of Bataan Death March POWs, 104;

INDEX 339

Bataan Death March stop of, 99, 100, 101, 103, 104–5; sun treatment punishment for Bataan Death March POWs, 104
San Francisco, California: Berry, K. L., awaiting transport to Philippines in, 51, 52; Presidio Letterman Hospital, Berry, K. L., in, 215
Santa Barbara, California: Japanese attack on oil refinery near, 85
Santos, Alfredo (General), 243
Sazawa, Hideo (Colonel), 179, 181
Schofield Barracks: Berry, K. L., at, 34; and Pearl Harbor, 34
scurvy, 77
Searle, Albert C. (Colonel), 215
Second Army Corps. *See* II Corps
2nd Philippine Constabulary, 77
2nd Texas Infantry. *See* Texas 2nd Infantry
segregation: and Texas, 252; US Army, 40. *See also* 24th Infantry
Segundo, Fidel, 63, 67, 69–74, 81, 82, 83
Selby, Wallace R. (Commodore), 157
Selective Service Director, Texas, 228, 229
Selleck, Clyde A. (Colonel): birthday gift for Berry, K. L., at Mukden camp, 205; Cheng Chia Tun camp, 196, 198
Semenov, Grigori (General), 28–29, 30; Berry, K. L., meeting with, 30; tea car, *31*
Shansi province, China, 44, 46
Sharp, William F. (Major General), 60
Sheppard, E. G. (Private), 157
Sherman Silver Purchase Act, 3
Sherry, Dean (Major), 164
Shirakawa, Gifu (Japan), 194
Shirakawa POW camp. *See* Berry, K. L., and Camp Shirakawa; Camp Shirakawa; Camp Shirakawa, Allied POWs in
Shivers, Robert Allan: Berry, K. L., choice as Texas representative to Eisenhower presidential inauguration, 237; Berry, K. L., reappointments as adjutant general of Texas, 231; as Texas governor, 231

Siberia, post-World War I, 24–32; and Baker, Newton, 26; Berry, K. L., military service in, 24, 27–30, 32, 34, 117, 148, 230; Bolsheviks, 24–25, 26, 27, 28, 30, 32; Cossacks, 24, 25, 26, 28, 30; Graves, William S., as commander in, 26–27, 30, 32; map, *25*; Mensheviks, 24; and Wilson, Woodrow, 26, 32, 42
Siberian Presidio Replacement Detachment No. 1, 24
Sigma Delta Psi Honorary Athletic Fraternity: Berry, K. L., 11; Littlefield, Clyde, 11
Silver Star, 223
Simson, Ivan, 134
Singapore, 118, 134, 182; Japanese invasion of, 79; Japanese siege of, 69
Sino-Japanese War (1894–1895), 42
Sixth Army, US: Bataan, recapture of, 200; Lingayen Gulf, invasion of, 200; Manila, attack on, 200
slavery: end of, 4; and Texas, 3
Sledge, Theodore J. (Colonel), 170–71, 175
Smith, Al, 39
Smith, Ross B. (Lieutenant Colonel), 53, 51
snails as food, 128–29
Southeast Asia: 1960s troubles in, 250
Spanish-American War, 4–5, 6, 14; military training with athletic training during, 33
Stalin, Joseph: and Yalta Conference, 200
Standing Sentinel, 257
Stansell, Joshua A. (Colonel): birthday gift for Berry, K. L., at Mukden camp, 205
starvation and malnutrition: and Bataan Death March POWs, 99, 104, 105; and Battle of Bataan, 77, 89, 90, 92; and Camp Cabanatuan POWs, 203; and Camp Karenko POWs, 141, 150; and Camp O'Donnell POWs, 110; and Camp Shirakawa POWs, 155, 156, 176; and Mukden camp POWs, 206
Steel, Charles L. (Colonel), 52; and Berry, K. L., 51; Camp Karenko, 130

Stevens, Luther R. (Brigadier General): Battle of Bataan, 95, 96; Battle of Bataan, after, 103; birthday gift for Berry, K. L., at Mukden camp, 205; Camp O'Donnell, 110; Wainwright Travelers 1950 gathering, 234, 235
Stickney, Henry H. (Colonel), 215
Stivers, Paul, 54
stock market collapse, 39
Stoudt, Daniel (Private), 111
Stowell, Allen L. (Colonel): birthday gift for Berry, K. L., at Mukden camp, 205; Camp Tarlac, 117
Subic Bay, 64, 213
Sumatra: Japanese invasion of, 79
Sutherland, Richard (Major General), 66–67, 75, 90
Suzuya Maru, 120, 121; Berry, K. L., dates/time on, 121; Mallonee, Richard C., description of, 121

Taft, Robert A., 236
Taft, William Howard, 8
Tai-Yuan-Fu, China, 46
Taiwan, 118, 121, 123, 151, 153, 183, 185, 186, 189; Berry, K. L., arrival, 122; Japanese control of, 42; map of POW camps in, 120; US air raids on, 188, 191, 200, 201. *See also specific cities and POW camps*
Takao, Taiwan, 120; *Nagaru Maru* arrival at, 121
Tamazato POW camp. *See* Camp Tamazato
Tank Group headquarters, 99
Tarlac City, Philippines, 111
Tarlac POW camp. *See* Berry, K. L., and Camp Tarlac; Camp Tarlac; Camp Tarlac, Allied POWs in
Tarpley, Tom, 54
Tenney, Lester: Bataan Death March, 99, 101–2, 104, 105; Camp O'Donnell, 109
Terrell Military College, 229
Texas: annexation of by United States, 4. *See also specific cities, towns, and counties*

Texas, governors of. *See* Daniels, Price; Jester, Beauford; Moody, Dan; Shivers, Robert Allan
Texas Air National Guard: Berry, K. L., and adjutant general, 228; federalization of during Korean War, 232
Texas Army National Guard, 228
Texas Distinguished Service Medal, 244
Texas Heritage Foundation, 244
Texas National Guard, 228; Berry, K. L., adjutant general role, 227; Berry, K. L., in, 11, 13; border protection by, 11; move to Fort Sam Houston, 16
Texas National Guard Officer Candidate Academy School, 244
Texas Normal College, 6
Texas 2nd Infantry: Berry, K. L., in, 12, 13–14; deactivation of, 16; on football team, 15–16, 33; protecting US-Mexico border, 13–15
Texas State Guard, 228
Texas State Guard Reserve Corps, 232
Texas Woman's University, 6
3rd Battalion: and Eliseo Rio, 63. *See also* Battle of Bataan, and Berry, K. L.
3rd Infantry, 63, 66–67, 70, 71, 72, 74, 81. *See also* Battle of Bataan, and Berry, K. L.
3rd Regiment, Philippine Army, 62, 63, 64, 66; engineer battalions, 66; headquarters, 64; in I Corps, 64; 155th Coast Defense two-gun battery, 66; and Rio, Eliseo, 62; self-propelled artillery battery, 66; 3rd Artillery battalion, 66; 295th Artillery battalion, 66. *See also* Battle of Bataan, and Berry, K. L.
31st Division, 64, 92
Thompson, John W. (Colonel), 215
Tientsin, China, 40, 41, 42, 43, 44, 45, 46; Berry, K. L., in, 213; Japanese occupation of, 46
Tokyo, Japan: US bombing of, 201
Townsend, Glen R. (Colonel), 86, 94
track. *See* University of Texas track
trains: POW camp transport: from

INDEX 341

Cheng Chia Tun camp to Mukden camp, 203; from San Fernando to Capas, 99, 100, 101, 103, 104–5
Traywick, Jesse T. (Colonel), 196
Treaty of Brest-Litovsk, 24
Truesdell, Karl (Colonel), 44
Truk (Caroline Islands): Allies' victory in, 180; US air raids on, 188
Truman, Harry S.: atomic bomb, approval to drop, 206; atomic bomb, first knowledge of, 206; Berry, K. L., Japan, calls to surrender, 206; and Korean War, 236; MacArthur, Douglas, firing of, 236; 1948 presidential victory, 236; US president, elevation to, 202
Tsuneyoshi, Yochio, 109
Tuol River (Philippines), 80
Turner, Charlie: Texas 2nd Infantry football team, 15; University of Texas football team, 15
21st Division, 92
21st Pursuit Squadron, 102
24th Infantry: and Berry, K. L., 39–40, 40; and Fort Benning construction, 39–40; and segregation in Army, 40
Twenty-One Demands agreement, 42
27th Bombardment Group, 101
27th Infantry, 34
26th Cavalry, 64, 70, 73, 77, 89
23rd Infantry, 34
"Two-Bowlers," 173

Ugi (Lieutenant), 114
Uhl, Frederick E. (General), 215
Uhrig, Jacob E. (Colonel): birthday gift for Berry, K. L., at Mukden camp, 205
United States–Mexico border: problems, 12; Texas 2nd Infantry protecting, 13–15
University of North Texas, 6
University of Texas: Berry, Eugene "Gene," at, 8; and Berry, K. L., 7, 8, 16, 226; Berry, K. L., return to as student, 35–37; *Cactus* (yearbook), 9, 10, 11; establishment of Forty Acres Club, 250; H. J. Lutcher Start Center for Physical Culture and Sport, 10, 35; and Jester, Beauford, 226. *See also* University of Texas football; University of Texas track; University of Texas wrestling
University of Texas football: Berry, Eugene "Gene," 8, 9; Berry, K. L., 8, 9, 9, 10, 33, 35–36, 36. *See also* Longhorns
University of Texas track: Berry, Eugene "Gene," 10; Berry, K. L., 8, 10, 36
University of Texas wrestling, 8, 10
University of Vermont, and Berry, K. L.: assistant professor of military science and tactics, 39, 40; freshman football coach, 39; varsity basketball coach, 39
Ura (Lieutenant), 114, 118
USAFFE: Roosevelt, Franklin D., creation of, 50
USS *Colbert*: mine explosion of, 215
USS *Coolidge*, 52–53
USS *Enterprise*: attack on Japanese at Wake Island from, 85
USS *Grant*, 41, 46
USS *Hornet*: bomber attack on *Ōryoku Maru*, 189–90
USS *Houston*: Japanese sinking of, 85
USS *Langley*: Japanese sinking of, 85
USS *Louisiana*, 52
USS *Relief*: 213–15; postwar transport of RAMPs home, 214–15
USS *Scott*, 52, 53
USS *Thomas*, 32

Vachon, Joseph P. (Colonel): birthday gift for Berry, K. L., at Mukden camp, 205
Vargas, Jesus (Lieutenant General): and Berry, K. L., Philippine Legion of Honor award, 242, 243, 244
Verkhne-Udinsk, Siberia, 25, 27, 28, 30, 32
Vietnam War: and Lyndon B. Johnson presidency, 252
Villa, Francisco "Pancho," 13, 14, 16, 18; United States–Mexico border town raids, 11, 13
violence against POWs: bayoneting by Japanese military, 103; beatings by

Japanese military, 103; beheadings by Japanese military, 103; and fellow POWs, 155–56; and POW camp guards, 103, 122, 142–44, 146, 148, 154, 201
Viyasan: island of, 63
Vladivostok, Siberia, 25, 27, 32, 45; Berry, K. L., leaving on USS *Thomas*, 32
Volckmann, Russell W. (Major), 95

Wainwright, Jonathan M. (Lieutenant General), 208; Battle of Bataan, 64–71, 73–77, 79, 82, 83, 86–90, 92, 93; Berry, K. L., appreciation letter to, 133; and Berry, K. L., Distinguished Service Medal ceremony, 224; and Berry, K. L., promotions, 131, 132, 222; Berry, K. L., and Texas Military District assignment, 226; and Brougher, William E., 96; Camp Karenko, 127–28, 131, 139, 140, 144; Camp Karenko, move from, 151; Camp Tarlac, 114; Cheng Chia Tun camp, 195; death of, 236; Galbraith, Nicoll F., and Distinguished Service Medal ceremony, 224; Japanese request to urge US and British authorities to treat Japanese POWs kindly, 144; Lazy K Ranch visit, 237–38, 238; and OSS rescue team, 210; postwar transport home, 211; surrender, 126, 131; Wainwright Travelers 1950 gathering, 234, 235; and War Plan Orange, 59. *See also* I Corps
Wainwright Travelers, The, 233, 234, 235, 236. *See also* Wainwright Travelers 1950 gathering
Wainwright Travelers 1950 gathering, 233–35, 234, 235; Berry, K. L., as speaker, 233; Lutes, Leroy, as speaker, 233; Wainwright, Jonathan M., as host, 233
Wakasugi, Jiro (2nd Lieutenant): and Camp Karenko, 124, 127, 130, 145; and Camp Shirakawa, 154
Wake Island: Japanese takeover of, 60, 62; USS *Enterprise* bomber attack on Japanese at, 85

Wakefield, Paul: Berry, K. L., and Texas Distinguished Service Medal, 244; cowbell owned by family of, 237
Walker, Bert. *See* Longhorns
Walker, Jack (Captain): Berry, K. L., letter of appreciation to, 204, 265; birthday gift for Berry, K. L., at Mukden camp, 205
Walter Reed Hospital, 220
War Plan Orange: and MacArthur, Douglas, 60, 61, 62
Ward, Frederick A. (Colonel), 161
Warsaw, Poland: Russian Army near, 200
water, lack of: Bataan Death March POWs, 102, 104, 105; Camp O'Donnell, 110, 111; Camp Shirakawa, 53, 154, 168–70; *Ōryoku Maru*, 187. *See also* dehydration
Weaver, James R. N. (Brigadier General): birthday gift for Berry, K. L., at Mukden camp, 205; Camp Shirakawa, 154, 158
Wetherby, Loren A. (Colonel): Bataan Death March, 104; and Battle of Bataan, 92; and Berry, K. L., 52
Whitehurst, Matthew S. (Master Sargeant), 161
Wilson, Ovid "Zero" (Lieutenant Colonel): and Berry, K. L., 53, 191; and Berry, K. L., Mukden camp, 204; birthday gift for Berry, K. L., at Mukden camp, 205; Fort McKinley, 54; Mukden camp, 191; and Parker, George M., 53
Wilson, Woodrow, 8, 11, 13, 14; Arizona National Guard, federalization of, 13; declaration of war against Central powers, 18; New Mexico National Guard, federalization of, 13; and Pershing, John, 16; and Siberia, 26, 32, 42; Texas National Guard, federalization of, 13; World War I, drafting of National Guard during, 19
Wimmer, Coke. *See* Longhorns
Wohlfeld, Mark (Captain), 101
Wolfhounds, 27, 28, 32

Wood, Leonard (Major General), 18
Wood, Stuart (Colonel): Camp Karenko, 146; Camp Tarlac, 117; Cheng Chia Tun camp, 196, 200
workers' rice, 138, 172, 173
World War I, 11, 16; end, 24; Germany, defeat of, 47; onset, 32; and Russian Empire, 24; United States entry into, 18, 23; Wilson, Woodrow, drafting of National Guard during, 19
World War II: D-Day, 182. *See also specific military battles, men, and POW camps*; atomic bombs; Bataan Death March; Bataan Peninsula; Churchill, Winston; Germany; Great Britain; Hitler, Adolph; I Corps; II Corps; Japan; Japanese military, and World War II; MacArthur, Douglas; North Africa, and World War II; Pearl Harbor; Philippines; Roosevelt, Franklin D.; Russia; Russian army; Stalin, Joseph; Taiwan; Truman, Harry S; Yalta Conference; World War II, US bombing raids
World War II, US bombing raids: on *Enoura Maru*, 191; on Japan, 182, 184, 201, 202, 206; on Keelung Harbor, 187; on Kyushu, 193; on Leyte, Philippines, 188; on Okinawa, 184, 187; on *Ōryoku Maru*, 189–90, *190*; on Taiwan, 184; on Tokyo, Japan, 201; on Truk Japanese base, 188. *See also* atomic bombs
Worthington, Josiah W. (Colonel), 176, 215
Wright-Patterson Army Air Corps Base (Ohio), 220
Wynkoop, Hueston R., 55

Yalta Conference, 200
Yamamoto, Isoroku (Admiral): and Pearl Harbor attack, 148; shot down by US fighter jets, 148
Yamamoto: US sinking of, 201
Yamanaka, Bob, 154
Yawata, Kyoto (Japan), 192
Young, Adlai C. (Colonel), 51, 52

Zimmermann, Arthur, 18

www.ingramcontent.com/pod-product-compliance
Lightning Source LLC
Chambersburg PA
CBHW060549080526
44585CB00013B/493